Statistics: Extracting Information from Data

Bryan R. Crissinger

Vine & Branch Publishing

Statistics: Extracting Information from Data

Published by Vine & Branch Publishing.
bryan@vineandbranchpublishing.com

The author and publisher welcome feedback on this first edition.

ISBN   978-1-7354231-0-4   paperback
ISBN   978-1-7354231-1-1   pdf ebook

Contents

Chapter 1: An Introduction to Data

In the summer of 2007 I took a trip to Yellowstone National Park. One of the most famous (perhaps *the* most famous?) geyser in the world is Old Faithful, so of course I had to check it out. On the morning of August 4th, I arrived at the Old Faithful viewing area at around 11:20am and was

greeted by the sign shown here. A large crowd had already gathered to witness one of Old Faithful's very predictable eruptions. The geyser did not disappoint, erupting in spectacular fashion at 11:34am, six minutes early, and as the crowd dispersed to check out the other thermal features in the area, I couldn't help but ponder how the park service knew that they'd be within about 10 minutes of their predicted time. As a data analyst, I had a pretty good idea: someone must have collected data on Old Faithful's eruptions and figured out a way to use the data on previous eruptions to predict the time until the next one. Yellowstone gets over 3 million visitors every year and since Old Faithful is one of the most popular features in

the park, you might imagine how important it is to give visitors a good idea of when the next eruption will happen. Tourists can then plan their visit accordingly. For example, if you get there an hour or so before the next eruption, you can walk around and see some of the other thermal features first or browse the visitor center and gift shop instead of waiting around at the Old Faithful site. As you can see, the skies were clear and the sun was beating down pretty harshly on the shadeless viewing area. Sure enough, a few minutes later, Old Faithful erupted in spectacular fashion. I didn't care whether the predicted time on the sign was spot on or not. All that mattered to me was that it was close enough.

Dear reader, I'd like to introduce you to my discipline – Statistics. **Statistics** is the science of extracting information from data to answer a question. That definition is one of my favorites because it's packed with meaning. Let's pull out the good stuff:

- "science": Statistics as a discipline is a science; it is not simply a branch of mathematics.

Yes, there are many math tools that statisticians and data analysts use, but the same is true for physicists, chemists, and biologists. However, we don't think of physics, chemistry, or biology as branches of mathematics. They are sciences. The same is true of statistics. You'll see later in the book that one of the techniques used is a rigorous application of the scientific method (Question → Hypothesis → Collect Data → Analyze Data → Conclusion).

> "Statistics is the grammar of science."
> – Karl Pearson

- "extracting": I love this word! It means that effort is required. The story doesn't simply appear out of the data on its own. It takes work to extract it and to make sure that the data are not misinterpreted.

Sometimes learning statistics will be difficult because it may be something you've never done before. Just like learning how to hit a tennis forehand with both accuracy *and* pace is difficult and takes lots of practice, you may find the same is true of learning statistics. However the payoff is well worth the effort. People with data analysis skills are very high demand these days.

> "Data do not give up their secrets easily. They must be tortured to confess."
> – Jeff Hopper

- "information" vs. "data": These two words are often used interchangeably, but they do not refer to the same thing.

Data (characteristics observed on objects) are everywhere. Your purchase of this book generated a bunch of data (price, time of purchase, location of purchase, method of payment, etc.). From financial data on publicly-traded companies to flight status data for an airline to class performance on standardized tests, it is easy to get your hands on data these days. Finding and accurately reporting the information in those data – the story that they tell – is something very different.

> "Your relevance as a data custodian is your ability to analyse and interpret it."
> – Wisdom Kwashie Mensah

- "to answer a question": The techniques and methods I'll show you in this book are designed to answer very practical questions on a wide variety of topics.

> "The best thing about being a statistician is that you get to play in everyone else's backyard."
> – John Tukey

There are four themes that you'll find throughout your study of statistics.

- <u>Variability</u>: Our world displays variability in many ways. If you stop and think about it, we're drowning in a sea of variability. People differ from one another on many characteristics. Your electricity bill is almost never the same month-to-month. Corporate revenues go up and down from quarter to quarter for a host of reasons. Understanding variability and how to measure it is fundamental to our discipline.

- <u>Uncertainty</u>: Predictions, forecasts, and estimates are common. "The forecasted high temperature today is 64 degrees." "XYZ Corporation beat analysts' estimates by $0.20 per share." "Next prediction for Old Faithful geyser is within 10 minutes of 11:40am." With each of these comes uncertainty. The actual high temperature today will likely not be exactly 64 degrees. The estimate of XYZ's earnings was off by $0.20 per share. Old Faithful didn't erupt at exactly 11:40am. Predictions, forecasts, and estimates are almost never exactly correct but we don't mind if they are "close enough." Statisticians are experts at quantifying and measuring how close is "close enough."

> "Essentially all [predictions] are wrong, but some are useful."
> – George Box

- Randomness: Would you believe that it's best to leave it up to chance? The best way to answer many important questions involving data is to use randomness in the data collection process. Now don't misunderstand me; by "random" I don't mean "haphazard." Think of randomness as structured uncertainty. When I toss a fair coin, the outcome is uncertain but there's some nice structure there: over many, many tosses, I'd get close to 50% heads and 50% tails. As you'll learn, data collected in a haphazard and non-random way (like a sidewalk survey) are much less useful than data obtained by using a random mechanism (like a coin toss).

- Inference: Webster's defines the verb "infer" this way: "to derive by reasoning; conclude or judge from premises or evidence."[1] The answers to many questions involve going beyond the data (evidence) at hand in a logical, reasoned way to make a conclusion or prediction. Here are three common types of statistical inference:

    o Predictive Inference: A predictive inference uses data collected on a bunch of objects to make a prediction or estimate about a new object. For example, you might want to estimate the market value of a piece of real estate, so you collect data on a bunch of properties and use those data to make the prediction.

    o Generalization Inference: A generalization inference is a conclusion about a large group of objects based on data collected on a smaller subset of those objects. Surveys are excellent examples of generalizations. Many public opinion surveys collect only a thousand or so responses but attempt to generalize the opinions of those thousand people to the entire U.S. adult population.

    o Causal Inference: A causal inference is a conclusion about the cause of a difference or relationship we observe in data collected on a set of objects. It attempts to answer the "why?" question. For example, a class of students who was taught using a new instructional method performed better than another class taught using the old method. Does that mean the new method is better? Can we conclude that the difference in the teaching methods was the cause of the higher performance? As you'll see, it depends on how the study was conducted.

Here are a few more examples to whet your appetite for what's to come.

Trisomy 18

When my wife and I went in for the big sonogram before Noah, our oldest child, was born, the doctor told us that the sonogram revealed some spots on his brain that could be an indication of a rare but very serious genetic disorder called trisomy 18 (T18). Babies with T18 seldom live past their first birthday if they are born alive at all. We followed up with a mom's blood test that successfully (the success rate is over 95%) identifies cases and non-cases of T18. Noah's test came back negative and he's fine, but if it had come back positive, what would you think? Do you think there's a good chance that a baby has T18 given a positive test result? The answer requires a careful understanding of the variability of the traits of "T18 status" and "test result" for a group of babies. Stay tuned.

Comparing Batteries

Have you ever wondered if brand-name batteries are worth the higher price? Let's say you went to your favorite store and purchased a sample of 10 brand-name batteries and 10 generic batteries. You test each one in a high-drain device with the following results.

Time-to-Failure (hours) Brand-Name	Time-to-Failure (hours) Generic
4.55	4.31
4.74	4.93
4.78	4.14
4.53	4.51
4.65	3.78
4.24	3.95
4.25	4.95
4.13	4.15
4.57	3.52
4.34	3.59
mean = 4.478	*mean* = 4.183

Notice the variability? Also notice that the brand-name batteries lasted 0.295 hours longer (4.478 – 4.183), on average, than the generic batteries. Does that mean the brand-name batteries are worth the extra price? The answer requires a *generalization inference* from only those 20 batteries to all batteries of each type at your store. You'd like to make a conclusion that goes beyond just those 20 batteries. The answer also depends on whether or not randomness was used to select the 20 batteries. If so, then the uncertainty introduced by the randomness must also be considered. Stay tuned.

House Value

Here are data collected on a sample of houses sold in the county where I live. More variability! How could you use these data to predict the selling price of a house having 2000 square feet that's currently on the market? The answer requires a fairly sophisticated accounting of the variabilities of the house sizes and selling prices, both separately and together. You might suspect that larger houses generally sell for higher prices. Quantifying that relationship and using it to predict the selling price of the new house is an example of *predictive inference*. The house hasn't sold yet so your prediction would carry some uncertainty with it. Stay tuned.

Square Feet (x)	Price (y)
2240	$275,000
2648	$315,000
1800	$260,000
1824	$299,900
1888	$255,000
1586	$249,900
1273	$219,900
2232	$249,900
1548	$230,000
1048	$180,000

Webpage Design

Does the location of the "Buy Now" button on a webpage affect sales of a product? You design an experiment where you randomly assign all visits to a product page to one of two versions of the

page: one where the "Buy Now" button is at the top and one where the "Buy Now" button is at the
bottom. For each visit, you determine whether the customer purchased the product or not and
tabulate the results.

	purchased	didn't purchase	total
top	75	435	510
bottom	104	418	522
total	179	853	1032

Note that 75/510 (14.7%) of customers who saw the "top" version purchased the product vs.
104/522 (19.9%) in the "bottom" group. Does that mean putting the button at the bottom caused
the higher purchase rate? The answer requires a *causal inference* about the effect of the button
location. Since there are many other reasons why customers decide to purchase or not, the button
effect must be isolated in order to make a valid causal inference. As you'll learn, randomness plays
a big role in making that happen but introduces uncertainty that must be dealt with. Stay tuned.

One of my goals in writing this book is that you'll come away with an understanding of how
statistics is not just math but a science. Here's an example of how we need to change the way we
think about certain math tools when we use them in a data context.

Police and Crime

Here is a plot of the number of police per 1000
people and the number of crimes per 1000 people for
several cities. The plot also includes the best-fitting
line. Its equation is

$$\hat{y} = 14.6x - 0.89.$$

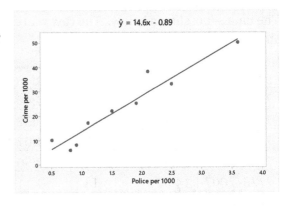

The slope of this line is $\frac{14.6}{1}$. A typical algebra I
interpretation of the slope would sound something
like this:

"Increasing the number of police by 1 per 1000 people would increase the number of crimes by
14.6 per 1000 people."

Is that correct? Of course not. There is variability around that line so a summary like the slope of
the best-fitting line is simply an estimate and there is uncertainty associated with it. More
importantly, the algebra I interpretation is an unjustified *causal inference*. It implies that the *reason
why* there are more crimes in some cities is that there are more police there, as if the mere presence
of police officers is the cause of crime. The common mathematical interpretation won't do.

So then how should we think about the slope of this line? We need to apply the concepts of
variability, uncertainty, and inference to the notion of "slope." Stay tuned and keep reading!

Data and Variables

Before we start to tackle the tasks of extracting information from data, we need to start with some basic terms.

Data are simply characteristics of objects. What makes data science so useful is that those objects can be just about anything, from people (we'd call them "subjects" instead of "objects"), to companies, to eruptions of a geyser. If we list each object as a row in a table and record the various characteristics of those objects in the columns of the table, we call the table a **data set**, that is, a collection of data on several objects. Each characteristic that we record in a column is often called a **variable**, because what we record in that column will very likely vary from object to object.

Example: Let's take the Old Faithful data set as an example.[2] The objects are the eruptions and each one gets its own row in the data set. There are several variables recorded for each eruption:

> Date: date of the eruption
> Time: time of the eruption using a 24-hour clock
> Predicted: a prediction of when the next eruption will happen (24-hour clock)
> Duration: how long the eruption lasted from start to finish (mm:ss)
> DIS: duration converted to the number of seconds
> DIM: duration converted to the number of minutes

The data set is shown below. The row representing the eruption that I saw shortly after arriving is shaded.

Date	Time	Predicted	Duration	DIS	DIM
7/28/2007	8:16	8:18	3:49	229	3.817
7/28/2007	9:45	9:46	3:32	212	3.533
7/28/2007	11:14	11:15	4:07	247	4.117
7/28/2007	12:48	12:44	3:50	230	3.833
7/28/2007	14:07	14:18	4:11	251	4.183
7/28/2007	15:35	15:37	3:37	217	3.617
7/28/2007	17:08	17:05	3:51	231	3.85
7/28/2007	18:41	18:38	3:53	233	3.883
7/29/2007	8:14	8:09	3:45	225	3.75
7/29/2007	9:42	9:44	4:10	250	4.167
7/29/2007	11:13	11:12	3:53	233	3.883
7/29/2007	12:43	12:43	3:51	231	3.85
7/29/2007	14:22	14:13	3:59	239	3.983
7/29/2007	15:45	15:52	4:07	247	4.117
7/29/2007	17:14	17:15	4:09	249	4.15
7/29/2007	18:54	18:44	3:27	207	3.45
7/30/2007	9:53	9:48	3:54	234	3.9
7/30/2007	11:15	11:23	3:51	231	3.85

Date	Time	Predicted	Duration	DIS	DIM
7/30/2007	12:45	12:45	3:39	219	3.65
7/30/2007	14:18	14:15	3:45	225	3.75
7/30/2007	15:53	15:48	3:55	235	3.917
7/30/2007	17:20	17:23	4:10	250	4.167
7/30/2007	18:53	18:50	3:47	227	3.783
7/31/2007	8:05	8:06	3:56	236	3.933
7/31/2007	9:38	9:35	3:58	238	3.967
7/31/2007	11:10	11:08	3:42	222	3.7
7/31/2007	12:40	12:40	3:55	235	3.917
7/31/2007	14:11	14:10	4:13	253	4.217
7/31/2007	15:44	15:41	3:59	239	3.983
7/31/2007	17:10	17:14	4:10	250	4.167
7/31/2007	18:47	18:40	3:52	232	3.867
8/1/2007	8:09	8:13	2:06	126	2.1
8/1/2007	9:14	9:14	4:51	291	4.85
8/1/2007	10:48	10:44	3:37	217	3.617
8/1/2007	12:15	12:18	3:35	215	3.583
8/1/2007	13:45	13:45	4:04	244	4.067
8/1/2007	15:22	15:15	3:55	235	3.917
8/1/2007	16:48	16:52	3:55	235	3.917
8/1/2007	18:23	18:18	3:41	221	3.683
8/2/2007	7:58	8:02	2:00	120	2
8/2/2007	9:06	9:03	4:12	252	4.2
8/2/2007	10:43	10:36	3:58	238	3.967
8/2/2007	12:13	12:13	4:08	248	4.133
8/2/2007	13:43	13:43	4:04	244	4.067
8/2/2007	15:12	15:13	3:38	218	3.633
8/2/2007	18:03	18:03	3:45	225	3.75
8/3/2007	9:12	9:11	3:58	238	3.967
8/3/2007	10:42	10:42	4:04	244	4.067
8/3/2007	12:15	12:12	2:04	124	2.067
8/3/2007	13:16	13:20	4:05	245	4.083
8/3/2007	15:00	14:46	3:43	223	3.717
8/3/2007	16:28	16:30	3:57	237	3.95
8/3/2007	18:03	17:58	3:39	219	3.65
8/4/2007	8:39	8:47	4:06	246	4.1
8/4/2007	10:10	10:09	3:44	224	3.733
8/4/2007	11:34	11:40	4:16	256	4.267
8/4/2007	13:07	13:04	3:55	235	3.917
8/4/2007	14:42	14:37	4:08	248	4.133

Date	Time	Predicted	Duration	DIS	DIM
8/4/2007	16:04	16:12	3:59	239	3.983
8/4/2007	17:34	17:34	3:31	211	3.517
8/4/2007	19:02	19:04	3:45	225	3.75

It is important to note that when we refer to a variable, we are referring to a characteristic of a single object, not a summary of data collected on many objects. For example, the *mean duration* is a very interesting summary (and we'll get there soon) but it is not a variable in this data set. Recording the mean duration for one eruption doesn't make any sense because the mean is a summary of the durations of several eruptions. The *duration* of the eruption is the variable. Whenever I need to define a variable in a data set, I find it helpful to ask, "What is the characteristic recorded on a *single object*?" In this example, the answer is, "its duration."

Example: Here is a data set showing the top 20 U.S. institutions of higher education ranked by tuition and fees.[3]

Institution	State	Highest Degree	Locale	Tuition	%FT Faculty
Aviator College of Aeronautical Science and Technology	FL	Associate's	41	74514	100
Columbia University in the City of New York	NY	Graduate	11	57208	46.9
University of Chicago	IL	Graduate	11	56034	82.4
Vassar College	NY	Bachelor's	21	55210	84.2
Trinity College	CT	Graduate	12	54770	75.2
Landmark College	VT	Bachelor's	42	54750	100
Jewish Theological Seminary of America	NY	Graduate	11	54684	52.9
Harvey Mudd College	CA	Bachelor's	21	54636	92
Franklin and Marshall College	PA	Bachelor's	13	54380	89.2
Tufts University	MA	Graduate	21	54318	63.5
Amherst College	MA	Bachelor's	21	54310	92.6
University of Southern California	CA	Graduate	11	54259	60.5
Reed College	OR	Graduate	11	54200	95.4
Sarah Lawrence College	NY	Graduate	21	54010	35.2
Bucknell University	PA	Graduate	32	53986	97.1
Colgate University	NY	Graduate	32	53980	100
Carnegie Mellon University	PA	Graduate	11	53910	94.9
Kenyon College	OH	Bachelor's	32	53560	95.1
Williams College	MA	Graduate	31	53550	83.8
Brandeis University	MA	Graduate	13	53537	66.7

The objects are the institutions; each row represents one institution. The column labels identify the variables recorded on each one:

Institution: institution name
State: state where the institution is located
Highest Degree: highest degree granted by the institution
Locale: code for location type

11 = city (large)	31 = town (fringe)
12 = city (midsize)	32 = town (distant)
13 = city (small)	33 = town (remote)
21 = suburb (large)	41 = rural (fringe)
22 = suburb (midsize)	42 = rural (distant)
23 = suburb (small)	43 = rural (remote)

Tuition: total yearly tuition and fees (in dollars)
%FT Faculty: percent of the faculty that are full-time

You may already have some questions of these data. For example, "What is the average tuition for these top 20 schools?" "What percent grant graduate degrees?" While you may already know how to compute these two common summaries, there is a fundamental difference between them. The difference is that calculating an average or mean requires one type of data and calculating a percentage requires another type. Understanding the difference between these two types of data or variables is very important in data science and statistics. Let's look at the two types.

Tuition is an example of a numeric variable; sometimes we say that the tuition data are numeric data. A **numeric variable** is a characteristic of an object that is either a measurement or a count. Both the tuition and age data for each of the 20 institutions are numeric variables as they measure some quantitative characteristic of the institution. Although not shown here, suppose we had listed the number of students enrolled at each institution. Since "number of students" is the result of a counting process, it is also a numeric variable.

Highest degree, on the other hand, is an example of a categorical variable; sometimes we say that the highest degree data are categorical data. A **categorical variable** is a characteristic of an object that classifies the object into one of several groups. We see three different categories of the highest degree variable in the 20 institutions (Associate's, Bachelor's, and Graduate). The state data are also categorical. The values recorded in the data set are not measurements or counts but are simply labels that classify or group the institutions by state.

What do you think of the locale variable? You might think it's numeric since you see numbers as data values. Be careful though: not every list of numbers is numeric data. Notice that these numbers just represent categories of location type and while there is some natural ordering from large cities to remote rural locations, there is no explicit measurement or count represented by the data values. So it's a categorical variable.

Example: Suppose we collect some data on graduates with a bachelor's degree in economics five years after graduation. We'd call the graduates "subjects" here instead of "objects" because they're people.

ID Number	GPA	Honors	JobGrad	Starting Salary	Sector
704653	3.92	summa cum laude	1	35000	government
706445	2.56	none	0	39500	public company
700995	3.56	cum laude	0	40000	private company
702112	3.81	magna cum laude	1	37500	government
701976	3.22	none	1	38000	government

Let's consider each variable in this data set and determine whether it is numeric or categorical. (Cover up the answers below if you want to try it yourself first!)

ID Number: Even though these are "numbers" in one sense, this variable is used simply to uniquely identify the graduates without using their names. Since the values do not represent measurements or counts, this is not a numeric variable. We could say it's categorical but because each graduate would have their own category, it's not a typical categorical variable either. Very often, unique identifiers such as ID numbers or names are used as neither numeric nor categorical variables. What makes these columns very useful is that we can use them to merge together data from difference sources. If we had another data set having different variables on these same graduates, it would be easy to combine the data for each graduate by adding more columns to the data set. Without a unique identifier, knowing which data goes with which graduate would be difficult. (Using names as identifiers is problematic for at least two reasons: 1) there can be two people with the same name and 2) there are privacy concerns.)

GPA: This is a <u>numeric</u> variable since it is a measure of academic achievement.

Honors: This is a <u>categorical</u> variable as it allows us to group the graduates together in several different honors categories. As you may know, these honors are often based on GPA. In this data set, graduates with GPAs of 3.9 or higher graduate with "summa cum laude" honors; those with GPAs between 3.7 and 3.9 graduate with "magna cum laude" honors; and those with GPAs between 3.5 and 3.7 graduate with "cum laude" honors. No honors are conferred upon those with GPAs below 3.5.

While not numeric, there is a natural ordering here with "summa cum laude" being the highest and "none" being the lowest. Categorical variables having this ordered property are further classified as ordinal. Categorical variables that do not have this ordered property (like sector in this data set) are further classified as nominal.

JobGrad: This variable indicates whether the graduate had a full-time job lined up by their graduation date (1 = yes; 0 = no). You may be tempted to classify JobGrad as numeric but resist the temptation! The 1s and 0s simply indicate groups or categories; they do not measure or count anything, so JobGrad is a <u>categorical</u> variable. By the way, using 1 and 0 to represent "yes" and "no" is a very common practice.

Starting Salary: Of course this is a <u>numeric</u> variable, as is any dollar amount. Notice that there are no commas used as the thousands separators as we often do when writing dollar amounts. When you start entering data in row/column format, it's a good idea to not use commas in numeric data.

Sector: This <u>categorical</u> variable shows into which sector of the economy the graduate's job is classified.

One mistake novice data scientists make is treating categorical data represented with digits (like ID number and JobGrad) as numeric, which brings me to the reason why all of this numeric/categorical stuff is important. We use one set of statistical tools to summarize numeric data and another set to summarize categorical data. Using a tool designed for numeric data to summarize categorical data won't work very well, just like using a hammer to drive a wood screw (instead of a screw driver) won't work very well. Using the right tool for the job is crucial. The first two sections in the next chapter illustrate the basic tools used to extract information from numeric and categorical data. (I know you just can't put this book down, but do it anyway and take a break before you start the next chapter.)

Chapter 1 Exercises

1. Determine whether the variable is numeric or categorical.

    a. percent of flights that are on-time in a month
    b. company sales (in dollars) in a quarter
    c. flight status (on-time or delayed)
    d. primary paint color of a car
    e. number of red cars in a parking lot

2. A college admissions office collects data on applicants in a large database. Shown here is a portion of the data set. Students' names, ID numbers, application dates, high school GPAs, highest SAT score, whether or not the student is a National Merit Scholar (0 = no, 1 = yes), and primary extra-curricular activity (0 = none, 1 = sports, 2 = instrumental music, 3 = choral music, 4 = theater, 5 = community service, etc.) are recorded.

Name	ID	ApplDate	HS GPA	SAT	NMS	ExtraCurr
John Doe	123456789	11/28	3.82	1160	0	1
Juanita Brown	234567890	12/04	3.92	1390	0	3
...	...	...	...	...	...	...
Julia Patel	901234567	2/14	4.00	1580	1	5

    a. What are the objects/subjects in this data set?
    b. List the variables and their types (numeric or categorical).
    c. Suppose the ExtraCurr variable is defined as the number of extra-curricular activities in which the student participates instead of the primary extra-curricular activity. Would ExtraCurr then be numeric or categorical?

Exercises 3 – 9: For each data description, identify the objects/subjects in the data set, the variables, and their types (numeric or categorical).

3. An appliance store takes inventory of its refrigerators. An employee walks through the store and back room and for each fridge writes down its make, model, price, size (cubic feet), EPA estimated annual energy cost (in dollars), and freezer type (top, bottom, side).

4. Dairy farmers have gone high-tech. In addition to ear tags identifying each cow in a herd, cows today are fitted with radio-frequency identification (RFID) tags. Together with software, each cow's milk production per day (in pounds) can be easily recorded and monitored.

5. Zillow.com provides data on properties currently for sale. A property listing shows the asking price, an estimate of the property's value, the size of the house (in square feet), number of bedrooms, number of bathrooms, heating type (oil, gas, heat pump, etc.), whether or not it has central air conditioning, and lot size (acres).

6. Many financial advisors strongly urge clients to participate in their company's 401(k) or 403(b) retirements plans. The reason is that by contributing a small percent of their paycheck to their own retirement account, the employer will often match a percentage of that contribution. It's free money! One company collected data on each employee in an effort to determine the participation rate in its 401(k) plan and whether males and females participate at different rates.

7. Trisomy 18 is a genetic abnormality where a baby has an extra copy of chromosome 18 in some or all cells in the body. (Down's syndrome occurs when a baby has an extra copy of chromosome 21). According to the Trisomy 18 Foundation, trisomy 18 occurs in about 1 in every 2500 pregnancies.[4] Because trisomy 18 causes serious developmental abnormalities, babies with trisomy 18 rarely live to their first birthday. A hospital keeps record of Trisomy 18 status as well as Down's syndrome status for each baby born.

8. A university wants to compare the number of burglaries reported on its campus last year to colleges and universities in its peer group. The university's office of institutional research compiles a list of the number of burglaries reported at each college/university.

9. In 1929 Edwin Hubble discovered that galaxies further away from us (in light years) are generally moving faster away (in km/sec) than galaxies closer to us.[5] This discovery along with others like it are generally regarded as evidence that the universe is expanding and that it had a beginning.

Exercises 10 – 12: For each data summary, identify the objects/subjects in the data set, the variables, and their types (numeric or categorical).

10. A survey of internet dating use was conducted using a random sample of 1957 U.S. adults. Among the results reported was that only 2.9% of the adults 65 and older in the sample had ever used internet dating while 26.8% of those 18 – 24 years old had ever used internet dating.[6]

11. In an asthma study), two treatments for asthma were compared. Patients were randomly-assigned to one of three treatment groups (salmeterol, triamcinolone, or placebo). After randomization but before the treatment period began, the average peak expiratory flow rates were 446.5 L/min, 443.9 L/min, and 459.4 L/min for the placebo, salmeterol, and triamcinolone groups, respectively.[7]

12. In a study of firefighting tasks aboard a Navy ship, 20 male and 20 female Navy personnel participated in a task where they were to attempt to start a pull-start gas-powered water pump (NOSC TR 818, Naval Architectural Research for Women Aboard Ship, RL Pepper and MD Phillips, Unclassified, Sept 1982). The average force for the female group was 12.25 foot-pounds and 19.60 foot-pounds for the male group.

13. The fraction of all adults in the U.S. who approve of current President's job performance is frequently estimated by public opinion polls. In one such poll, 46% of a random sample of 1500 U.S. adults approved of the President's job. Determine whether each of the statements below is true or false. If false, give a reason why.

    a. The subjects are the U.S. Presidents.
    b. The variable is the percent who approve of the current President's job performance.
    c. The variable is categorical.

Notes

[1] Webster's Encyclopedic Unabridged Dictionary of the English Language (1989), dilithium Press, Ltd., New York.

[2] Old Faithful Visitor Center log books (www.geyserstudy.org/ofvclogs.aspx), transcribed by Lynn Stephens. Many thanks to the park rangers, volunteers, and Lynn for collecting and compiling these data.

[3] U.S. Department of Education College Scorecard (as of 5/21/2019).

[4] www.trisomy18.org.

[5] Hubble, E. (1929). "A Relation Between Distance and Radial Velocity Among Extra-Galactic Nebulae," *Proceedings of the National Academy of Sciences*, 15 (3), 168 – 173.

[6] Based on a survey conducted June 10 – July 12, 2015, Pew Research Center.

[7] Lazarus, S. C., Boushey, H. A., et al. (2001). "Long-Acting β_2-Agonist Monotherapy vs Continued Therapy with Inhaled Corticosteroids in Patients with Persistent Asthma: A Randomized Controlled Trial," *Journal of the American Medical Association*, 285 (20), 2583 – 2593.

Chapter 2: Summarizing Data

2.1 Summarizing One Categorical Variable

In this section, we'll look at a few basic tools used to summarize data obtained on one categorical variable. Since categorical data consist of category or group labels, the first step is to start counting. We'd like to know how many objects are in each category and how those numbers compare.

Example: Here are the highest degree data for the top 20 institutions by tuition.[1] Even for such a small data set it takes a few seconds to see that the most common highest degree granted by these institutions is a graduate degree. It is much more useful to extract that information and present it to the reader in the form of a **frequency table** as shown below.

Highest Degree	Count	%
Associate's	1	5
Bachelor's	6	30
Graduate	13	65
Total	20	100

The frequency table simply lists all of the degrees found in the data set and shows how many of each there are. The percent (%) column shows the frequency (Count) as a percentage of the total. For example, the 8 companies in the banking industry are $\frac{8}{20} * 100 = 40\%$ of the 20 companies, at least twice as many as in any other industry.

Institution	Highest Degree
Aviator College of Aeronautical Science and Technology	Associate's
Columbia University in the City of New York	Graduate
University of Chicago	Graduate
Vassar College	Bachelor's
Trinity College	Graduate
Landmark College	Bachelor's
Jewish Theological Seminary of America	Graduate
Harvey Mudd College	Bachelor's
Franklin and Marshall College	Bachelor's
Tufts University	Graduate
Amherst College	Bachelor's
University of Southern California	Graduate
Reed College	Graduate
Sarah Lawrence College	Graduate
Bucknell University	Graduate
Colgate University	Graduate
Carnegie Mellon University	Graduate
Kenyon College	Bachelor's
Williams College	Graduate
Brandeis University	Graduate

Resist the urge to try to generalize these figures and make statements like, "About 5% of institutions of higher education grant Associate's degrees" or "Over twice as many institutions of higher education grant graduate degrees compared to Bachelor's degrees." The data we have here are not a representative sample from all institutions of higher education; they certainly were not randomly obtained! In fact, they are the top 20 in a ranking based on tuition, a very special group indeed. The summaries we obtain therefore are not to be generalized to any larger group of institutions.

Once we have the frequency table, it is natural to represent the percentages visually in either a **bar chart** or **pie chart**. In the bar chart shown here, the height of the bar over each category shows the percent in that category. The categories are arranged in alphabetical order, but the ordering is not necessarily that important. The chart makes it very easy to see how frequent each degree is relative to the other degrees.

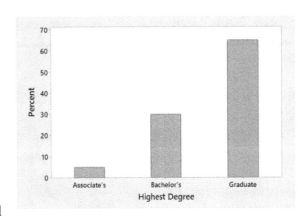

In the pie chart here, the percentages are displayed as wedges of a disk or "slices of a pie." To determine how big each slice is, we simply take the corresponding percentage out of 360°. For example, the Associate's slice takes up 5% of 360° or 0.05(360°) = 18°. The Bachelor's slice takes up the next 0.30(360°) = 108°, and the Graduate slice takes up the remaining 0.65(360°) = 234°.

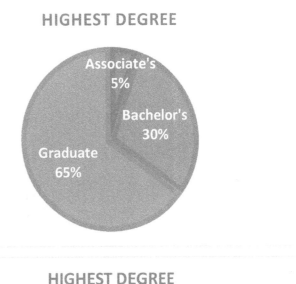

One note about pie charts: I don't like three-dimensional (3D) ones. In every 3D pie chart I've seen, the thickness of the pie chart doesn't represent anything meaningful and serves only to introduce perspective which can distort the relative sizes of the slices. Here is the same pie chart in 3D.

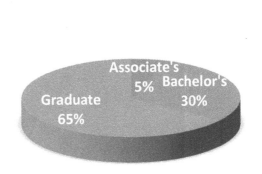

Notice how the slices that appear closer also appear larger than the ones that appear farther away. The Graduate slice appears larger and the Bachelor's slice appears smaller in the 3D chart compared to the 2D one. Of course you might say, "I wouldn't be fooled because I'd just look at the actual numbers shown." Good for you! The problem is that in some pie charts, the actual percentages are not shown and you have to visually compare the slices. As you can see, in a 3D graph, that visual comparison can be compromised by the unnecessary perspective. *A good graph should convey the information in the data accurately, without distortion, and within a fraction of a*

second. While details such as the percent labels are very helpful, the reader shouldn't have to look at those too much to decipher the message. An accurate message should be apparent almost immediately.

Example: Consider the data set of all U.S. colleges and universities.[2] Are these schools equally spread out across the states or are there some states that are home to more schools than others? Since this question concerns the categorical variable State, we can use a frequency table to extract that information from the data and use a bar chart to visually display the information.

Here is a bar chart of the numbers of schools in each U.S. state and territory. The states are ordered by the frequency so it's very easy to visually rank them. This type of bar chart where the categories are ordered from highest to lowest frequency is called a **Pareto diagram** named for Vilfredo Pareto (1848 – 1923), an Italian economist.

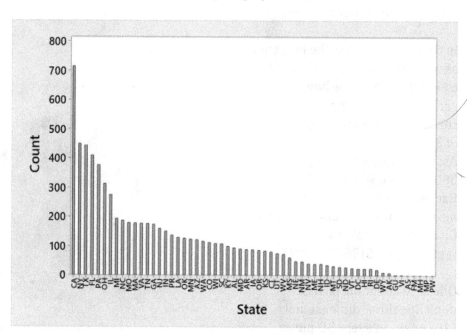

The bar chart easily extracts for us the information that there are substantially more schools in California than in the other states. Perhaps that's not so surprising given that California is the most populous state.

The pie chart also shows California as home to substantially more top schools than the other states, but because there are so many states, many of the slices are very small and the labels so close together that the names are not easily legible. For this reason, bar charts are preferred over pie charts when there are many categories.

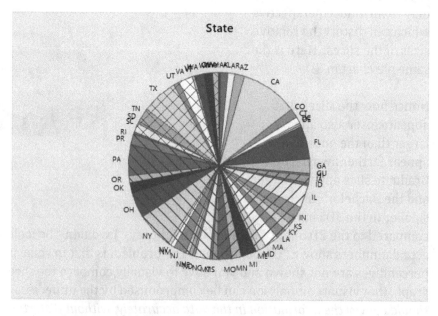

Section 2.1 Exercises

1. The data set below shows the marital status for each of a sample of people (S = single, M = married, P = separated, D = divorced, W = widowed).

S	S	M	S	P	M	M	M	D	D	W
S	P	M	M	S	S	M	D	S	S	S
P	D	S	S	M	S	S	S	S	M	M
M	S	S	D	P	P	D				

    a. Make a frequency table.
    b. What percent of these people are unmarried (single, divorced, or widowed)?
    c. Make a bar chart.
    d. Make a pie chart.

2. Refer to exercise 2 in chapter 1 showing data for college applicants. For which of the variables in that data set would it be appropriate to make a bar chart?

3. According to the Centers for Disease Control there were 2,744,248 deaths in the U.S. in 2016. The frequency table below shows a summary of the data.[3]

Cause of Death	Count
Heart disease	635,260
Cancer	598,038
Accident	161,374
Chronic lower respiratory disease	154,596
Cerebrovascular disease	142,142
Alzheimer's disease	116,103
Diabetes	80,058
Influenza and pneumonia	51,537
Nephritis, nephrotic syndrome and nephrosis	50,046
Suicide	44,965
Other	???

    a. Identify the objects/subjects and the variable represented by the frequency table.
    b. What would the data set look like?
    c. What percent of deaths in 2016 were from
        i.   heart disease?
        ii. heart disease and cancer combined?
        iii. causes other than the top 10 shown?
    d. What proportion of deaths from both heart disease and cancer were from cancer?
    e. Create an appropriate graph.

4. In an August 2017 study of TV-watching, the Pew Research Center surveyed a random sample of 1,893 U.S. adults. One question they asked was, "How do you usually watch television these days?" The pie chart shown here displays the results.[4]

    a. A few adults reported that they watch TV some other way or that they don't watch TV. How many adults were like this?
    b. Make a frequency table that includes the frequency of each category. Note: Frequencies must be whole numbers.
    c. Why would a 3-dimensional pie chart not be the best representation of these data?

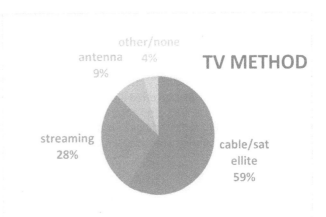

5. In a 2001 study of treatments for asthma published in the Journal of the American Medical Association, 164 volunteer patients that met the entrance criteria were randomly assigned to one of three treatments (salmeterol, triamcinolone, or placebo). A partial frequency table summarizing one of the variables is shown here.[5]

Ethnicity	Count	%
White	115	
Asian or Pacific Islander		
Black	24	
Hispanic	13	
Other	3	
Total		

   a. Complete the frequency table.
   b. Identify the subjects/objects represented by the frequency table.
   c. Identify the variable represented by the frequency table.

6. Mutual funds are convenient ways for individuals to invest money. Fidelity Investments is a large investment company that gives individuals access to thousands of mutual funds. One way to classify a mutual fund is by what kinds of assets it invests in; two popular asset classes are stocks (equities) and bonds. At the time I'm writing this problem, there are 2,977 U.S. equity funds, 1,980 taxable bond funds, 1,623 asset allocation funds, 1,507 international equity funds, 909 municipal bond funds, 675 sector equity funds, and 772 other types of funds available at Fidelity Investments.[6]

Asset Class	Count	%
U.S. equity	2977	28.5
Taxable bond	1980	19.0
Asset allocation	1623	0.155
International equity	1507	14.4
Municipal bond	909	8.7
Sector equity	675	6.5
Other	772	7.3925
Total	10,443	100.0

   a. Identify the subjects/objects and the variable represented by the frequency table.
   b. What would the data set look like?
   c. Find two "mistakes" in the frequency table.
   d. Consider the bar chart shown here. Is it true that there are over twice as many municipal bond funds as sector equity funds? Why or why not?

7. The status of each aircraft during February 2019 was observed and summarized by the U.S. Department of Transportation, Bureau of Transportation Statistics.[7] The frequency table shown here summarizes the data.

Arrival Status	Count	%
On time	396,837	74.4
Air carrier delay	31,505	5.9
Weather delay	4,356	0.8
National Aviation System delay	40,374	7.6
Security delay	220	0.04
Aircraft arriving late	43,022	8.1
Cancelled	15,255	2.9
Diverted	1,606	0.3
Total	533,175	

   a. Identify the subjects/objects and the variable represented by the frequency table.
   b. Add up the percentages. Why don't you get exactly 100.0%?
   c. Calculate the percentages to 2 decimal places. Then add them up.
   d. Make another frequency table for "reason for not on time" summarizing the data for only the delayed/cancelled/diverted flights. Your percents should add to 100.

8. In a survey conducted by Schoen Consulting, 1,350 U.S. adults were randomly-selected and interviewed about the Holocaust. Responses to the question, "Have you ever seen or heard the word Holocaust before?" are tabulated here.[8]

Response	Count	%
Yes, I have definitely heard about the Holocaust		89
Yes, I think I've heard about the Holocaust		
No, I don't think I've heard about the Holocaust		3
No, I definitely have not heard about the Holocaust		1
Total		

a. Identify the subjects/objects and the variable represented by the frequency table.
b. Complete the frequency table.
c. Make a graph.
d. How many people surveyed either didn't think they'd heard about or definitely had not heard about the Holocaust?

2.2 Summarizing One Numeric Variable

Because numeric data are either measurements or counts and not categories or group indicators, we need to use different methods to summarize them. The relevant information we want to extract from numeric variables is how the data values are distributed across the range of possibilities.

Example: Here are 28 randomly selected heights (inches) out of 202 students who reported their height on a survey at the beginning of a statistics course.

62.0 62.5 63.0 64.0 64.0 65.0 65.0 65.0 65.0 65.0 66.0 67.0 67.0
67.0 67.0 67.0 68.0 68.0 68.0 69.0 69.0 69.0 70.0 71.0 72.0 72.0
72.0 76.0

Note that in the case of data on a single variable, we can list the values like this instead of in a column to save space. Who knows, it might have saved you a few cents on this book! I've put the heights in order from low to high as well.

Some questions we might have about these heights are:
- What is a typical height?
- How do heights vary around the typical height?
- Are there any unusually high or low ones?

We can answer these questions fairly easily by constructing a dot plot. A **dot plot** is a graph of a set of numeric data values where each dot represents a data value placed on a number line. If data values are repeated, the dots are stacked on top of each other. The dot plot for the random sample of heights is shown below.

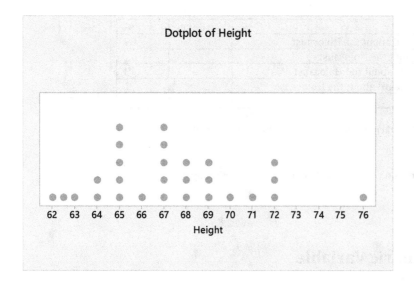

The graph allows us to extract some interesting information from these data that were not readily apparent in the list of numbers. We see that a typical height in this sample is somewhere around 67" and that there is a somewhat unusually tall student (compared to the rest) at 76". The concentration of heights is greatest around 67" (from about 65" to 69") indicated by the higher columns of dots and then tapers off on either side of that so that there aren't that many short or tall students.

Since these 28 heights are a random sample from the 202 heights, we know that they are representative of the 202 heights in the "fairness" sense. That is, the random sample of 28 allowed each student the same chance of being part of the sample. No groups of students (males, females, basketball players, racehorse jockeys, etc.) were over- or under-represented by the sampling process. Even so, we might wonder how close these heights as a group would be to the group of 202 heights. How do we know that we didn't pick the taller students by accident, that is, just by chance? The answer to this question will come later, so be patient.

One of the disadvantages of dotplots is that they do not work well for large data sets; we soon run out of space at the top if there are too many data values repeated. Dotplots also don't work well for data sets that don't have many repeated values. In those cases, the dots may not stack up much at all making it harder to see where the data tend to cluster. A graph that avoids these problems is the histogram. A **histogram** shows the frequency of data values in each of several intervals on the number line. Here's how it works:

- Break up the range of the data values between the lowest and highest value into several intervals of equal width. One rule of thumb is to use between 5 and 20 intervals.
- Count how many data values are in each interval.
- Use a bar above each interval to represent the count (or %) in the interval.

Example: Here are the 28 randomly selected heights (inches) again.

62.0	62.5	63.0	64.0	64.0	65.0	65.0	65.0	65.0	65.0	66.0	67.0	67.0
67.0	67.0	67.0	68.0	68.0	68.0	69.0	69.0	69.0	70.0	71.0	72.0	72.0
72.0	76.0											

Since the heights range from 62.0" to 76.0", let's define intervals 2" wide starting at 62" and ending at 76". The lower endpoint of each interval is included but not the upper

Height Interval	Count	%
[62,64)	3	10.7
[64,66)	7	25.0
[66,68)	6	21.4
[68,70)	6	21.4
[70,72)	2	7.1
[72,74)	3	10.7
[74,76)	0	0.0
[76,78)	1	3.6
Total	28	100.0

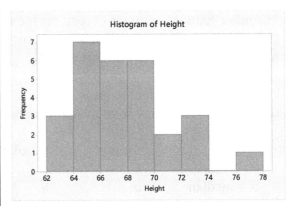

endpoint. Note that there are no gaps between the intervals; we don't want anyone's height to fall through the cracks! There are 3 heights in the interval [62,64), 7 heights in the interval [64,66), etc. The histogram is a graphical representation of the table.

The histogram shows some of the same features as the dotplot, including the somewhat unusually taller student. One difference is that now we don't know that student's height exactly, just that it's somewhere between 76" and 78".

Remember the questions we might have about a distribution of data values?
- What is a typical value?
- How do the values vary around the typical one?
- Are there any unusually high or low ones?

Dotplots and histograms can give us graphical answers to these questions, but we often want more formal, numeric answers. In the next few pages, we'll look at some very common numeric summaries that help to answer these questions about a distribution of numeric data values.

Measures of Center

Measures of center can help to answer the question, "What is a typical value?" They help us to locate the center of the distribution. Two very common measures of center are the mean and the median.

Suppose we denote a set of n numeric data values as $x_1, x_2, \ldots, x_n$. The **mean** (or average) is just the sum of the data values divided by n:

$$mean = \frac{x_1 + x_2 + x_3 + \cdots + x_n}{n}$$

It is also the "balance point" in the distribution. If you imagine the histogram or dotplot as a seesaw and where the dots or bars were made of some solid material so that taller stacks of dots or taller bars were heavier, the mean is where you would put the fulcrum of the seesaw to balance it.

The other common measure of center is the **median**, defined as the middle value in the ordered list. If there is an even number of data values such that there is no middle value, the median is defined to be the mean of the two middle values.

The notation we use for the mean and median depends upon whether the data were obtained from the entire population or from a sample from the population:

	Sample	Population
mean	$\bar{x}$	μ
median	m	η

In many cases, sample summaries are denoted with Arabic letters and population summaries are denoted with Greek letters.

The following example illustrates these two measures of center and highlights their differences.

Example: In a survey at the beginning of a statistics course, students reported how fast they've ever driven a car (in miles per hour). Here are those fastest speeds for 30 randomly selected students followed by a histogram.

90	88	85	92	100	70	80	85	90	100
110	70	90	90	90	70	80	90	110	90
85	90	90	110	160	70	115	90	20000	90

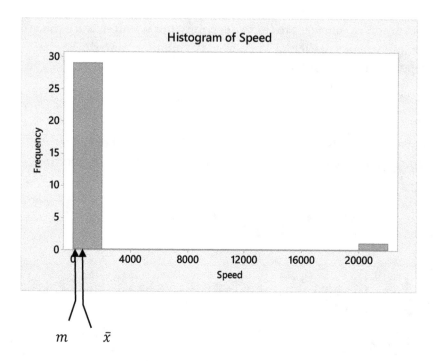

The mean fastest speed is

$$\bar{x} = \frac{90+88+85+\cdots+90}{30} = 755.\bar{6} \ mph.$$

Note that the mean is very much influenced by the impossibly high fastest speed of 20,000 mph. (I think that student just wanted the attention but they made my point for me!)

To get the median we must first put the speeds in order from low to high:

70	70	70	70	80	80	85	85	85	88
90	90	90	90	<u>90</u>	<u>90</u>	90	90	90	90
90	92	100	100	110	110	110	115	160	20000

There are 30 speeds so the median is the mean of the 15th and 16th speeds. They are underlined in the list above:

$$m = \frac{90+90}{2} = 90 \ mph.$$

This is a much more reasonable value to report as typical because it is not influenced by the 20,000 mph value. In fact, the median would have been 90 mph no matter how high the highest number (or how low the lowest number) would have been. The median is much more resistant to unusual values than is the mean.

Measures of Variability

Variability exists in any real data set; the values are not all the same! Quantifying the amount of variability in a data set is one of the "bread and butter" tools of statistics. In this section, we'll focus on how to do just that.

Example: In a survey at the beginning of a statistics course, students responded to the question, "How many people do you think live in Canada?" There were two forms of this question depending upon which of two statements preceded the question. Students saw either, "The U.S. population is about 319 million," or "The Australian population is about 23 million." Students were shown one of these two statements at random. Here are the responses (in millions) for 30 randomly selected students grouped by which statement they saw first (US or AUS).

US	1.5	200	500	200	275	300	250	30	200	150
	57	250	150	10						

AUS	20	245	35	50	20	20	20	10	24	15
	10	100	50	35	28					

At a quick glance it's pretty easy to see that the responses for students who got the US form of the question tended to give higher responses than students who got the Australia form. What is not so clear is that the US responses also varied much more. Here are histograms of the responses on the same scale.

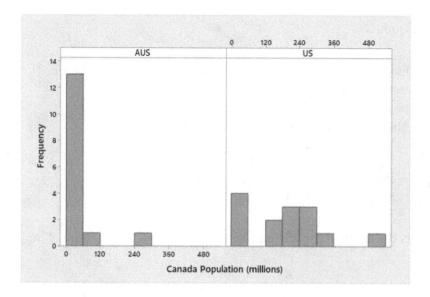

Notice the obviously higher center of the US distribution of responses. Notice also that the US responses are spread out more around their center compared to the AUS responses which are "clumped up" more around their center. This histogram shows this higher variability with lower frequencies of values in more places along the horizontal axis for the US responses. Over half of the AUS responses are in the lowest interval, a concentration that we don't see in any of the intervals in the US histogram.

One simple measure of variability is the **range**, the difference between the largest and smallest values. The ranges of Canada responses for both forms are

US: $range = 500 - 1.5 = 498.5$
AUS: $range = 245 - 10 = 235$

You can think of the range as measuring the length of the "footprint" that the distribution leaves on the horizontal axis. The range of US responses is over twice as big as that for the AUS responses and therefore it picks up on the higher degree of variability in the US distribution.

The problem with the range is that it is defined by the extremes and ignores what is happening in the rest of the distribution. A more sensitive measure of variability is the standard deviation, a very useful value in statistics. The **standard deviation** is a measure of the *average distance the data values are from their mean*. This definition provides a clue as to how we compute the standard deviation. For data obtained from the entire population of interest, the standard deviation is

$$\sigma = \sqrt{\frac{(x_1-\mu)^2+(x_2-\mu)^2+(x_3-\mu)^2+\cdots+(x_N-\mu)^2}{N}}$$

where N is the size of the population. The differences $(x_i - \mu)$ measure how far each data value is from the mean. These differences are the "deviations." Since the mean is the balance point in a distribution, the sum of these differences will always be 0, so we square the differences before we add them; in terms of measuring variability, we don't care whether a distance is positive or negative, that is, whether a data value is above or below the mean. Dividing by how many distances we have gives an "average squared distance" and then taking the square root gets us back to an "average distance."

For data obtained from a sample, the standard deviation is computed a bit differently:

$$s = \sqrt{\frac{(x_1-\bar{x})^2+(x_2-\bar{x})^2+(x_3-\bar{x})^2+\cdots+(x_n-\bar{x})^2}{n-1}}$$

where n is the sample size and we use the sample mean $\bar{x}$ instead of the population mean μ. The biggest difference is in the denominator where we divide by $n - 1$, which is a bit unusual since we're computing an "average distance." When have you ever computed an average by "adding them up and dividing by *one less than how many*?" The short answer at this point is that using $n - 1$ makes the sample standard deviation s a slightly better estimate of the population standard deviation σ if your purpose is generalization. The reason? Data from a sample are almost always closer to their own mean ($\bar{x}$) than the mean of the whole population (μ). Using n in the denominator of s tends to underestimate σ, so if that's your purpose, you'd tend to be a little low in your estimate and think that the population data are a bit less variable than they really are.

Example: Consider the sample of Canada responses grouped by form.

US	1.5	200	500	200	275	300	250	30	200	150
	57	250	150	10						

AUS	20	245	35	50	20	20	20	10	24	15
	10	100	50	35	28					

Let's calculate and interpret the standard deviations for both groups of responses. We'll call them s_{US} and s_{AUS}. First find the sample mean for each group, $\bar{x}_{US}$ and $\bar{x}_{AUS}$.

$$\bar{x}_{US} = \frac{1.5+200+500+\cdots+10}{14} = \frac{2573.5}{14} \approx 183.8214286 \quad \text{(don't round this off too much!)}$$

$$\bar{x}_{AUS} = \frac{20+245+35+\cdots+28}{15} = \frac{682}{15} = 45.4\bar{6}$$

Then plug these in to the standard deviation formula to find the average distance from the mean for each distribution:

$$s_{US} = \sqrt{\frac{(1.5-183.8214286)^2+(200-183.8214286)^2+(500-183.8214286)^2+\cdots+(10-183.8214286)^2}{14-1}} \approx 135.0$$

$$s_{AUS} = \sqrt{\frac{(20-45.4\bar{6})^2+(245-45.4\bar{6})^2+(35-45.4\bar{6})^2+\cdots+(28-45.4\bar{6})^2}{15-1}} \approx 59.7$$

The average distance from the mean in the US distribution of Canada responses is about 135.0 million; some values are closer to the mean than that (150 and 200, for example) and some are farther from the mean than that (30, for example). Compare 135.0 million to the average distance from the mean in the AUS distribution of Canada responses, 59.7 million, and you can see the much smaller variability there.

A close cousin to the standard deviation is the **variance**, the average squared distance from the mean, calculated as $(standard\ deviation)^2$. In this example, the two sample variances are

$$s_{US}^2 = 135.0^2 \approx 18,216.3$$

$$s_{AU}^2 = 59.7^2 \approx 3,560.8$$

Although we use the variance in many kinds of calculations, the number itself is difficult to interpret because the units of variance are the square of the units of the data values. Do you know what "18,216.3 square millions" means? I don't.

Here is a summary of notation differences for standard deviation and variance.

	Sample	Population
standard deviation	s	σ
variance	s^2	σ^2

Before we leave this subsection, look back again at the differences in the distributions of Canada responses between the two forms (US and AUS). Are you wondering why students got different versions of the question? Since students were assigned the version of the question at random, we can attribute any differences we see to either chance or the different versions themselves. Of course it is possible that the different "anchoring" numbers used in the two forms had nothing to do with students' responses. It is possible that those who would have guessed a higher number anyway just happened to be assigned to the US form and those who would have guessed a lower number anyway just happened to be assigned to the AUS form (although it turns out the probability of that happening is very low). It is also possible that the difference in the anchoring numbers was responsible for the generally higher and more variable responses in the US group and generally lower and less variable responses in the AUS group. The punch line is that it is not a good idea to provide any kind of extra information that may influence a person's response to a survey question. In a jury trial, that would be called "leading the witness" and would not be allowed by the judge.

By the way, in 2014 the population of Canada was about 35.5 million.[9]

Shape

Shown below are two samples of data. Here are a few summaries:

	$\bar{x}$	m	s
Sample A	5.01	4.88	5.03
Sample B	4.96	3.34	5.08

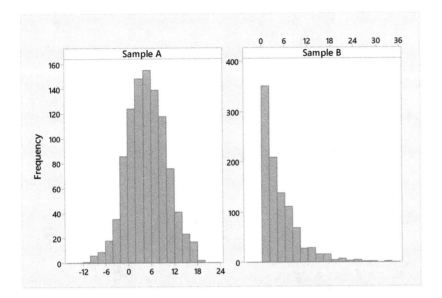

You'll notice that the means and standard deviations are quite similar for both samples while the median for sample A is a good bit higher than the median for sample B. The most striking difference between these two distributions, however, is shown in the histograms by where the values cluster. In sample A, the higher concentrations of values are in the middle and then the frequencies taper off fairly symmetrically toward the extremes. In sample B, values are concentrated more heavily on the low extreme and then taper off asymmetrically toward the high end. Another way to describe the difference is with respect to where the "peaks" are (places of highest frequencies) in the distributions. The peak in sample A is around 5, very close to the mean, but in sample B, the peak is in the interval [0,2), on the low extreme.

The difference we've been discussing is neither a difference in center nor a difference in variability, but a difference in *shape*. The **shape** of a distribution is a description of the visual impression we get from a histogram or dotplot.

Shown below are several important shapes. The table gives a description of each shape and an example of a variable with that shape that corresponds to the histogram.

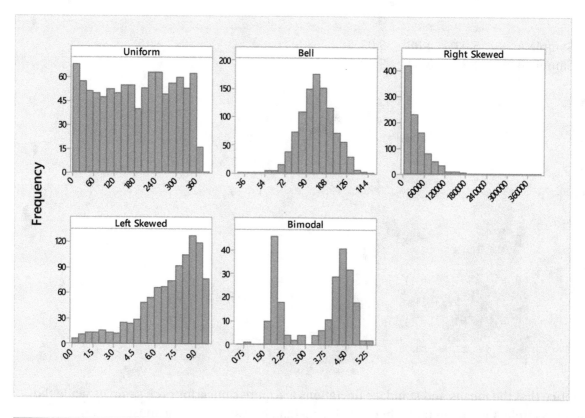

Shape	Description	Example
Uniform	no clear peak in the distribution; values do not tend to cluster anywhere	day of the year of a person's birth (1 – 365)
Bell	clear peak in the middle of the distribution; values become less frequent toward each extreme in a fairly symmetric way	intelligence quotient (IQ) scores
Right Skewed	clear peak on the low end of the distribution; values become less frequent toward the high end	personal incomes
Left Skewed	clear peak on the high end of the distribution; values become less frequent toward the low end	scores on a fairly easy quiz out of 10 points
Bimodal	two clear peaks in the distribution	duration times (minutes) of Old Faithful geyser

Let's look at some of the examples more closely. In distributions that are symmetric, the mean (balance point) and median (middle number) will be fairly close to each other. In the distribution of IQ scores shown here, the mean and median are

$$mean = 99.9$$
$$median = 99.6.$$

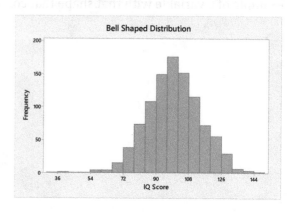

Values on the high end tend to be well-balanced by values on the low end and so their effects on where the balance point is cancel out.

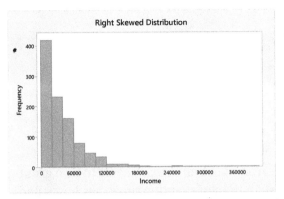

In a right skewed distribution, there are no values on the low end that are low enough to balance out the leverage of the values on the high end and so the balance point must generally be higher than the middle number. In this distribution of incomes, for example, the mean and median are

$$mean = \$37{,}136$$
$$median = \$26{,}789.$$

In a left skewed distribution, there are no values on the high end that are high enough to balance out the leverage of the values on the low end and so the balance point must generally be lower than the middle number. In this distribution of quiz scores, for example, the mean and median are

$$mean = 7.0$$
$$median = 7.6.$$

Here is a distribution of eruption times (in minutes) of Old Faithful, collected in August of 1985. The mean and median are

$$mean = 3.5$$
$$median = 4.1$$

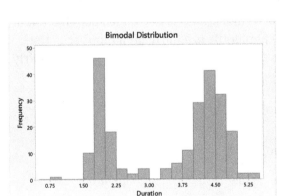

but neither one does a good job of identifying a typical duration time. This is because the distribution's shape is *bimodal*. A **bimodal** distribution has two peaks in the distribution; it has two places where there is a high frequency of values. (The technical definition of a *mode* is *the most frequent data value* but a more useful definition is *a location of high frequency*.) Here we see there are many eruptions clustered around 2 minutes or so and another cluster around 4.5 minutes. Without this description of the distribution's shape, the numeric summaries of center and variability don't make nearly as much sense and could even be misleading!

Some food for thought: Compare the distribution of Old Faithful durations in August of 1985[10] to the one in late July/early August of 2007 (see the data shown at the beginning of this chapter). Both distributions are shown below on the same horizontal scale.

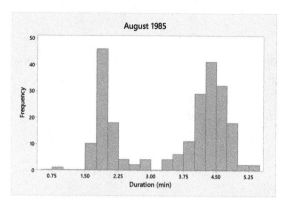

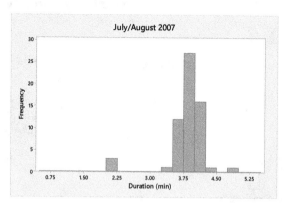

Notice that the bimodality in the distribution has disappeared; there no longer seems to be a group of shorter eruption times. Whether this change in the distribution of eruption times has happened slowly over time is difficult to say with just these two "snapshots" but it certainly begs further questions: Is this a real change? If so, why has the distribution changed? As you see here, sometimes extracting information from data to answer one question leads to several other questions.

Percentiles

So far we've learned about some ways to answer questions about the typical value in a distribution and questions about how much variability there is in a distribution. In order to answer questions about whether there are any unusually high or low values in the distribution, we need to clearly define what we mean by saying a value is "far from typical." The percentiles in a distribution of data are one way to do this. The pth **percentile** in a set of numeric data is the value such that about $p\%$ of the data values are below it. For example, the 90th percentile is a fairly high number in the distribution; about 90% of the values are less than it. On the other hand, the 20th percentile is fairly low; only about 20% of the values are less than it.

To find the pth percentile in a data set, follow these steps[11]:

- Put the data values in order from low to high.
- Find the address (k) of the percentile as $k = \frac{p}{100}(n + 1)$, where n is the number of values in the data set.
- The pth percentile is then the kth value in the ordered list. You may need to interpolate if k is not an integer.

Example: Consider the durations (in seconds this time) of the set of 61 Old Faithful eruptions shown at the beginning of this chapter. We'll find the 10th, 25th, 50th, and 75th percentiles to illustrate the method.

229	212	247	230	251	217	231	233	225	250	233	231	239
247	249	207	234	231	219	225	235	250	227	236	238	222
235	253	239	250	232	126	291	217	215	244	235	235	221
120	252	238	248	244	218	225	238	244	124	245	223	237

219   246   224   256   235   248   239   211   225

First, put the data values in order from low to high.

120	124	126	207	211	212	215	217	217	218	219	219	221
222	223	224	225	225	225	225	227	229	230	231	231	231
232	233	233	234	235	235	235	235	235	236	237	238	238
238	239	239	239	244	244	244	245	246	247	247	248	248
249	250	250	250	251	252	253	256	291				

The address of the 50th percentile is $k = \frac{50}{100}(61 + 1) = 31$. That means the 50th percentile is the 31st number in the ordered list: 235 seconds. Notice that we've just found the median! This approach matches how we find the median of a set of data but gives us a general approach to finding any percentile, not just the 50th.

The address of the 25th percentile is $k = \frac{25}{100}(61 + 1) = 15.5$. That means the 25th percentile is the 15.5th number in the ordered list. "But wait," you say, "there is no 15.5th number!" So let's interpolate between the 15th and 16th numbers:

> 15th number: 223
> 16th number: 224.

To interpolate, follow these steps:

- Find the distance (difference) between the two numbers.
- Multiply the distance by the fractional part of the address.
- Add the result to the lower of the two numbers.

Here, the distance between the two numbers is $224 - 223 = 1$. Then we multiply that distance by 0.5, the fractional part of the address of 15.5: $1(0.5) = 0.5$. Finally we add the result to the lower number: $223 + 0.5 = 223.5$ and take 223.5 as the 25th percentile.

The address of the 75th percentile is $k = \frac{75}{100}(61 + 1) = 46.5$. That means the 75th percentile is the 46.5th number in the ordered list, so we interpolate again:

> 46th number: 244
> 47th number: 245
> Distance between them: $245 - 244 = 1$
> Multiply by fractional part of address: $1(0.5) = 0.5$
> Add to lower number: $244 + 0.5 = 244.5$

Finally, the address of the 10th percentile is $k = \frac{10}{100}(61 + 1) = 6.2$. That means the 10th percentile is the 6.2nd number in the ordered list, so we interpolate again:

6th number: 212
7th number: 215
Distance between them: $215 - 212 = 3$
Multiply by fractional part of address: $3(0.2) = 0.6$
Add to lower number: $212 + 0.6 = 212.6$

Note where these percentiles are in the distribution of duration times:

10th percentile = 212.6 seconds
25th percentile = 223.5 seconds
50th percentile = 235 seconds
75th percentile = 245.5 seconds

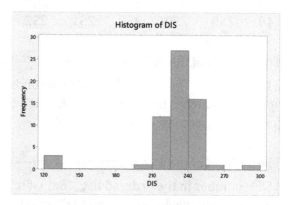

The 10th percentile is fairly low; only about 10% of the times are lower than 212.6 seconds. The 75th percentile is somewhat higher than typical; only about 25% of the times are higher than 245.5 seconds.

The 25th, 50th, and 75th percentiles are also known as the first quartile ($Q1$), second quartile ($Q2$), and third quartile ($Q3$), respectively; they divide the set of numbers into quarters where there are roughly the same number of values in each quarter.

We can use the quartiles for more than just to locate the middle number and the 25th and 75th percentiles. Another use is to measure variability. The middle 50% of the data values are located between $Q1$ and $Q3$; the difference is called the interquartile range (IQR):

$$IQR = Q3 - Q1$$

You can think of the IQR as the footprint of the middle 50% of the data values. The more variable the middle 50% of the data values are, the larger the distance between $Q1$ and $Q3$ and, hence, the larger the IQR.

Example: Consider the two distributions of Canada population guesses, one for students who got the U.S. form (US) and one for students who got the Australia form (AUS). We saw that the US distribution had more variability than the AUS distribution in terms of both the range and the standard deviation. Here are both data sets, sorted from low to high. Let's find the IQR for each.

US	1.5	10	30	57	150	150	200	200	200
	250	250	275	300	500				

AUS	10	10	15	20	20	20	20	24	28
	35	35	50	50	100	245			

US ($n = 14$)	AUS ($n = 15$)
Find $Q1$ (25th percentile): • Address $k = \frac{25}{100}(14 + 1) = 3.75$ • 3rd number = 30; 4th number = 57 Distance: $57 - 30 = 27$ Multiply: $27(0.75) = 20.25$ Add: $Q1 = 30 + 20.25 = 50.25$ Find $Q3$ (75th percentile): • Address $k = \frac{75}{100}(14 + 1) = 11.25$ • 11th number = 250; 12th number = 275 Distance: $275 - 250 = 25$ Multiply: $25(0.25) = 6.25$ Add: $Q3 = 250 + 6.25 = 256.25$ $IQR = Q3 - Q1 = 256.25 - 50.25 = 206$	Find $Q1$ (25th percentile): • Address $k = \frac{25}{100}(15 + 1) = 4$ • Find 4th number in the ordered list: $Q1 = 20$ Find $Q3$ (75th percentile): • Address $k = \frac{75}{100}(15 + 1) = 12$ • Find 12th number in the ordered list: $Q3 = 50$ $IQR = Q3 - Q1 = 50 - 20 = 30$

The interquartile range is more than six times as big for US than for AUS, meaning the middle 50% of the Canada values is much more spread out in the US group than in the AUS group. Here is a comparison of the measures of variability for these two groups.

	s	s^2	range	IQR
US	135.0	18,216.3	498.5	206
AUS	57.0	3,560.8	235.0	30

On every measure, the US group exhibits more variability in Canada guesses than the AUS group.

Z-scores

Another very convenient way to define clearly what we mean by "far from typical" is by using the value of the standard deviation as a ruler. The **z-score** for a data value is a number that measures how many standard deviations the value is from its mean.

$$z = \frac{value - mean}{SD}$$

If a data value has a z-score of 0, we know the data value is the same as the mean; it is "perfectly typical" in that sense. A data value that is above the mean will have a positive z-score; values below the mean have negative z-scores. The absolute value of the z-score can be a good indication of how far from typical a data value is. Values having z-scores that are large in absolute value are far from typical. Of course, if the mean is not a good measure of what is typical (in a bimodal distribution, for example), then the z-score would only tell us how many standard deviations a value is from the mean and not how far it is from a typical value.

Using the standard deviation in this way is important because knowing the raw "distance from the mean" doesn't necessarily tell us what we need to know. As we'll see in the next example, *the variability of the distribution determines what is far from typical and what is not.*

Example: Shown here are the distributions of the percent of faculty that are full-time for schools whose top degree is the Bachelor's and those whose top degree is a graduate degree. The data come from the set of top 20 schools ranked by tuition.[12]

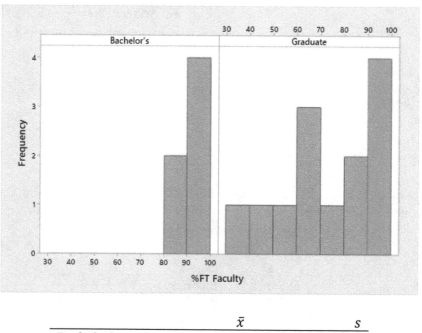

	$\bar{x}$	s
Bachelor's	92.2	5.34
Graduate	73.4	21.01

Since the full-time faculty rates for the Bachelor's schools are much less variable than those for the Graduate-degree schools, a Bachelor's school with 72.2% full-time faculty (20 percentage points below average) would look much more unusual than a school granting graduate degrees with 53.4% full-time faculty (also 20 percentage points below average). Even though both schools would have below-average full-time faculty rates by about the same amount, the smaller variability among Bachelor's schools rates would make that difference stand out much more than it would in the more variable distribution of rates among schools granting graduate degrees. The z-scores for each example are

Bachelor's school with 72.2% full-time faculty:

$$z = \frac{72.2 - 92.2}{5.34} = \frac{-20}{5.34} = -3.75$$

This school would have a full-time faculty rate that is 3.75 standard deviations below the mean.

Graduate degree school with 53.4% full-time faculty:

$$z = \frac{53.4 - 73.4}{21.01} = \frac{-20}{21.01} = -0.95$$

This school would have a full-time faculty rate that is 0.95 standard deviations below the mean.

Both z-scores are negative. That means both schools have below-average full-time faculty rates. But notice how much larger in absolute value the z-score is for the Bachelor's school's rate. That's telling us the 72.2% is much lower compared to the other Bachelor's schools' full-time rates than is the 53.4% compared to the other graduate degree schools' rates.

Outliers

Sometimes it is the unusually large or small values that are the interesting features in a data set. In the random sample of fastest speeds driven, the unusually large value of 20,000 mph stands out like a green flamingo! In some data sets, there are large or small values that do not stand out as prominently. In this section, we'll look at two more formal definitions of what is "unusually large" or "unusually small." Data values identified by these methods are called **outliers**. Both methods use an upper and a lower boundary, beyond which any data values would be flagged as outliers. The difference between the methods is how the boundaries are calculated.

One method locates the boundaries, called fences, as one and a half $IQRs$ from the first and third quartiles:

$$lower\ fence = Q1 - 1.5(IQR)$$
$$upper\ fence = Q3 + 1.5(IQR)$$

Data values above the upper fence or below the lower fence are considered outliers.

Example: Consider the durations (in seconds) of the set of 61 Old Faithful eruptions shown at the beginning of this chapter. Here they area again in order from lowest to highest.

120	124	126	207	211	212	215	217	217	218	219	219	221
222	223	224	225	225	225	225	227	229	230	231	231	231
232	233	233	234	235	235	235	235	235	236	237	238	238
238	239	239	239	244	244	244	245	246	247	247	248	248
249	250	250	250	251	252	253	256	291				

In a previous example we found

$$Q1 = 223.5$$
$$Q3 = 245.5$$

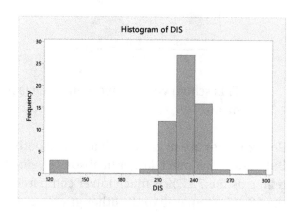

so the interquartile range is

$$IQR = Q3 - Q1 = 245.5 - 223.5 = 22.$$

One and a half $IQRs$ is 33 seconds so to find the lower and upper fences, locate the values that are 33 seconds below the first quartile and 33 seconds above the third quartile:

$$lower\ fence = Q1 - 1.5(IQR) = 223.5 - 1.5(22) = 223.5 - 33 = 190.5$$
$$upper\ fence = Q3 + 1.5(IQR) = 245.5 + 1.5(22) = 245.5 + 33 = 278.5.$$

Note that there is one duration that is higher than the upper fence (291) and there are three values below the lower fence (120, 124, 126). These four values would be considered outliers according to this method. Notice that they stand out in the histogram as well.

Another method to locate the boundaries that define what values are outliers uses the z-score concept. Remember that a z-score for a data value tells how many standard deviations a data value is from the mean.

Shown below are four of the distribution shapes you saw earlier: a uniform distribution of birthdays, a bell-shaped distribution of IQ scores, a right skewed distribution of incomes, and the bimodal distribution of Old Faithful eruption durations. The means and standard deviations for each data set are also shown below. The symbols μ and σ are used as the mean and standard deviation for the IQ scores, indicating that those scores were obtained from every person in the population of interest. The symbols $\bar{x}$ and s are used as the mean and standard deviation for the other variables, which means those scores were obtained from a sample of persons/eruptions.

Variable	Mean	SD
Birthday	$\bar{x} = 184.33$	$s = 108.11$
IQ Score	$\mu = 99.88$	$\sigma = 14.51$
Income	$\bar{x} = 37,136$	$s = 37,300$
Duration	$\bar{x} = 3.48$	$s = 1.21$

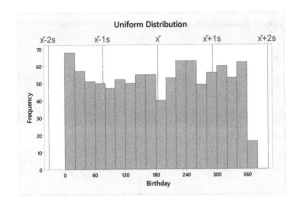

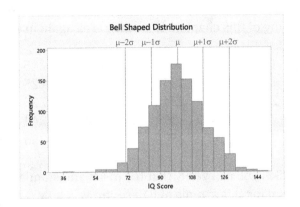

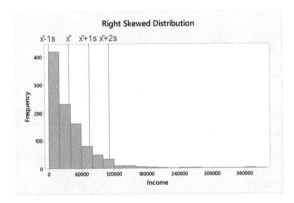

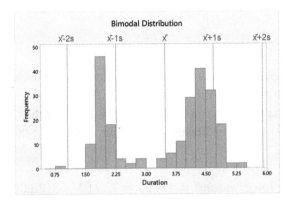

In the histograms, the location of the mean is shown along with places in the distribution that are 1- or 2-standard deviations away from the mean. For example, in the sample of birthdays, the mean is shown at 184.33, 1-standard deviation below the mean (184.33 – 108.11 = 76.22), 2-standard deviations below the mean (184.33 – 2(108.11) = -31.89), 1-standard deviation above the mean (184.33 + 108.11 = 292.44) and 2-standard deviations above the mean (184.33 + 2(108.11) = 400.85).

Because the scale of each of these variables is very different, "far from the mean" means (sorry about all the "means…" I don't mean to be so mean) something different for IQ scores (30 to 40 or so points above or below average) than it does for incomes (around $100,000 or so above average). However, notice that once we translate these distances from the mean into z-scores (the number of standard deviations from the mean), a fairly consistent pattern is shown: *you don't find many data values more than 2 standard deviations away from the mean in either direction.* In the distribution of birthdays and eruption durations, all or almost all of the data values are within 2 standard deviations from the mean. Even in the distributions of income and IQ scores, few values are more than 2 standard deviations from the mean.

In terms of z-scores, we'll call a data value an outlier if it is more than 2 standard deviations away from the mean:

$$lower\ bound = mean - 2SD$$
$$upper\ bound = mean + 2SD$$

Example: Consider the durations (in seconds) of the set of 61 Old Faithful eruptions shown at the beginning of this chapter. Here they area again in order from lowest to highest.

120	124	126	207	211	212	215	217	217	218	219	219	221
222	223	224	225	225	225	225	227	229	230	231	231	231
232	233	233	234	235	235	235	235	235	236	237	238	238
238	239	239	239	244	244	244	245	246	247	247	248	248
249	250	250	250	251	252	253	256	291				

If you like, you can verify that the mean and standard deviation for this sample of 61 eruptions is

$$\bar{x} = 229.51$$
$$s = 28.00$$

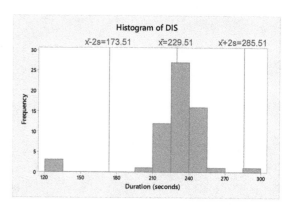

although it will take you a little while to do by hand. In the histogram, the mean duration is shown along with durations that would be exactly 2 standard deviations (about 56 seconds) above and below the mean:

$$lower\ bound = \bar{x} - 2s = 229.51 - 2(28.00) = 173.51$$
$$upper\ bound = \bar{x} + 2s = 229.51 + 2(28.00) = 285.51.$$

Note that there is one duration that is higher than the upper bound (291) and there are three values below the lower bound (120, 124, 126). These four values would be considered outliers according to this method. Notice that they stand out in the histogram as well. These are the same four outliers identified by the lower and upper fences, but that is just by coincidence; there is no rule that says both methods must identify the same outliers.

Boxplots

I'll close this section with one of my favorite graphs for numeric data, the boxplot. Boxplots were invented by John Tukey (1915 – 2000), a very influential statistician who was the first person to use the word "software" to refer to a computer program.[13] The reason for not mentioning boxplots earlier along with dotplots and histograms is that we didn't have all the necessary background for the boxplot's features. A boxplot is a graphical representation of a distribution that shows the 5-number summary (*minimum, Q1, median, Q3, maximum*) as well as any outliers according to the upper and lower fences. It is a one-dimensional graph, meaning you need only a single axis instead of two axes (a vertical one and a horizontal one) for histograms and dotplots.

To construct a boxplot, follow these steps:

Draw a box starting at $Q1$ and ending at $Q3$.
Draw a line inside the box at the median.
Indicate any values outside the fences with a special symbol like an asterisk.

Draw lines (called "whiskers") extending from each end of the box to the most extreme data values just inside the fences.

Example: Consider the set of Canada population guesses in the random sample of 30 students.

Here they are in order.

1.5	10	10	10	15	20	20	20	20	24
28	30	35	35	50	50	57	100	150	150
200	200	200	245	250	250	275	300	500	

The 5-number summary is

$minimum = 1.5$
$Q1 = 20$
$m = 50$
$Q3 = 200$
$maximum = 500.$ (Can you verify $Q1$, m, and $Q3$ on your own?)

To get the fences, we need the interquartile range: $IQR = Q3 - Q1 = 200 - 20 = 180$. The fences are

$lower\ fence = Q1 - 1.5IQR = 20 - 1.5(180) = -250$
$upper\ fence = Q3 + 1.5IQR = 200 + 1.5(180) = 470.$

The lower whisker extends from the lower end of the box down to 1.5 (the lowest value above the lower fence). There are no values below the lower fence of –250 so there are no low outliers. The upper whisker extends from the upper end of the box up to 300 (the highest value below the upper fence). There is one value (500) above the upper fence of 470 giving one high outlier. The boxplot is shown below annotated to show where the important features are. Note: The fences are not

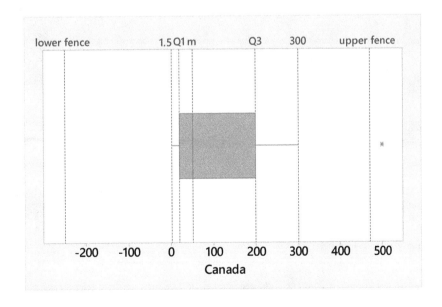

shown explicitly in the boxplot; they are used only to determine where outliers are, if any. The whiskers do not extend to the fences but to the highest and lowest values within the fences.

The figure below shows a histogram of the Canada data with the boxplot directly below it on the same scale.

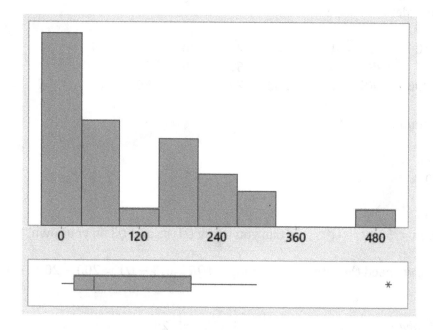

Example: Here are side-by-side boxplots of the Canada data, one for each form (US or AUS).

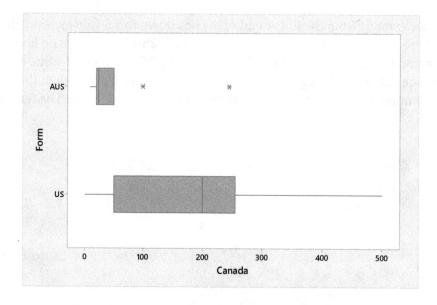

Notice how the boxplots make it very easy to compare the main features of each distribution of Canada responses: The US distribution of population guesses has a much higher center and much greater variability than the AUS distribution. Two guesses in the AUS distribution are considered unusually high according to the fences.

The box part of the boxplot shows where the middle 50% of the data values are located and also how variable they are in terms of the IQR since the $IQR = Q3 - Q1$ measures the length of the box. This middle 50% is situated at a higher spot on the scale and is also more spread out in the US distribution than the middle 50% in the AUS distribution. In this way the box part of a boxplot can be used to assess both center and variability.

Boxplots can also give you some indication of shape. In the figure below, five distributions are shown with different shapes. The boxplots are shown underneath and on the same scale as the histogram so you can see how the boxplots reflect the shape of the distribution:

Shape	Boxplot Features
Uniform	symmetric; median close to the middle of the box; few, if any, outliers; lengths of whiskers close to half a box length
Bell	symmetric; median close to the middle of the box; whiskers longer than half a box length
Right Skewed	median closer to $Q1$; higher whisker longer than lower whisker; outliers, if any, on the high side
Left Skewed	median closer to $Q3$; lower whisker longer than higher whisker; outliers, if any, on the low side
Bimodal	median closer to the "group" with more data values; whiskers close to half a box length

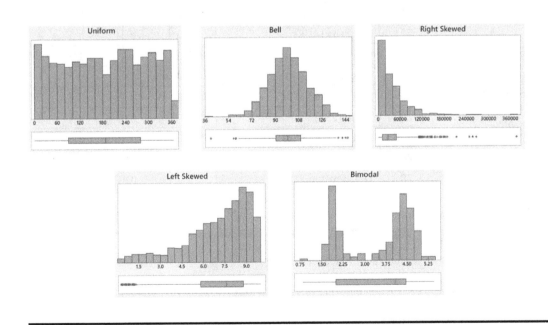

Section 2.2 Exercises

1. Shown here are medal counts for each country with at least one medal from the 2018 Winter Olympic Games held in Pyeongchang, South Korea.

   a. Make a dotplot of the number of gold medals.
   b. Make a histogram of the number of gold medals.
   c. Find the mean and median number of gold medals awarded per country.
   d. What notation would you use for the mean and median?

2. Here are the annual salaries (dollars) for all employees in a small family business.

   25,000
   28,000
   32,000
   33,000
   35,000
   50,000

   a. Would you consider these data a sample or a population data set?
   b. Make a histogram of the salaries.
   c. Compare the mean and median salaries.
   d. If the highest salary was changed to $40,000, what effect would that have on the mean? On the median?

3. What shape would you expect data on each of the following variables to have? Explain your answer.

   a. weights of newborn baby boys at a local hospital
   b. time you have to wait for the "walk" sign at a certain intersection
   c. amount won for all lottery tickets sold at a convenience store last month
   d. number of menstrual cycles last year for students at a local college

4. What shape would you expect data on each of the following variables to have? Explain your answer.

   a. number of emails sent to the teacher for each student in a class
   b. scores on the SAT last year
   c. ages of attendees at a middle school swim meet
   d. number of nights per month that college students set an alarm to wake up

Rank	Country	Gold	Silver	Bronze	Total
1	Norway (NOR)	14	14	11	39
2	Germany (GER)	14	10	7	31
3	Canada (CAN)	11	8	10	29
4	United States (USA)	9	8	6	23
5	Netherlands (NED)	8	6	6	20
6	Sweden (SWE)	7	6	1	14
7	South Korea (KOR)	5	8	4	17
8	Switzerland (SUI)	5	6	4	15
9	France (FRA)	5	4	6	15
10	Austria (AUT)	5	3	6	14
11	Japan (JPN)	4	5	4	13
12	Italy (ITA)	3	2	5	10
13	Olympic Athletes from Russia (OAR)	2	6	9	17
14	Czech Republic (CZE)	2	2	3	7
15	Belarus (BLR)	2	1	0	3
16	China (CHN)	1	6	2	9
17	Slovakia (SVK)	1	2	0	3
18	Finland (FIN)	1	1	4	6
19	Great Britain (GBR)	1	0	4	5
20	Poland (POL)	1	0	1	2
21	Hungary (HUN)	1	0	0	1
21	Ukraine (UKR)	1	0	0	1
23	Australia (AUS)	0	2	1	3
24	Slovenia (SLO)	0	1	1	2
25	Belgium (BEL)	0	1	0	1
26	New Zealand (NZL)	0	0	2	2
26	Spain (ESP)	0	0	2	2
28	Kazakhstan (KAZ)	0	0	1	1
28	Latvia (LAT)	0	0	1	1
28	Liechtenstein (LIE)	0	0	1	1

e.   date in November (e.g. 25) on which Thanksgiving fell during the last 70 years

5.   Shown here is a histogram of the number of deaths from several large earthquakes in the U.S.[14]

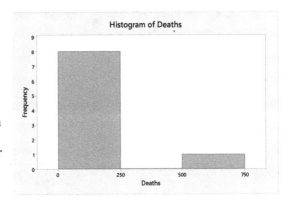

   a.   Identify the objects/subjects and the variable represented by the histogram.
   b.   How many earthquakes are in the data set?
   c.   What's wrong with the histogram?
   d.   You'll find the data in a table in section 2.4. Use the data to make a better histogram.
   e.   How would the mean and median compare? Verify your comparison by calculating both.

6.   Refer to the earthquake data shown in a table in section 2.4.

   a.   Make a histogram of the earthquake magnitudes.
   b.   Compare the mean and median magnitude. Does the comparison make sense based on the shape? Explain.

7.   There are 10 men and 30 women in a choir. The mean height of the men is 70 inches and the mean height of the women is 65 inches. Find the mean height of all 40 singers.

8.   Shown here are histograms of data on nine variables for the top 50 Major League Baseball players ranked by number of career home runs.[15] The variables are: G = games, AB = at bats, HR = home runs, RBI = runs batted in, BB = walks, SO = strikeouts, SB = stolen bases, CS = caught stealing, BA = batting average.

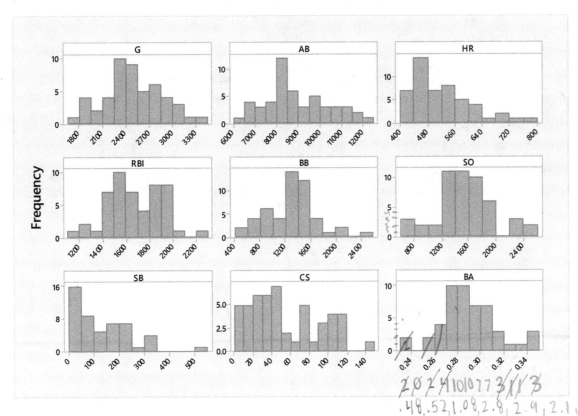

   a.   What is the shape of the distribution of the number of home runs for these 50 players?
   b.   What percent of these players have fewer than 1200 career strikeouts?
   c.   How would the mean and median compare in the distribution of batting averages?
   d.   While you can't calculate it exactly, say as much as you can about the median number of stolen bases.

e. How would you describe the shape of the distribution of the number of times caught stealing? What do you think is the explanation for this shape?

f. Is it surprising that many of the top 50 home run leaders of all time had relatively few career stolen bases? Explain.

g. What questions do you have after looking at these pictures?

9. Refer to the data on the top 20 schools ranked by tuition from chapter 1.

a. Guess the shape of the distribution of tuition.

b. Make a histogram of the tuition data. Was your guess correct?

c. Without calculating anything, how do you think the mean tuition would change if the tuition for Aviator College of Aeronautical Science and Technology is removed?

d. Without calculating anything, how do you think the median tuition would change if the tuition for Aviator College of Aeronautical Science and Technology is removed?

e. Verify your answers to parts c. and d. by calculating the mean and median tuition with and without the tuition amount for Aviator College of Aeronautical Science and Technology.

10. Suppose you changed the smallest value in a data set to something else. Under what circumstances would the median be unaffected? Under what circumstances could the median change?

11. Shown here is a histogram of a random sample of credit card transaction amounts from the set of all such amounts for a school district in a recent fiscal year.[16]

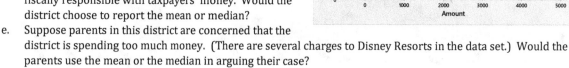

a. How many intervals were used?

b. What is the width of each interval?

c. What proportion of the sample of transactions were returns/refunds?

d. Suppose the school district wants to report a measure of center that will assure constituents that the district is fiscally responsible with taxpayers' money. Would the district choose to report the mean or median?

e. Suppose parents in this district are concerned that the district is spending too much money. (There are several charges to Disney Resorts in the data set.) Would the parents use the mean or the median in arguing their case?

12. A 401(k) account is a personal retirement account to which employers and employees can contribute on a pre-tax basis. Earnings in the account are also tax-deferred. A January 2020 article posted at nerdwallet.com reported that the average balance for all 401(k) accounts at Fidelity Investments was $103,700.[17]

a. How do you think the median account balance compares to this figure? Explain.

b. The article goes on to say, "If that still seems high, consider that averages tend to be skewed by outliers, and in this case, that number is being propped up by those rare millionaires." Comment on the statistical terms used in this sentence.

c. The next line is: "The median, which represents the middle balance between highs and lows, is just $24,500." Is this a good interpretation of the median? Explain.

13. Consider these data sets.

Data set I:	0	5	10
Data set II:	5	5	5
Data set III:	4	5	6

a. Answer these questions without calculating anything: Which data set will have the largest standard deviation? Which will have the smallest variance?

b. Calculate s^2, s, and the range for each data set.

c. Add 3 to each value in data set III. Then recalculate $\bar{x}$, s^2, s, and the range. How did each of these summaries change?

14. Answer true or false and explain your answer.

   a. In a data set where the range = 0, the standard deviation = 0.
   b. In a data set where the standard deviation = 0, the range = 0.

15. Refer to the 2018 Winter Olympic medals data from exercise 1. Make dotplots of the number of silver medals and the number of bronze medals on the same scale, one right above the other. Use the dotplots to compare the centers, shapes, and variabilities of two distributions. Be sure your answer is in context.

16. Here are the number of fouls committed and the points scored for the players on a basketball team during one year.

   a. Make dotplots of the number of fouls and points. Be sure you use the same scale for both.
   b. Compare the centers, shapes, and variabilities of the two distributions.
   c. Make a dotplot of the player numbers. Is there any useful information you can extract from that graph? Explain.
   d. Change the number of fouls for one player so that the ranges of fouls and points are the same.
   e. Change the number of fouls for one player so that the standard deviations of fouls and points are the same.

Player	Fouls	Points
1	1	0
2	22	75
3	3	3
4	14	62
5	4	9
6	10	20
7	5	29
8	22	56
9	3	5
10	3	2
11	24	56

17. Make up two data sets that have the same range but different means.

18. Make up two data sets that have the same mean but different ranges.

19. Make up two data sets that have the same mean and range but different standard deviations.

20. Stock screeners are strategies used to select a portfolio of stocks in which to invest. Different screeners use different criteria (value vs. growth, large vs. small companies, etc.). The data shown here are the returns for a particular month for all stock screeners.[18]

-7.47	-6.52	-3.43	-3.26	-2.28	-1.84	-1.74	-1.59	-1.53	-1.46	-0.40	-0.24
-0.20	-0.04	0.65	1.40	1.56	1.66	1.70	1.75	1.78	2.02	2.17	2.21
2.21	2.40	2.44	2.58	2.99	3.22	3.28	3.45	3.45	3.47	3.47	3.47
3.57	3.64	3.74	3.94	4.08	4.10	4.19	4.48	4.55	4.65	4.84	4.86
4.86	4.99	4.99	5.74	5.75	6.24	6.37	6.53	7.04	7.48	7.70	8.51
8.53	8.68	10.65									

   a. The return on an investment is the percentage change in the value of the investment during a certain time period. A return of 0 means the value didn't change. What proportion of these stock screeners lost money during this particular month?
   b. Make a histogram.
   c. Why would a histogram be better than a dotplot for these data?
   d. Describe the distribution in terms of center, shape, and variability.
   e. Calculate and interpret the standard deviation.

21. Consider the number of menstrual cycles last year for each student at a local college.

   a. If half are male and half are female, what would you guess the mean to be?
   b. What would a mean of 9 imply?

22. The median of the data set (x, y, 22, 10) is 30. Give values of x and y that make this work.

23. Consider these summaries of a data set: $n = 50, \bar{x} = 40$, and $s = 7.2$. I'm considering adding one of the following pairs of values to the data set:

    0 and 80   40 and 40   38 and 42   30 and 50   – 1 and – 2

    a. Which pair(s) will increase s?
    b. Which pair(s) will decrease s?

24. Consider the four distributions shown here.

    a. Without calculating anything, which distribution do you think has the highest standard deviation? the lowest?
    b. Calculate s for each distribution.

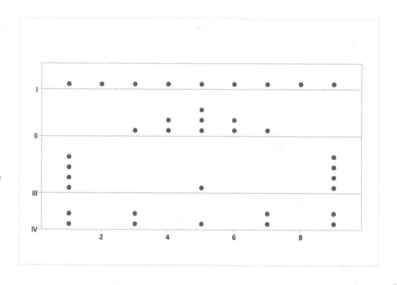

25. Suppose that in a particular administration of the Scholastic Aptitude Test (SAT), the mean and standard deviation of the math scores were 515 and 116 respectively.

    a. A score of 680 is at the 91st percentile for the math section. What does that mean?
    b. Calculate and interpret the z-score for a student who scores a 480 on the math section.
    c. The lowest and highest possible SAT math scores are 200 and 800. Are there any possible scores that would be considered outliers according to the z-score criterion? If so, identify them.
    d. The 25th and 75th percentiles are 440 and 600, respectively. Are there any possible scores that would be considered outliers according to the fences? If so, identify them.

26. Refer to question 25 about SAT math scores. The z-scores for three students' SAT math scores are:

    Charles: -0.99   Indie: 0.73   Francisco: 1.59

    a. Which students scored above the mean?
    b. Find the SAT math scores for these three students.

27. Refer to question 16 about the data for the basketball team.

    a. Find the 25th, 50th, and 75th percentiles for the number of points scored.
    b. Find the 25th, 50th, and 75th percentiles for the number of fouls committed.
    c. Calculate the IQR for both distributions.
    d. Calculate and interpret the z-score for the number of fouls committed by player 11.
    e. Change the points for player 2 from 75 to 100. Then recalculate the mean, standard deviation, median, range, and IQR. Which of these summaries are resistant to extreme values? Which are not?

28. The body can absorb alcohol so that the blood alcohol concentration (BAC) decreases by a certain number of percentage points per hour. For example, if someone's BAC goes from 0.10% to 0.08% in one hour, their rate would be 0.02 percentage points per hour. Of course this rate can vary. Suppose the standard deviation of hourly BAC reduction is 0.003 percentage points.

    a. A person is observed to have a BAC that decreased by 0.021 percentage points in an hour, corresponding to a z-score = 2. Interpret the z-score for this decrease.
    b. What is the mean decrease in BAC per hour?

c.   A person's BAC is currently 0.110%. The legal limit for driving is 0.08%. Should this person plan to drive home in an hour? Explain.

29. Refer to question 20 about the stock screener data.

a.   Find the 10th and 90th percentiles.
b.   Calculate and interpret the z-score for a stock screener that had a 5.75% return for that particular month.
c.   Calculate and interpret the z-score for a stock screener that lost 3% for that particular month.

30. An investor wants to invest money in the bond market. Her top considerations in deciding where to invest are rate of return (higher numbers are better) and maturity. Since she would like access to the funds within the next few years, lower maturities are better. She considers rate of return and maturity of equal importance. The rates of return and maturities of two investments are shown below. The mean and standard deviations of rate of return and maturity for all bond investments are also shown.

	Investment A	Investment B	Mean	Standard Deviation
Rate of Return	8.2%	6.8%	6%	1.5%
Maturity	4.1 years	2.4 years	4.5 years	6.0 years

a.   Calculate and interpret the z-score for the maturity for Investment B.
b.   If she can choose only one investment, which should she choose? Justify your answer.

31. Here is a boxplot of the 50 U.S. state populations as of the 2010 Census.[19]

a.   Approximately, what was the median population?
b.   Would the mean be higher, lower, or about the same as the median? Explain.
c.   Can you guess the states whose populations are outliers?

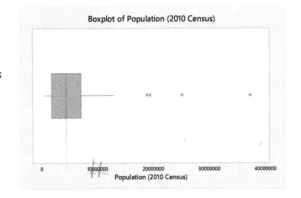

32. Shown here are the maximum wind speeds for the Atlantic storms of the 2018 hurricane season.[20]

a.   Find $Q1$, $Q2$, and $Q3$.
b.   Calculate the fences and identify any outliers according to the fences.
c.   Make a boxplot.
d.   What does the boxplot suggest about the shape of this distribution?
e.   Calculate and interpret the z-score for tropical depression Eleven. Note: this storm wasn't strong enough to be named.
f.   Which values are outliers according to the z-score criterion?

Name	MaxWind (mph)	Name	MaxWind (mph)
Alberto	65	Isaac	75
Beryl	80	Joyce	45
Chris	105	Eleven	35
Debby	50	Kirk	60
Ernesto	45	Leslie	90
Florence	140	Michael	155
Gordon	70	Nadine	65
Helene	110	Oscar	105

33. Refer to question 20 about the stock screener data.

a.   Calculate the fences.
b.   Identify any outliers according to the fences.
c.   Make a boxplot.
d.   Calculate the z-scores for any outliers you identified by the fences. Would these values be considered outliers according to the z-score criterion?

34. Make a boxplot using the number of points and number of fouls using the data from the basketball team in question 15. Put the boxplots side-by-side on the same scale.

35. Here is a boxplot of the number of Atlantic storms (hurricanes, tropical storms, and tropical depressions) for the years 2000 through 2018.[21]

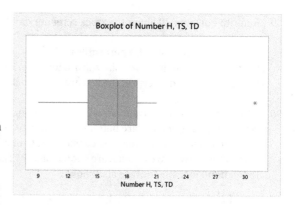

   a. What are the objects in this data set?
   b. What is the variable?
   c. What was the median number of storms?
   d. Fill in the blank: About 75% of the years had fewer than _____ storms.
   e. Fill in the blank: About half of the years had between _____ and _____ storms.
   f. Fill in the blank: About _____% of the years had more than 14 storms.
   g. Which (if any) of these is the upper fence? 19, 21, 31

36. Refer to question 11 about the credit card transaction data for a school district. Here are the 50 randomly-selected transaction amounts in order.

-135.00	-61.32	4.11	14.95	15.10	22.85	25.00	25.00	25.00	33.33	35.70
35.74	37.94	37.98	40.00	43.70	44.63	51.40	55.00	56.72	60.00	60.00
60.00	67.34	79.80	79.95	90.00	95.00	126.97	140.50	148.75	150.08	150.08
178.88	345.00	349.95	373.20	468.95	494.00	500.00	504.00	589.00	625.00	812.46
819.00	845.00	845.00	861.43	3500.00	4865.00					

   a. Find the 5-number summary.
   b. Calculate the fences and identify any outliers according to the fences.
   c. Make a boxplot.
   d. What does the boxplot suggest about the shape of this distribution?
   e. Calculate and interpret the z-score for the $500.00 amount.
   f. Which values are outliers according to the z-score criterion?
   g. How do the mean and standard deviation suggest that this distribution of transaction amounts is right-skewed?

2.3 Comparing Groups

One fairly common use of statistical methods and data science is to use data to answer questions about how groups compare. For example, doctors might like to know whether patients who take medication B do better than those who take medication A so they compare clinical outcomes of those two groups of patients. A company might like to know whether changes to a webpage result in more sales so it compares results of people visiting different versions of a particular webpage. In this section we'll use the tools we learned about in the previous section to compare groups.

Comparing Groups using Numeric Data

To compare groups using numeric data, we can use all of the summary tools appropriate for numeric data, for example, measures of center (mean, median), measures of variability (range, IQR, standard deviation), dotplots, histograms, and boxplots. Dotplots, histograms, and boxplots must be on the same scale. When using boxplots, use side-by-side boxplots so that you don't have to repeat the scale several times. It is very easy to stack a bunch of boxplots one above the other on the same horizontal scale. You can alternatively draw boxplots on a vertical scale and stack a bunch of them side-by-side.

Example: Between my undergraduate work in math and statistics and my graduate work in statistics, I worked for two years as a data analyst. One of the studies I worked on was a multi-center clinical trial of two treatments for asthma (salmeterol and triamcinolone).[22] Volunteer patients that met the entrance criteria were randomly assigned to one of three treatments (salmeterol, triamcinolone, or placebo). After the randomization but before the treatment period (referred to as "baseline"), data were collected on a bunch of variables, including the primary outcome variable for the study: Peak Expiratory Flow Rate in the morning (AMPEFR). Higher numbers mean better lung function. If the randomization worked, we shouldn't have seen much difference between the distributions of baseline AMPEFR among the three treatment groups.

Here are some numeric comparisons of the AMPEFR data:

	Placebo	Salmeterol	Triamcinolone
n	56	54	54
$\bar{x}$	446.5	443.9	459.4
s	101.4	112.3	107.0

Notice that there are some differences among the three groups on AMPEFR. Does that mean the randomization didn't work as intended? Was the salmeterol group at a disadvantage because it had sicker patients in it to start? Or were these differences typical for randomly assigning patients to the three treatments? The answer is that the differences in the sample means are typical for random assignment of 164 patients but you'll have to wait for the details behind that answer for a while. This is another good example of when extracting information from data to answer a question leads to more questions!

"Wait!" you say. "You can't compare these groups because the sample sizes aren't the same!" The authors used the mean and standard deviation as summaries which are both averages and take the differences in sample sizes into account. The mean is the average AMPEFR and the standard deviation is the average distance from that mean AMPEFR. You don't have to worry about their values being distorted by different sample sizes. All of the summary numbers and graphs for numeric data you've learned about so far are designed to measure or illustrate features of the distribution of numbers, regardless of how many there are. You *do* have to be careful about sample size differences when dealing with categorical data (more on that later).

You also have to be wary when using numeric variables such as the number of events when objects could have very different exposures.

Example: Shown below are the number of burglaries reported in 2014 at three institutions of higher learning.[23]

Institution	Number of Burglaries
A	27
B	26
C	13

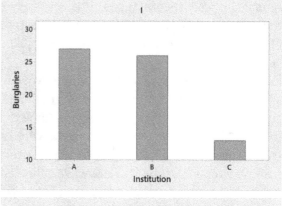

Data like these are often displayed using a "chart with bars." Here are three such charts. Which is the best one? Notice though that, unlike the bar chart described earlier in the chapter, this chart is not showing category frequencies or percentages. Instead the charts here show the values of a numeric variable that is a frequency (charts I and II) or rate (chart III) of the number of events (burglaries) for each object (institution). Notice how school C looks good compared to schools A and B; there were fewer burglaries there! If you're at institution C, you like the first two charts, especially the chart I.

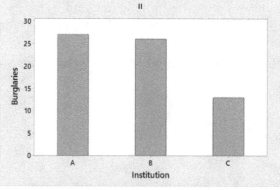

I hope you didn't choose chart I as the best one. Notice that both charts I and II show the number of burglaries at each school but because the vertical axis in chart I does not start at 0, the differences between the schools are exaggerated. School C looks lower than it actually is. Chart I makes it look like school B has about 4 times as many burglaries as school C, but that's not true. Of the two, chart II

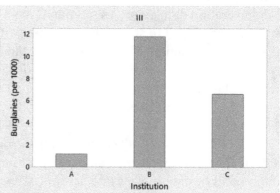

is better because the relative numbers of burglaries is accurately reflected by the visual bar heights.

But what if school C had fewer students on campus? Comparing schools on the number of burglaries is unfair unless the institutions have similar numbers of students. Places with more people tend to have higher crime counts just because there is more opportunity for crime to happen. You are generally more exposed to crime where there are more people compared to where there are fewer people. So let's adjust these numbers and compare the schools based on a burglary *rate*, say, the number of burglaries per 1,000 students.

Institution	Number of Burglaries	Number of Students	Burglary Rate (per 1000)
A	27	12,757	1.19
B	26	1,131	11.77
C	13	1,199	6.60

The burglary rates per 1,000 students tell a completely different story. Institution A is very large compared to institutions B and C, so while 27 burglaries seems relatively high, it is very small for the number of students there. Institution C no longer looks as good. School A has the smallest burglary rate by far and since the rate per 1,000 students is a much fairer way to compare these schools of different sizes, chart III is the best. It summarizes the data without distorting the message. While charts I and II are technically correct (the bar heights are where they should be), they distort the comparison among the schools. Remember, *a good graph should convey the information in the data accurately, without distortion, and within a fraction of a second.*

Notice that these data are not "grouped" in the usual way; we are comparing the objects (institutions) themselves, not groups of institutions. In that sense this example differs from the asthma example where it was the treatment groups of objects (subjects) we compared, not the subjects themselves. Nevertheless, it illustrates the need for caution when using variables that are counts of the number of events.

Example: How do students in a relationship compare to students who are not in a relationship with respect to GPA? To try to answer this question, here are histograms of reported GPA for those in a relationship and those not in a relationship for 206 students in an introductory statistics course.

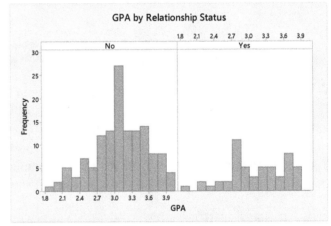

Notice that the two distributions of GPA have similar centers ($\bar{x}_{No} = 3.09$, $\bar{x}_{Yes} = 3.10$). There are fewer students in a relationship and their GPAs are a bit more variable ($s_{No} = 0.46$, $s_{Yes} = 0.49$, $IQR_{No} = 0.65$, $IQR_{Yes} = 0.86$). The distribution of GPA for students not in a relationship is fairly bell-shaped.

Could we use these data to say that, "being in a relationship or not does not affect a student's GPA?" Whoa, not so fast! That statement is pretty strong. First, it sounds like a generalization to all students from data based on our set of 206 students. These data were obtained from only one introductory statistics course at one university, so generalizing to all students is certainly not appropriate. Second, the statement implies cause/effect ("...does not *affect* a student's GPA"), or in this case, no cause/effect. Establishing whether or not a causal connection exists between two variables (in this case GPA and relationship status) is very difficult and should not be stated or implied without understanding the statistical foundations for making such a statement. Suppose being in a relationship has a detrimental effect on GPA in general but we just happened to observe a set of students who are in a relationship and are also generally better students than average. In that case, we would not be able to see the effect that being in a relationship has on GPA. Sounds like we need to answer more questions, doesn't it?

Example: Do females tend to have more Facebook friends than males? Here are boxplots of the reported number of Facebook friends for males and females for 206 students in an introductory statistics course.

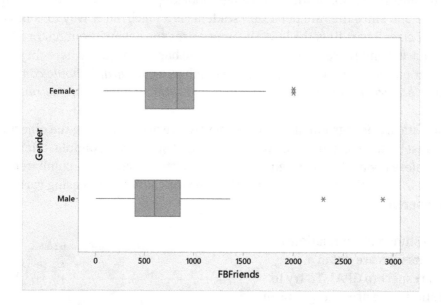

While the highest number of Facebook friends came from males, males generally had fewer Facebook friends than females ($m_F = 825, m_M = 600$). The male group also had slightly lower variability in the number of Facebook friends ($IQR_F = 500, IQR_M = 456$).

You might be tempted to use these data to say females have more Facebook friends, on average, than males. Remember where these data came from: one statistics course at one university.

Generalizing to all males and females is certainly not appropriate. These students may be different in terms of their Facebook usage than the general population. For one, their ages are typical of college-aged students: mostly between 18 and 22 with a few exceptions (see the boxplot shown here). Lest you think I forgot to draw a line inside the box at the median, it turns out that both the median and $Q1$ are 19 for the age data.

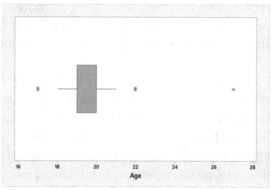

It could certainly be the case that females generally have more Facebook friends than males, but we could not establish that based on these data.

Comparing Groups using Categorical Data

To compare groups using categorical data, we can use all of the summary tools appropriate for categorical data, primarily percentages and graphs like bar charts. Here's where you have to be careful about different group sizes. It is important that we compare groups using the percentages in each category instead of the frequencies so that the comparisons are not distorted by differing group sizes.

Example: The following data are based on a Pew Research Center report about online dating usage.[24] In a random sample of U.S. adults, people were asked whether they've ever used an online dating site or app. The frequencies and percentages of responses in each age group are shown below in a two-way frequency table or **contingency table**. A contingency table is a two-dimensional array of frequencies where the rows represent categories of one variable and the columns represent categories of another variable. You can think of comparing groups as a problem where there are two categorical variables involved: a grouping variable and the categorical variable used to compare the groups. Let's let the rows represent the age groups and the columns represent people's response to the question about internet dating.

Each cell of the contingency table shows the frequency in that cell and the row percentage which expresses the frequency as a percentage of the row total. For example, 52 of the 194 adults in the sample aged 18 – 24 years used internet dating.

		Ever Used Internet Dating?		
		Yes	No	
	18 - 24	52 (26.8%)	142 (73.2%)	194
	25 - 34	60 (21.9%)	214 (78.1%)	274
Age Group	35 – 44	49 (14.8%)	186 (85.2%)	330
	45 – 54	43 (18.3%)	287 (81.7%)	235
	55 – 64	49 (11.9%)	362 (88.1%)	411
	65+	15 (2.9%)	498 (97.1%)	513
		268 (13.7%)	1689 (86.3%)	1957

Notice how using just the frequencies to compare the groups would be a bit misleading. While it is technically true that the 25 – 34 year old group had the most internet daters, on a percentage basis the 18 – 24 year old group actually had the most because there were more internet daters compared to the group size in that age group than in any other age group. Think of it this way: if each age group had the exactly the same *percentage* of internet daters, say 20%, then there would still be differences in the *frequencies* of internet daters because of the different age group sizes. *Unless the group sizes are quite similar, it is better to compare groups using percentages instead of frequencies.*

Example: Consider the asthma study where 164 patients were randomly assigned to one of three treatment groups (placebo, salmeterol, or triamcinolone).[25] The randomization process is designed to balance the treatments with respect to all other potential variables related to lung function. Shown below are summaries of the ethnicity variable for each of the treatment groups in an effort to see whether randomization balanced the treatment groups with respect to ethnicity. Because ethnicity is a categorical variable, we'll use a contingency table to summarize the data numerically and both a bar chart and stacked bar chart for a graphical summary.

In the contingency table below, the frequencies in each ethnicity are reported for each treatment group as well as overall. The row percentages are shown in each cell using the row total as the denominator. These row percentages are also used to construct the graphs. In the segmented bar chart, the width of a segment is the percentage in that ethnicity.

		Ethnicity					
		Asian	Black	Hispanic	White	Other	
Treatment Group	Placebo	4 (7.1%)	9 (16.1%)	5 (8.9%)	0 (0.0%)	38 (67.9%)	56
	Salmeterol	2 (3.7%)	7 (13.0%)	3 (5.6%)	1 (1.9%)	41 (75.9%)	54
	Triamcinolone	3 (5.6%)	8 (14.8%)	5 (9.3%)	2 (3.7%)	36 (66.7%)	54
		9 (5.5%)	24 (14.6%)	13 (7.9%)	3 (1.8%)	115 (70.1%)	164

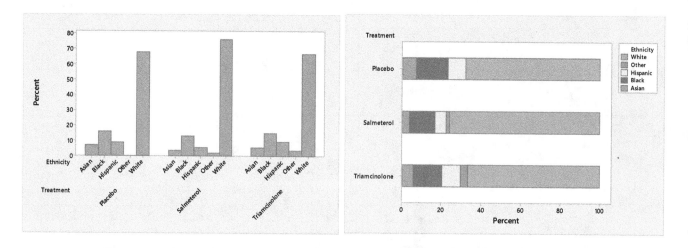

Nine of the 56 placebo patients were black (16.1%), which is a bit higher than the percentage of black patients in the other two treatment groups. There are some differences in the percentages in each ethnicity across the treatment groups, so we might wonder if randomization did the job of balancing with respect to ethnicity. On the other hand, it would be unreasonable to expect exactly the same ethnicity pattern for the three groups. It turns out that the differences in the ethnicity patterns we see across the three treatment groups are not unusual at all for random assignment. We can feel comfortable that the randomization worked to balance the groups with respect to ethnicity.

In these group comparisons where two categorical variables are involved, percentages must be interpreted with great care because their interpretation depends on the denominator used to compute them. Consider the internet dating example. The cell frequency of 52 for the 18 – 24 year-old internet daters can be expressed as a percentage in at least 3 different ways! Here they are with how they would be interpreted in words:

Row percent: 100*(52/194) = 26.8%
About 26.8% of the 18 – 24 year-olds in the sample used internet dating.

Column percent: 100*(52/268) = 19.4%
About 19.4% of the internet daters in the sample were 18 – 24 years old.

Table percent: 100*(52/1957) = 2.7%
About 2.7% of the adults in the sample were 18 – 24 year-olds who used internet dating.

Since that example was focused on comparing the age groups, the row percentages are the most relevant. The following example illustrates using percentages and rates which are computed using different denominators.

Example: Trisomy 18 (T18) is a genetic abnormality where a baby has an extra copy of chromosome 18 in some or all cells in the body. (Down's syndrome occurs when a baby has an extra copy of chromosome 21). According to the Trisomy 18 Foundation, T18 occurs in about 1 in every 2500 pregnancies.[26] Because T18 causes serious developmental abnormalities, babies with T18 rarely live to their first birthday. A certain genetic test on the mother's blood is positive for T18 in 99% of cases and negative in 97.6% of non-cases. Based on these numbers, if a pregnant mother tests positive, how likely do you think it is that her baby has T18?

Data to the rescue! Imagine a population of 10,000,000 pregnant women. Observe two categorical variables on each woman: 1) whether or not her baby has T18 and 2) the result of the mother's blood test for T18. The data would look like this:

Woman	T18 Status	Test Result
1	No	Negative
2	Yes	Positive
...	...	...
10,000,000	No	Negative

Let's apply the rates and percentages to a population of 10,000,000 pregnant women. Using the 1/2500 figure, we'd expect $\frac{1}{2500}(10,000,000) = 4,000$ cases of T18; the remaining 9,996,000 pregnancies would not be cases of T18. Now, using the 99% positive test rate for cases, 99% of those 4,000 cases would test positive: 0.99(4,000) = 396. By subtraction, 40 would falsely test negative. Finally, using the 97.6% negative test rate for non-cases, 97.6% of the 9,996,000 non-cases would test negative: 0.976(9,996,000) = 9,756,096. By subtraction, 239,904 would falsely test positive. These numbers allow us to fill in the contingency table as shown below. The frequencies in each "combination of categories" or "contingency" are shown along with the *row percentages* which use the row totals as the denominator. The row and column totals are also shown.

	Test Result		
		Positive	Negative	
T18 Status	Yes	3,960 (99.0%)	40 (1.0%)	4,000
	No	239,904 (2.4%)	9,756,096 (97.6%)	9,996,000
		243,864	9,756,136	10,000,000

While the vast majority of cases and non-cases are correctly identified by this test, there are still a very few number of cases that are not detected (negative test for a T18 baby) and a not-so-small number of non-cases for which the test is positive.

To answer our question then, of those 243,864 pregnant mothers who test positive, only 3,960 of them or 3,960/243,864 = 1.6% actually have babies with T18. It is *not* likely that a pregnant mother who tests positive has a baby with T18! You say, "How can this be? The test is correct for 99% of cases and 97.6% of non-cases." It's because the disorder is so rare so that 99.96% (2499/2500) of pregnancies are non-cases AND the test is not perfect. Even a small error rate in the test for the non-cases (2.4%) gives a large number of false positives compared to the number of true positives. The success rates are reported as *row percentages* but the practical answer to a mother's question about what a positive test result means must be answered using the *column percentages* that use the column totals as the denominator. It's all about the denominator.

Section 2.3 Exercises

1. Shown here are the distributions of the percent of faculty that are full-time for schools whose top degree is the Bachelor's and those whose top degree is a graduate degree. The data come from the set of top 20 schools ranked by tuition that you'll find in chapter 1.

	$\bar{x}$	s
Bachelor's	92.2	5.34
Graduate	73.4	21.01

    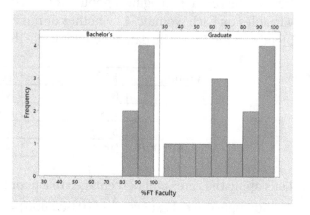

    a. Compare the centers, shapes, and variabilities of the two distributions. Be sure your answer is in context.
    b. Verify the two standard deviations using the data.

2. Refer to exercise 32 in section 2.2. Here are boxplots of the wind speeds of the 2018 Atlantic storms by type: H = hurricane, TS = tropical storm/depression.

    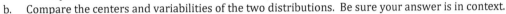

    a. Determine whether each of these statements is true or false. If false, give a reason why.
        i.   All of the hurricanes had higher maximum wind speeds than any of the tropical storms/depressions.
        ii. The line inside the boxes shows the mean wind speed.
        iii. The lower fence for the hurricanes is about 75 mph.
    b. Compare the centers and variabilities of the two distributions. Be sure your answer is in context.
    c. Does it make sense that there is no overlap between these two distributions? Explain.

3. Here are side-by-side boxplots of the credit card transaction amounts by fiscal period (month of the fiscal year) for a school district.

    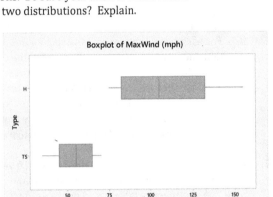

    a. Which fiscal period had the largest median transaction amount?
    b. Which fiscal period had the smallest IQR?
    c. Which fiscal period had the smallest range?

4. A certain state gave a math test to 11th and 12th grade students. For each school, the percent of students who scored at a "proficient" level was recorded. The percentages of students' scores classified as proficient are shown below.[27]

11th Grade:	42.27	30.33	41.94	54.76	29.41	8.33	41.46	49.55	11.36	38.39
	20.00	13.48	39.86	17.46	20.39	18.02	22.86	33.33	50.00	19.44
	16.98	17.39	34.09	23.08	31.73	15.84	8.25	40.88	29.41	46.74
	45.78	68.06	28.13	9.71	38.66	16.56	24.84	24.76	65.52	22.41
	40.27	12.40								

12th Grade:	25.00	33.33	11.76	25.00	25.00	50.00	66.67	33.33	18.18	17.39
	8.70	5.26	20.00	20.00						

    a. Make histograms of the two distributions, one above the other on the same scale.
    b. Make side-by-side boxplots of the two distributions on the same scale.
    c. Compare the centers, shapes, and variabilities of the two distributions. Comment on any unusual values. Be sure your response is in context.
    d. Some schools did not participate in the testing, some did not report their results, and for those schools who did, not all students participated. How does knowing this change the information you can extract from these data?

5. A 401(k) account is a personal retirement account to which employers and employees can contribute on a pre-tax basis. Earnings in the account are also tax-deferred. An article posted at nerdwallet.com reported the mean and median balances for all 401(k) accounts at Fidelity Investments by age group. The age groups were: 20-29, 30-39, 40-49, 50-59, 60-69. The following are the mean and median pairs for the age groups.[28]

    ($36,000 | $102,700)
    ($11,800 | $4,300)
    ($62,000 | $195,500)
    ($16,500 | $42,400)
    ($174,100 | $60,900)

a. For each pair, decide which number is the mean and which is the median.

b. Match each pair with the correct age group.

6. Consider the asthma study example from this section. Here are some numeric comparisons of the three treatment groups at baseline using data on forced expiratory volume (in liters) in one second (FEV1). FEV1 measures how much air a person can force out of their lungs in one second. The higher the number, the healthier their lung function. For healthy men and women, typical FEV1 values are over 4 liters and over 3 liters, respectively.

	Placebo	Salmeterol	Triamcinolone
n	56	54	54
$\bar{x}$	3.066	3.105	3.213
s	0.623	0.786	0.646

a. Each treatment group was comprised of both men and women. Give two different reasons why the means aren't around 3.5 (the average of 4 and 3 for healthy men and women).

b. For each treatment group, construct an interval that should capture most of the FEV1 values.

c. Consider moving someone's FEV1 from the triamcinolone group to the placebo group. Is it possible that both means would increase as a result? Why or why not?

d. Suppose Dietrich is a subject in the salmeterol group and Miriam is a subject in the triamcinolone group. Give FEV1 values for both Dietrich and Miriam that satisfy both these conditions:
   - Dietrich's FEV1 is farther away from the mean (in liters) than Miriam's FEV1.
   - Miriam's FEV1 has a higher z-score in absolute value than Dietrich's FEV1.

7. In a survey of U.S. parents of children under 18 years old, the Pew Research Center asked a representative sample of parents, "Do you think you spend too much time with your child/children, too little time, or about the right amount of time?" A contingency table based on the study is shown below.[29]

	too much	right amount	too little	
Mothers	70	579	221	876
Fathers	19	472	446	937

a. What are the variables in the data set? Are they numeric or categorical?

b. What percent of the respondents were fathers?

c. What percent of the respondents were fathers who thought they spent too little time with their child/children?

d. What percent of the fathers thought they spent too little time with their child/children?

e. Compare the mothers and fathers using the percent in each time category.

f. In what situation would it be okay to use the frequencies to compare mothers and fathers?

8. The U.S. Department of Education publishes data on crimes at college and university campuses. Shown below are the numbers of burglaries at several schools for a recent year.[30]

School	Number of Burglaries	Total Enrollment
Azusa Pacific University (main campus)	29	9,926
Ohio State University (main campus)	49	59,837
Benedict College	10	2,090

a. Make a graph that compares the three schools on the number of burglaries.

b. Calculate the number of burglaries per 1000 students for each school: burglaries / (enrollment/1000).

c. Make another graph that compares the three schools on the number of burglaries per 1000 students.

d. Which graph would Benedict College *want* to publish? Which graph *should* Benedict College publish?

9. The contingency table below summarizes the delivery status and the mother's level of prenatal care received during the pregnancy in the group of all twin births during 1995 – 1997.[31]

		Delivery Status				
		Preterm (induced)	Preterm (cesarean)	Preterm (natural)	Term or Post-term	
Mother's Level of Prenatal Care	Intensive	3455	14,857	14,889	28,489	61,690
	Adequate	7754	37,888	42,833	64,902	153,377
	Less than adequate	1738	10,234	13,374	38,098	63,444
		12,947	62,979	71,096	131,489	278,511

a.   What percent of mothers received at least adequate prenatal care during their pregnancy?
b.   What percent of twin births were preterm and involved mothers who had less than adequate prenatal care?
c.   What percent of twin births involved a cesarean section procedure?
d.   Compare the percent of preterm births among the three groups of mothers.
e.   Make a graph that presents the contingency table information.
f.   Table 1 of the article shows that in 1995 there were 96,736 twin births, in 1996 there were 100,750 twin births, and in 1997 there were 104,137 twin births. Add these three numbers and compare to the number of twin births summarized by the table above. Why do you think there is a discrepancy?

10. Tuberculosis (TB) is a potentially serious bacterial disease that affects the lungs. In the U.S., TB is rare. A common test for TB involves injecting a small amount of a liquid called tuberculin under the skin of a person's forearm. If the injection site becomes raised or hard in a few days, the test result is considered "positive" for TB. Otherwise the test result is considered "negative" for TB. The table below summarizes data on a large population of people for whom we know both their actual TB status and the outcome of the TB test.

		Test Result		
		Positive	Negative	
TB Status	Yes	*True* 94	*False* 6	100
	No	*False* 23,988	*True* 175,912	199,900
		24,082	175,918	200,000

a.   What percent of this population actually has TB?
b.   What percent of those who have TB tested positive?
c.   What percent of those who don't have TB tested negative?
d.   What percent of those who tested positive actually have TB?
e.   Why are your answers to parts b. and d. so different?

11. Pancreatic cancer is a rare but aggressive cancer that occurs in an estimated 2 per 10,000 people. Suppose a diagnostic test for pancreatic cancer was 95% accurate for people with pancreatic cancer. (This is called the *sensitivity* of the test.) Suppose that it was 99% accurate for people without pancreatic cancer. (This is called the *specificity* of the test.) What percent of people who test positive actually have pancreatic cancer?

12. A player's batting average in baseball is calculated by dividing the number of hits by the number of at-bats. Hits are classified as singles, doubles, triples, and homeruns. Plate appearances at which a player walks are not counted as at-bats. Shown below are frequency tables of plate appearance outcomes for two different players for two consecutive years.

	Last Year		This Year	
	Player A	Player B	Player A	Player B
Walk	81	2	1	67
Strikeout	91	21	14	140
Ground/Fly Out	249	27	39	218
Single	96	9	15	91
Double	47	9	2	18
Triple	5	0	0	6
Homerun	32	4	0	21
Total	601	72	71	561

a. Consider the frequencies for player A this year. What are the objects in the data set? What is the variable summarized by one of the columns in the frequency table? Sketch what the data set looks like.

b. How many at-bats did each player have each year? How many hits?

c. Calculate the batting average for each year for each player. How does player A compare to player B this year? Last year?

d. Calculate the combined batting average for each player across both years. How does player A compare to player B overall?

e. What you observed in parts c. and d. is an example of Simpson's Paradox. Make a graph that illustrates this contradiction. Why did it happen?

13. In a study of how much help should be provided to students doing accounting homework, researchers randomly assigned a group of volunteer students to three groups.[32] All students received the same reading on a topic not covered in class and took a pretest (10 multiple choice questions) on the material. Then students worked a homework problem. One group of students received no assistance on the homework, another group received check figures at several steps, and the third group received the complete solution. After working the homework problem, students took a posttest on the same topic. The posttest contained the same 10 questions from the pretest with the choices reordered. The improvement in test scores (posttest – pretest) was recorded for each student. Some summaries of the score differences are shown here.

	no assistance	check figures	complete solution
n	30	25	20
$\bar{x}$	2.43	2.72	1.95
m	3	2	2
min	-3	0	-5
max	7	7	8

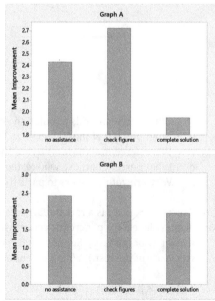

a. Your hypothesis is that providing the complete solution is detrimental to learning this material for students like these. Which graph better supports your hypothesis?

b. Your hypothesis is that there isn't much difference in the effects of the three solution treatments on learning this material for students like these. Which graph better supports your hypothesis?

c. Which graph *should* you present, regardless of your hypothesis about providing homework solutions? Why?

d. Interpret the minimums for each group.

e. Compare the centers and variabilities of the three distributions of improvement scores.

f. What clues are there about the shapes of the three distributions of improvement scores?

14. The U.S. Department of Education publishes all kinds of data on colleges and universities. Shown below are the median debt (in thousands of dollars) for students upon entering loan repayment for a random sample of institutions.[33] The data are grouped by the predominant degree granted.

Certificate:	9.5	5.5	9.8	4.8	9.5	12.0	9.0	9.4	9.5	3.5
	7.5	9.7	9.5	9.5	9.5	4.8	11.9	9.5	11.3	9.5
	11.2	12.4								

Associate's:	9.5	5.8	6.3	9.6	4.0	9.9	6.8	9.0	4.0	15.7
	3.5									

Bachelor's:	15.6	17.0	12.6	12.0	17.3	20.1	15.0	16.0	19.8

Other:	9.5	8.0	17.3	9.8

    a. What object/subject does each number above represent?
    b. If you were to reformat the data so that each object/subject got its own row, how many rows and columns would the data set have?
    c. Make side-by-side boxplots of the distributions of median debts.
    d. Compare the centers and variabilities of the four distributions.

15. A real estate firm wants to compare two appraisers (Jill and Anya) that it tends to use for its real estate transactions. To do so, it randomly selects 5 properties currently for sale and has both Jill and Anya appraise each house. The appraised values are shown below.

Property	Value (Jill)	Value (Anya)
1	$175,600	$181,650
2	$94,800	$96,800
3	$227,300	$236,500
4	$419,650	$423,850
5	$275,200	$279,900

    a. Find the mean, median, standard deviation, and IQR for both sets of appraised values.
    b. Make two side-by-side boxplots. Do the data convince you that Jill and Anya appraise differently?
    c. For each property calculate the difference in appraised values (Jill – Anya).
    d. Find the mean, median, standard deviation, and IQR for these differences.
    e. Make a boxplot of the differences. Do these data convince you that Jill and Anya appraise differently?
    f. Explain to the layperson what's going on here.

2.4 Relationships Between Two Numeric Variables

In addition to comparing groups, another common question requiring the extraction of information from data is about the relationship between two numeric variables. For example, Edwin Hubble discovered a striking relationship between two variables observed on galaxies: distance from us and recessional velocity. Suppose Southwest Airlines wants to track the percentage of flights that arrive on time. One variable would be the on-time proportion (number of on-time flights / number of total flights) and the other variable would be time.

Scatterplots

A **scatterplot** is a plot of (x, y) coordinates where the x- and y-coordinates are the values of numeric variables observed on a set of objects.

Example: In the 1920s the astronomer Edwin Hubble collected data on galaxies ("nebulae" as he called them). Two of the variables he observed were distance (how far the galaxy is from Earth in megaparsecs) and velocity (how fast the galaxy is moving away from Earth in km/sec). A megaparsec (Mpc) is one million parsecs; a parsec is about 3.26 light years. Shown below are Hubble's data on 24 galaxies.[34]

Galaxy	1	2	3	4	5	6	7	8
Distance (Mpc)	0.032	0.034	0.214	0.263	0.275	0.275	0.45	0.5
Velocity (km/sec)	170	290	-130	-70	-185	-220	200	290

Galaxy	9	10	11	12	13	14	15	16
Distance (Mpc)	0.5	0.63	0.8	0.9	0.9	0.9	0.9	1
Velocity (km/sec)	270	200	300	-30	650	150	500	920

Galaxy	17	18	19	20	21	22	23	24
Distance (Mpc)	1.1	1.1	1.4	1.7	2	2	2	2
Velocity (km/sec)	450	500	500	960	500	850	800	1090

If we treat distance as x and velocity as y, we can plot the ordered pairs using a scatterplot (shown here).

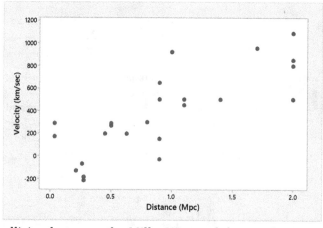

Notice the tendency for galaxies that are farther from us to be moving faster away from us than galaxies closer to us. This observation is evidence that the universe is expanding and was breaking news to the astronomers of Hubble's day. Also notice that there are some galaxies relatively close to us that have negative velocities, meaning they are moving *toward* us. Does that mean a collision between the Milky Way and those galaxies is imminent? Not at all; those galaxies are light years away from us (literally).

When describing the features of a single distribution of numeric data, it's important to highlight the center, shape, variability, and any outliers. In a similar way, there are four features of a scatterplot showing the relationship between two numeric variables that are worth noting:

- Form: Is the general pattern of scatter <u>linear</u> or <u>non-linear</u>.
- Direction: Do higher x's generally go with higher y's (<u>positive</u> direction) or do higher x's generally go with lower y's (<u>negative</u> direction)?
- Strength: How closely do the points follow the form? (<u>strong</u>, <u>moderate</u>, or <u>weak</u>)
- Outliers: Are there any points that look unusual?

Here are some examples that illustrate the features.

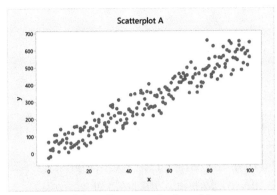

Strong, positive, linear relationship

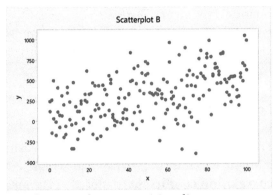

Weak-to-moderate, positive, linear relationship, with a couple possible outliers

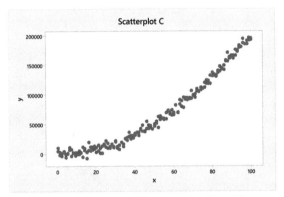

Strong, positive, non-linear relationship

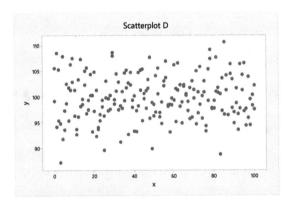

Weak-to-no relationship

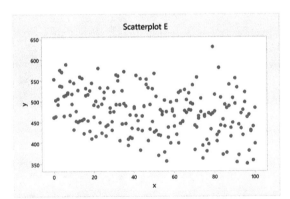

Weak, negative, linear relationship

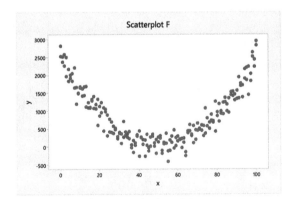

Strong non-linear relationship

When constructing an x,y scatterplot, be sure to scale both axes to match the range of the data. You want to "zoom in" as much as possible. In fact, "zooming out," especially on the y-axis, can distort the features of the relationship. The exception to this rule is a time series plot where it's best to begin the y-axis at 0. More on that later.

Here's scatterplot B again, but with the y-axis limits set too far beyond the range of the y-data. Notice the difference? Might you be tempted to call this relationship "strong" instead? Or perhaps you might have said that there was "no relationship" between x and Y? Zooming out in the y-scale has distorted the relationship.

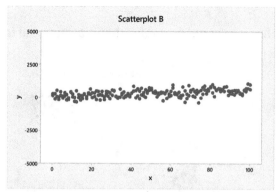

Correlation

There are several common summary measures used to quantify the relationship between two numeric variables. Perhaps none is more commonly-used than the correlation coefficient. The **correlation coefficient** (denoted r) is a measure of both the *strength* and *direction* of a *linear* relationship between two numeric variables. Here's how it's calculated along with a summary of important things to know about correlation.

Correlation Coefficient (r)

$$r = \frac{(x_1-\bar{x})(y_1-\bar{y})+(x_2-\bar{x})(y_2-\bar{y})+\cdots(x_n-\bar{x})(y_n-\bar{y})}{(n-1)s_x s_y}$$

- Appropriate only for <u>linear</u> relationships between two <u>numeric</u> variables.
- $-1 \leq r \leq 1$
- The sign of r indicates direction: positive or negative.
- Values close to -1 or 1 indicate a strong linear relationship.
- Values close to 0 indicate a weak linear relationship.
- Doesn't depend on which variable is the x-variable and which is the y-variable.

The real action here is the sum of all the products of differences between the x-data values and their mean and the y-data values and their mean. Let's illustrate with an example.

Example: Shown here are data collected on both the selling price and the square feet of living space for a sample of houses sold in the county where I live.[35] Describe the relationship between house size and selling price and calculate the correlation coefficient.

Square Feet	Price
2240	$275,000
2648	$315,000
1800	$260,000
1824	$299,900
1888	$255,000
1586	$249,900
1273	$219,900
2232	$249,900
1548	$230,000
1048	$180,000

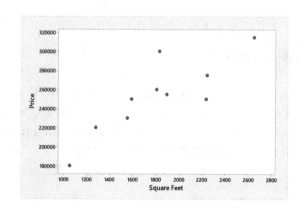

Let's use house size in square feet as the x-variable and price as the y-variable. The scatterplot shows a moderate-to-strong, positive, linear relationship between size and selling price for these houses. In order to calculate the correlation coefficient, we first need some basic summary measures:

$$\bar{x} = \frac{2240+2648+\cdots+1048}{10} = 1808.7 \qquad\qquad \bar{y} = \frac{275000+315000+\cdots+180000}{10} = 253,460$$

$$s_x = \sqrt{\frac{(2240-1808.7)^2+(2648-1808.7)^2+\cdots+(1048-1808.7)^2}{10-1}} = 478.3565$$

$$s_y = \sqrt{\frac{(275000-253460)^2+(315000-253460)^2+\cdots+(180000-253460)^2}{10-1}} = 38803.3847$$

Now we have all the pieces to calculate the correlation coefficient:

$$r = \frac{(2240-1808.7)(275000-253460)+(2648-1808.7)(315000-253460)+\cdots+(1048-1808.7)(180000-253460)}{(10-1)(478.3565)(38803.3847)} = 0.844$$

Note that r indicates that the linear relationship between house size and selling price is positive (it's a positive number) and rather strong (fairly close to 1).

Here's the scatterplot again divided up into four quadrants by two lines: a vertical line at $\bar{x} = 1808.7$ and a horizontal line at $\bar{y} = 253,460$. Consider each quadrant for a minute.

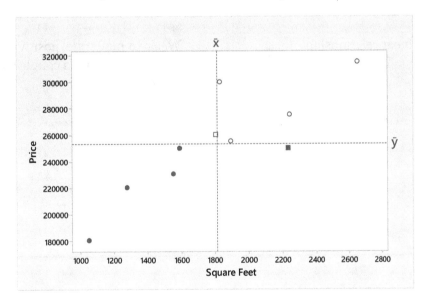

Upper Right : Points in this quadrant are both above the mean x and above the mean y. Therefore the sign of the products $(x_i - \bar{x})(y_i - \bar{y})$ will be $(+)(+) = +$.

Upper Left: Points in this quadrant are below the mean x and above the mean y. Therefore the sign of the products $(x_i - \bar{x})(y_i - \bar{y})$ will be $(-)(+) = -$.

Lower Left: Points in this quadrant are both below the mean x and below the mean y. Therefore the sign of the products $(x_i - \bar{x})(y_i - \bar{y})$ will be $(-)(-) = +$.

Lower Right: Points in this quadrant are above the mean x and below the mean y. Therefore the sign of the products $(x_i - \bar{x})(y_i - \bar{y})$ will be $(+)(-) = -$.

If a scatterplot shows a positive direction, there will be more points in the upper right and lower left quadrants which will contribute more positive products to the correlation, and so the correlation will be positive. For scatterplots with a negative direction, there will be more points in the upper left and lower right quadrants which will contribute more negative products to the correlation, and so the correlation will be negative.

Here are those six examples again with their correlation coefficients.

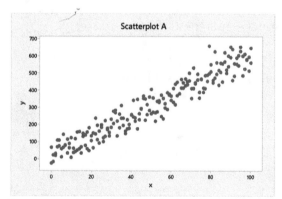

Strong, positive, linear relationship
$r = 0.958$

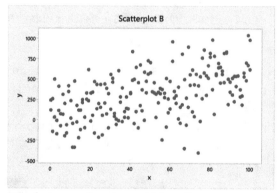

Weak-to-moderate, positive, linear relationship, with a couple stray points
$r = 0.518$

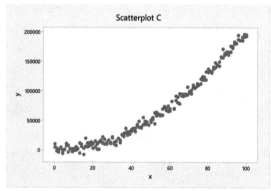

Strong, positive, non-linear relationship
$r = 0.965$

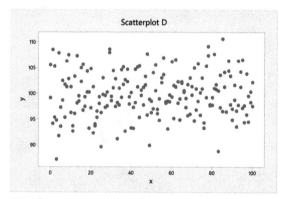

Weak-to-no relationship
$r = 0.050$

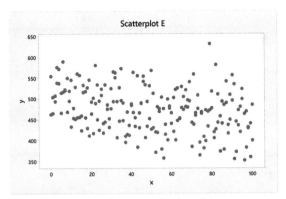

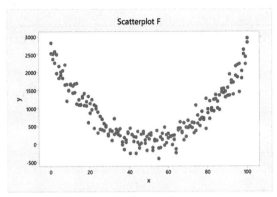

Weak, negative, linear relationship
$r = -0.316$

Strong non-linear relationship
$r = -0.010$

Notice that the correlation cannot pick up on any non-linearity in the relationship between x and y. For example in scatterplots C and F, the correlation value doesn't tell you about the curved form of the relationship. In fact, you might be tempted to conclude that there is little-to-no relationship between x and y based only on the correlation for scatterplot F when there is a very clear, strong, non-linear relationship! Beware of using the correlation by itself, without a scatterplot to go with it.

Here are a few examples of what happens to the correlation for a set of points that are co-linear.

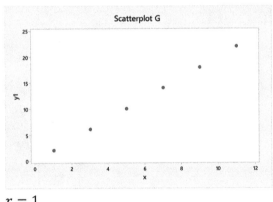

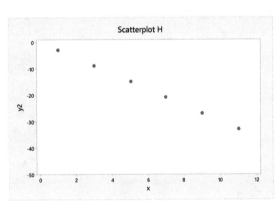

$r = 1$

$r = -1$

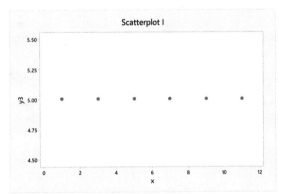

r is undefined

Note that for a set of points that are co-linear on a horizontal line, the correlation is undefined. Because there is no variability in the y-coordinates, $s_y = 0$, and division by 0 is undefined. Fortunately this situation (as well as any other set of perfectly co-linear points) rarely happens in practice with real data.

Beware of the effects of outliers on the correlation, especially for small sample sizes!

Example: Here are data on several major earthquakes in the U.S.[36] For each, the magnitude (on the Richter scale) and number of deaths attributable to the earthquake are shown. Two scatterplots are shown; one which excludes the San Francisco quake of 1906.

Date	Location	Magnitude	Deaths
8/31/1886	Charleston, SC	6.9	60
4/18/1906	San Francisco, CA	7.9	700
6/29/1925	California	6.8	13
3/10/1933	Long Beach, CA	6.4	120
2/9/1971	San Fernando Valley, CA	6.6	58
10/17/1989	San Francisco, CA	6.9	63
6/28/1992	Yucca Valley, CA	7.3	3
1/17/1994	California	6.7	57
8/23/2011	Virginia	5.9	0

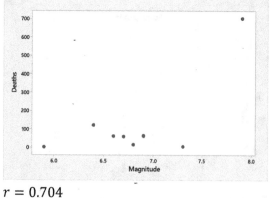

$r = 0.704$

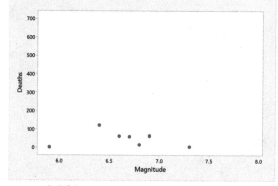

$r = -0.081$

The 1906 San Francisco earthquake's data point is highly unusual (because of the unusually large number of deaths) and so has a huge influence on the correlation for this small data set. Might you be tempted to conclude that there is a fairly strong positive relationship between the magnitude of the quake and the number of deaths if you had only the correlation coefficient? Who would argue? Stronger quakes kill more people, right? Perhaps if all else were equal. But of course in these data collected over more than a century, lots of things changed, like building codes in California. Notice that the correlation is very close to 0 when that point is removed. Be very careful not to take summary statistics too seriously that are highly dependent on only a few data values.

Prediction

In some cases, the primary purpose of the analysis of data on two numeric variables is *predictive*, that is, to predict the value of one variable (the y-variable) for an object whose x-variable value is known to us. How do we do this? We use data on a bunch of objects where we know the values of both. One fairly simple way to do that is to find the *line of best-fit* to a scatterplot and then use the equation of that line to predict y for a given value of x. The equation of the line is typically written in slope-intercept form:

$$\hat{y} = mx + b$$

where m is the slope of the line and b is the y-intercept. The y wears a "hat" because we don't want to imply that knowing x will give us the exact value of y for any object. In fact, the line of best fit often doesn't go through any of the points! So we'll refer to $\hat{y}$ as the **predicted value** of y.

Finding the equation of the best-fitting line in a scatterplot boils down to finding the slope (m) and the y-intercept (b). You can eye-ball any old line and you might get rather close to the best-fitting one, but to find *the* line that fits *the best* takes some effort.

Part of the problem is to first say exactly what we mean by "best-fitting." Among several contenders, we'll consider what's known as the least-squares criterion. The least-squares criterion says this: find the equation of the line that minimizes the sum of the squared vertical distances between the points and the line. The **least-squares line** is the line in slope-intercept form that accomplishes this.

Shown below are two lines drawn on the scatterplot of selling price vs. house size in square feet for the sample of houses recently sold. I drew the one of the left and the least-squares line is the one on the right. For each point the vertical distance to the line is shown. It's pretty clear that my line is not as good as the least-squares line because the distances from the points to the line are generally longer. The least-squares line gets closer to the points in that sense, even though it doesn't actually go through any of them.

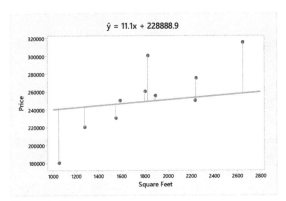

Bryan's line

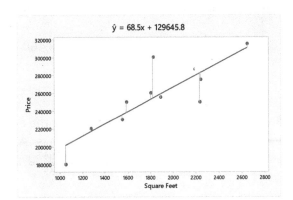

Least-Squares line

You can't do any better than the least-squares line in a vertical distance sense. It's the best! "But Bryan," you say, "how did you find the equation of the least-squares line?" If you're smelling an optimization problem here, your sniffer is working well. If you're sensing that there's some fairly intense math involved, right again! My hunch is that you don't really care much about the math and just want the bottom line. It's okay, I won't take it personally.

Here's how you get the slope and y-intercept of the least-squares line along with their interpretations:

Least-Squares Line $\hat{y} = mx + b$

Slope: $m = r\dfrac{s_y}{s_x}$ y-intercept: $b = \bar{y} - m\bar{x}$

- The slope is the predicted difference in y for an object with an x of one higher.
- The y-intercept is the predicted y for an object with an x of 0.

Example: Here are the data again for a sample of houses sold in the county where I live. Find the least-squares line and interpret the slope and y-intercept. Then use the line to predict the selling price of a house having 2000 square feet.

Square Feet (x)	Price (y)
2240	$275,000
2648	$315,000
1800	$260,000
1824	$299,900
1888	$255,000
1586	$249,900
1273	$219,900
2232	$249,900
1548	$230,000
1048	$180,000

We've already done most of the work including calculating the correlation coefficient:

$$\bar{x} = 1808.7 \qquad \bar{y} = 253{,}460 \qquad s_x = 478.3565 \qquad s_y = 38803.3847$$

$$r = 0.84389028$$

So just put these pieces together to get the equation of the least-squares line:

$$m = r\frac{s_y}{s_x} = 0.84389028\,\frac{38803.3847}{478.3565} = 68.4548$$

$$b = \bar{y} - m\bar{x} = 253460 - 68.4548(1808.7) = 129645.78996.$$

The least-squares line is

$$\hat{y} = 68.5x + 129645.8.$$

The interpretation of the slope (68.5) is:

"A house having a size of 1 square foot more would be predicted to sell for $68.5 more."

The interpretation of the y-intercept (129645.8) is:

"A house having 0 square feet would be predicted to sell for $129,646."

To predict the selling price of a house with 2000 square feet, just plug in 2000 for x in the least-squares line.

$$\hat{y} = 68.4548(2000) + 129645.78996$$
$$= \$266,555$$

In the scatterplot, the least-squares line shows the predicted location of a point for a house having 2000 square feet.

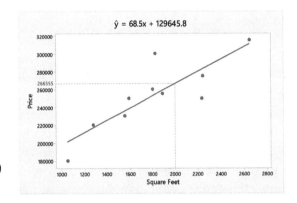

The prediction procedure is fairly straight-forward, especially since we didn't get into all the math involved. However, it is very easy to misuse and misinterpret these numbers! The following examples will illustrate some very common pitfalls when analyzing data on two numeric variables.

Extrapolation

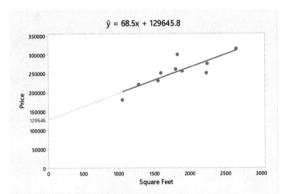

You might have wondered about the interpretation of the y-intercept in the house price example above. "Why would a house with no living space be predicted to sell for $129,646?" Good question. You'll notice that 0 square feet is well below the range of house sizes we have in the data set. If we'd extend the line down that far, $129,646 is where it would cross the y-axis. However, without data on houses having between 0 and 1000 square feet, we have no assurance that the pattern continues in that way. For the same reason, don't use the least-squares line based on these houses to predict the selling price for a 5000 square foot house since 5000 is well above the range of house sizes in the data set.

Extrapolation can occur in another way. Suppose I asked you to predict the selling price of a 2000 square foot house in New York City. Would you use the least-squares line to come up with $266,555? I hope not! The line is based on houses in my home county (which is very rural) and would certainly not describe well the relationship between house size and house price in New York City. Don't use the line to do prediction for an object that comes from a different group of objects than the ones on which the equation of the line is based.

Causal Inferences

You may have heard the warning, "Association is not necessarily causation." Just because you see a relationship or association between two variables doesn't necessarily mean that one is causing changes in the other. This is particularly relevant when you're dealing with data on two numeric variables. It is very tempting to make cause/effect conclusions based on what you see in the data.

Example: Here is a scatterplot showing the relationship between the number of people per television and the life expectancy for 40 countries.[37] The points for the U.S. (1.3, 75.5) and Ethiopia (503.0, 51.5) are labeled. The plot shows a fairly strong, negative, non-linear relationship between the number of people per TV and life expectancy. Countries having fewer people per TV (more TV's per person) generally have longer life expectancy. However, the data do not suggest that decreasing the number of people per TV in a country (perhaps by

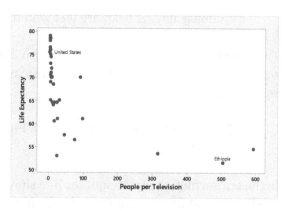

importing more TVs) is a way to increase life expectancy. Of course, the presence/absence of TVs has no effect on life expectancy. Very likely, the relationship you see here is caused by another variable (perhaps like socio-economic status) that causes both more TVs per person (because people have more disposable income) and longer life expectancy (those people have access to better health care). The real cause for what's seen in the data is not the number of people per TV so adding TVs won't affect life expectancy. You probably wouldn't have jumped to that conclusion because it doesn't make sense.

Now behold a different scatterplot, this time of life expectancy versus the number of people per physician and the same strong, negative, non-linear relationship in the data. The points for the U.S. (404, 75.5) and Ethiopia are shown (36660, 51.5). Countries having fewer people per physician (more physicians per person) generally have longer life expectancy. So decreasing the number of people per physician in a country (by adding more physicians) is a way to increase life expectancy, right? You might

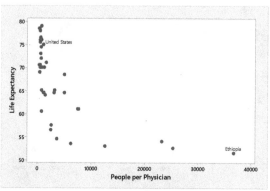

be very tempted to jump to that conclusion because it seems to make perfect sense. The answer is "No," however, for the same reason as for adding TVs. There is potentially another variable (perhaps education) that causes both more physicians per person (access to better education produces more physicians) and longer life expectancy (the better-educated make better lifestyle choices).

This causal fallacy is pervasive in our culture. If you start watching for it, you'll find examples of it almost daily. I believe there are at least a couple reasons for this.

1) We human beings are curious by nature and like to know the answer to the "why" question. When we see evidence of a relationship, association, or difference, we want to know why it exists. If we can come up with a reason easily (like adding physicians would give people better healthcare and prolong their life), it's easy to jump to that conclusion. The problem is that it doesn't matter whether we can easily think of a reason or not; the fact remains that the data often cannot provide convincing evidence of such a theory.

2) Most of us are formally taught since our first algebra course (or perhaps even earlier) the same way to interpret the slope of a line: "rise over run." A line having slope 5/3 means that "increasing x by 3 increases y by 5." And if x and y are the only variables in the universe (they are when you're doing algebra I problems), that's how it has to work. However, our universe is very complex. There are many variables that may affect y; x may or may not have anything to do with it. However, most of us have not been formally trained in multi-variable relationships. So it's very tempting for us to apply our two-dimensional thinking to multi-dimensional situations. We're like the ant who experiences the three-dimensional world through a largely two-dimensional experience. Our information is quite often very incomplete.

Example: Here is a data set for several cities that includes the number of police per 1000 people and the number of crimes per 1000 people for each city. The scatterplot shows a strong, positive, linear relationship between these two variables for these cities. Also shown is the equation of the least-squares line. Interpret the slope of the line.

Police per 1000 (x)	Crime per 1000 (y)
1.5	22
1.1	17
0.9	8
0.8	6
0.5	10
3.6	50
2.1	38
2.5	33
1.9	25

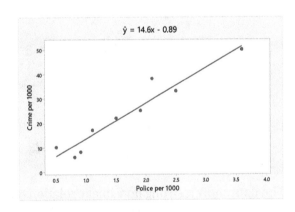

The best interpretation of the slope is:

"A city having 1 more police officer per 1000 people is predicted to have 14.6 more crimes per 1000 people."

There are several ways to misinterpret what 14.6 means. Here's are a few examples of an algebra I slope interpretation misapplied to data science:

"Increasing the number of police by 1 per 1000 people is predicted to increase the number of crimes by 14.6 per 1000 people."

"For every additional police officer per 1000 people, there is a predicted increase of 14.6 crimes per 1000 people."

Whoa! Do you think adding police officers to a city will *cause* an increase in crimes? That's what the above interpretations imply. Do you think that the way to decrease crime is to fire police officers? That's what the above interpretations imply. What we're likely seeing here is not a cause-effect relationship between the number of police officers and the number of crimes but a result of the fact that larger cities tend to have more of both!

Be careful not to imply that changes in x *cause* changes in y, unless the data were collected in a way that permits causal inferences. If the x-values were randomly assigned to objects, then a causal inference may be supported by the data.

Additive Amount (mg)	Weight (lbs)
0	4.63
0.1	4.45
0.2	4.96
0.3	4.79
0.4	4.97
0.5	5.03
0.6	5.17
0.7	5.09
0.8	5.28
0.9	5.1
1.0	5.33
1.1	5.34
1.2	5.45
1.3	5.29
1.4	5.46
1.5	5.41
1.6	5.29
1.7	4.97
1.8	5.23
1.9	5.17
2.0	5.11
2.1	4.97
2.2	4.89
2.3	4.82
2.4	4.65
2.5	4.75

Example: The manufacturer of chicken feed wants to determine the optimal amount of a nutritional supplement that is added to the product. Too little additive won't result in the desired effect but too much may be detrimental to growth. In an experiment, 26 chickens of the same breed, age, and size were each given the same amount of chicken feed but with a different amount of the nutritional additive. The additive amounts (from 0mg to 2.5mg) were randomly assigned to the chickens. The data are shown here.

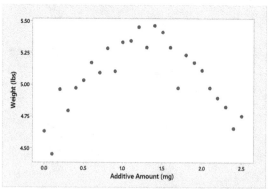

The data show a strong non-linear relationship between the amount of the additive and weight. Randomly assigning the additive amounts to the chickens effectively isolates its effect on weight. Chickens that would have grown more regardless of the additive amount were equally-likely to get a small, moderate, or large amount. So either the chickens that would have grown more anyway just happened to get a moderate amount of the additive at random (by the way the probability of that is $6.3x10^{-10}$) or the additive has an effect on weight. It looks like the optimum amount is around 1.3mg.

When the x-values are randomly assigned, then causal interpretations are permitted. Here are some examples:

"Increasing the additive amount up to 1.3mg tends to increase weight."

"For each additional 0.1mg, weight tends to increase until around the 1.3mg level. After that, additional increases in additive amount tend to be detrimental to weight."

One disadvantage of the tightly-controlled environment of an experiment like this is that the results hold only for chickens like the ones in the study. Other studies would need to be conducted to see if the 1.3mg optimum additive level applies to chickens of different breeds or ages.

Deterministic Language

One final caution when using the least-squares method is to always use non-deterministic language when reporting predictions or interpreting the features of the line. Our predictions are just that: *predictions*. George Box, a famous statistician, once said, "Essentially all models are wrong, but some are useful." If you use the least-squares line to predict y, your prediction is almost surely not going to be the actual y-value. When interpreting the slope of the line, the object with the x of one higher will almost surely not have a y that differs by exactly by m units from the object with the x of one lower. So it's best to use these words and phrases: "predicted," "estimated," "tends to," and "we expect."

Here are some good (non-deterministic) and bad (deterministic) ways to use the least-squares line in the police and crime example. Note that the modifiers "about" and "approximately" alone are not sufficient.

	Good (non-deterministic language)	Bad (deterministic language)
Slope Interpretation	"A city having 1 more police officer per 1000 people *is predicted to have* 14.6 more crimes per 1000 people."	"A city having 1 more police officer per 1000 people *will have* 14.6 more crimes per 1000 people."
	"A city having 1 more police officer per 1000 people *is estimated to have* 14.6 more crimes per 1000 people."	"A city having 1 more police officer per 1000 people *will have about* 14.6 more crimes per 1000 people."
	"A city having 1 more police officer per 1000 people *tends to have* 14.6 more crimes per 1000 people."	"A city having 1 more police officer per 1000 people *will have approximately* 14.6 more crimes per 1000 people."
	"*We expect* a city having 1 more police officer per 1000 people *to have* 14.6 more crimes per 1000 people."	
Prediction	"A city with 3.0 police officers per 1000 people *is predicted to have* 42.91 crimes per 1000 people."	"A city with 3.0 police officers per 1000 people *will have* 42.91 crimes per 1000 people."
	"A city with 3.0 police officers per 1000 people *is estimated to have* 42.91 crimes per 1000 people."	"A city with 3.0 police officers per 1000 people *will have about* 42.91 crimes per 1000 people."

	Good (non-deterministic language)	Bad (deterministic language)
	"A city with 3.0 police officers per 1000 people *tends to have* 42.91 crimes per 1000 people." "A city with 3.0 police officers per 1000 people *is expected to have* 42.91 crimes per 1000 people."	"A city with 3.0 police officers per 1000 people *will have approximately* 42.91 crimes per 1000 people."

Time Series

A **time series** is a sequence of numeric values collected over time. A plot of these values (on the vertical axis) versus time (on the horizontal axis) is called a **time series plot**. In this special kind of scatterplot it's common practice to connect the points with lines.

Example: The United States Bureau of Transportation Statistics collects data on on-time arrival performance for the major airlines. Shown below is a partial listing of data collected on Southwest Airlines flights arriving at Baltimore/Washington International (BWI) airport from June 2003 to September 2015.[38]

Year	Month	Arrivals	OnTime	OnTime%
2003	6	4481	3988	89.0
2003	7	4658	3929	84.3
...	...	...	...	...
2015	9	5590	4735	84.7

In addition to year and month indicators, the other variables are:

> Arrivals: total number of arriving flights that month
> OnTime: number of arriving flights that arrived on-time (less than 15 minutes late)
> OnTime%: percentage of on-time arrivals out of the total number of arrivals

The time series plot of the percentage of arriving flights that were on-time is shown below along with a histogram (turned on its side) on the same scale. Notice the very subtle decreasing trend in on-time performance that a histogram of the on-time percentages would not reveal. You'll also notice the substantial variability of on-time performance from month to month that could not be observed in the histogram. For these reasons, it's often better to use a time series plot instead of a histogram for data collected over time so that we can see changes over time. A histogram is appropriate to show the distribution of data collected at the same time. You can think of histograms, boxplots, and dotplots as photographs and time series plots as movies.

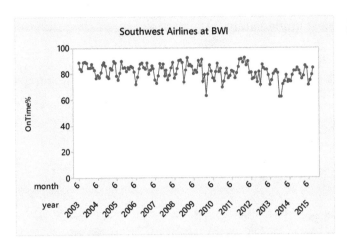

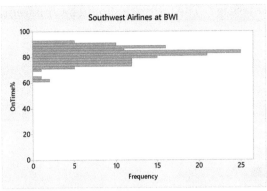

Here's a comparison of on-time performance for Southwest Airlines and American Airlines arrivals at BWI. At this airport, Southwest generally had better on-time performance throughout the time period.

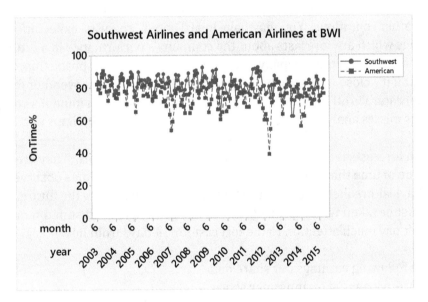

The pattern in a time series can be easily distorted by not starting the vertical axis scale at 0. This "zooming in" on the time series can make changes over time look bigger than they really are. The plots below show sales for a certain company over time. In the plot on the left, notice how sales appear to have doubled from 2015 to 2016 when, in fact, the increase from about $19 million to $21 million is only about 10.5%. The plot on the right where the sales axis starts at 0 shows that this increase is much less dramatic than the plot on the right makes it appear. Of course you might say, "I wouldn't be fooled because I'd just look at the numbers on the scale and realize that the increase is only about 10%." Good for you! The problem is that many people would not be as astute as you and would be taken in by the visual impression of the plot. Remember, *a good graph should convey the information in the data accurately and within a fraction of a second.* While the details such as axis scales are necessary, the reader shouldn't have to look at those too much to decipher the message. The message should be apparent almost immediately.

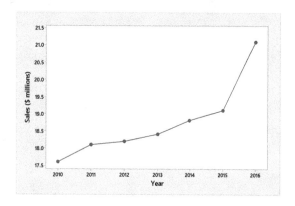 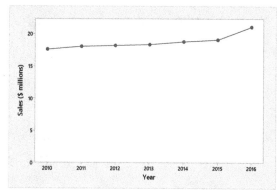

Forecasting

The term **forecasting** is often used to refer to the process of predicting where a time series is headed in the future. Weather forecasts are made by using current and historical weather data to predict future weather conditions. You may have heard about "analysts' expectations" for a company's earnings which are forecasts about the company's performance in a future time period based on data collected about the company and the economy. Like any prediction, a forecast is usually wrong, but if it's close enough, who cares? Would you care if the weather forecast for today's high temperature is off by 2 degrees? Probably not. Would you mind if a company's quarterly earnings misses analysts' expectations by $0.01 per share? Perhaps not.

The difficulty with forecasts is that they are, by definition, extrapolations. They are predictions using a future value of time that is, of course, beyond the range of the times observed in the time series. This isn't a deal-breaker as long as the forecast is not too far into the future. However, long-term forecasts must be taken with a grain of salt. I put a good deal of trust in tomorrow's weather forecast but I don't pay much attention to the forecast for 14 days from now.

Example: Use the following earnings per share data for Facebook to forecast annual earnings per share for 2019.[39]

Year	Earnings Per Share ($)
2015	1.31
2016	3.56
2017	5.49
2018	7.65

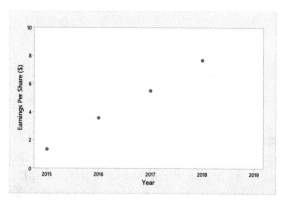

This very strong, positive, linear trend lends itself nicely to forecasting. First, find the correlation and the equation of the least-squares line:

$$\bar{x} = \frac{2015+2016+2017+2018}{4} = 2016.5 \qquad \bar{y} = \frac{1.31+3.56+5.49+7.65}{4} = 4.5025$$

$$s_x = \sqrt{\frac{(2015-2016.5)^2+\cdots+(2018-2016.5)^2}{4-1}} = 1.29099$$

$$s_y = \sqrt{\frac{(1.31-4.5025)^2+\cdots+(7.65-4.5025)^2}{4-1}} = 2.70569$$

$$r = \frac{(2015-2016.5)(1.31-4.5025)+\cdots+(2018-2016.5)(7.65-4.5025)}{(4-1)(1.29099)(2.70569)} = 0.99960948$$

$$m = r\frac{s_y}{s_x} = 0.99960948\frac{2.70569}{1.29099} = 2.095$$

$$b = \bar{y} - m\bar{x} = 4.5025 - 2.095(2016.5) = -4220.065$$

The least-squares line is:

$$\hat{y} = 2.095x - 4220.065.$$

To forecast earnings per share for 2019, simply plug in 2019 for x:

$$\hat{y} = 2.095(2019) - 4220.065 = 9.74$$

and the forecast for 2019 is $9.74 per share.

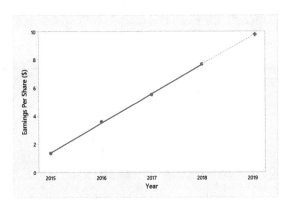

For a time series observed more frequently than a year, you'll need to create a time index to use as the x-variable. Start the index at 1 and increment by 1. Here's an example.

Example: Use the following earnings per share data for Ford Motor Company[40] to forecast quarterly earnings per share for the first quarter of 2019.

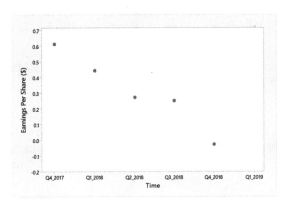

Year (y)	Quarter	Time Index (x)	Earnings Per Share ($)
2017	4	1	0.61
2018	1	2	0.44
2018	2	3	0.27
2018	3	4	0.25
2018	4	5	-0.03

Here's the equation of the least-squares line. I'll leave it as an exercise for you to verify it.

$$\hat{y} = -0.147x + 0.749$$

To forecast earnings per share for the first quarter of 2019, plug in 6 (the time index for Q1 2019) for x:

$$\hat{y} = -0.147(6) + 0.749 = -0.133$$

and the forecast for first quarter 2019 is –$0.133 per share.

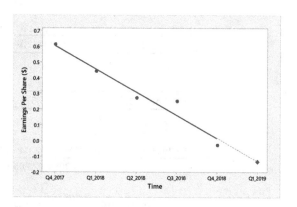

One implicit assumption you make when forecasting using this method is that the behavior of the time series will continue into the future as it has in the past. Sometimes, however, there is an unpredictable shock to the system that changes the behavior of the time series. In that case, forecasts for even the near-term can be very wrong. Those who remember the financial crash of 2008 will understand this very well.

Section 2.4 Exercises

1. Each part below lists a set of objects and two numeric variables observed on each object. Describe the form and direction of the relationship between the two variables that you would expect a scatterplot to show. Also describe the strength of the relationship if you can.

    a. people: height in inches, height in meters
    b. companies: number of employees, monthly amount paid in salaries
    c. college students: height, time to complete a crossword puzzle
    d. people: age, weight able to bench press

2. Each part below lists a set of objects and two numeric variables observed on each object. Describe the form and direction of the relationship between the two variables that you would expect a scatterplot to show. Also describe the strength of the relationship if you can.

    a. bowling balls dropped from the 20 tallest buildings in the world: height, speed at the bottom
    b. college seniors: SAT score, GPA
    c. vehicles: weight, miles per gallon
    d. mountains: height, number of people who summit per year

3. Shown here are four scatterplots. For each one, describe the form, direction, and strength of the relationship between the two variables.

4. Match these correlations to the four scatterplots in problem 3: -0.605, 0.156, 0.664, 0.988

5. A real estate firm randomly selects 5 properties currently for sale and has two appraisers (Jill and Anya) appraise each house. The appraised values are shown below.

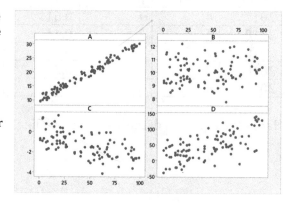

Property	Value (Jill)	Value (Anya)
1	$175,600	$181,650
2	$94,800	$96,800

Property	Value (Jill)	Value (Anya)
3	$227,300	$236,500
4	$419,650	$423,850
5	$275,200	$279,900

235,510   243,740

a. Make a scatterplot and describe the relationship between Jill's and Anya's appraised values.
b. Calculate the correlation coefficient.
c. Do the results of your analysis surprise you? Explain.

6. Here are the career number of stolen bases (SB), the number of times caught stealing (CS), and career batting average (BA) for the top 10 Major League Baseball players ranked by number of career home runs (HR)[41]:

Player	HR	SB	CS	BA
Barry Bonds	762	514	141	0.298
Hank Aaron	755	240	73	0.305
Babe Ruth	714	123	117	0.342
Alex Rodriguez	696	329	76	0.295
Willie Mays	660	338	103	0.302
Albert Pujols	638	111	41	0.301
Ken Griffey Jr.	630	184	69	0.284
Jim Thome	612	19	20	0.276
Sammy Sosa	609	234	107	0.273
Frank Robinson	586	204	77	0.294

a. Make a scatterplot and describe the relationship between the number of stolen bases and the number of times caught stealing for these players.
b. Calculate the correlation coefficient.
c. Which player seems to be unusual compared to the rest? Explain.
d. Remove the data point for the unusual player and recalculate the correlation coefficient. How did it change?

7. Refer to the data on the top 10 all-time home-run hitters in exercise 6.

a. Find the equation of the least-squares line. Let x = number of stolen bases and y = number of times caught stealing.
b. Interpret the slope of the line in context.
c. Predict the number of times caught stealing for a player having 400 career stolen bases.
d. As of the end of the 2018 season, Ricky Henderson held the MLB record for career stolen bases at 1,406. Using the least-squares line to predict the number of times Henderson was caught stealing is very questionable. Why?

8. Refer to the property appraisals data in exercise 5.

a. Find the equation of the least-squares line. Let x = Jill's value and y = Anya's value.
b. Interpret the slope of the line in context.
c. Predict Anya's appraisal value for a property that Jill appraised for $350,000.
d. Your boss wants you to predict Anya's appraisal value for a property that Jill has appraised for $1.5 million. What should you do?

9. In a study of the relationship between political advertising and voter turnout, researchers observed the number of presidential political impressions (advertisements) per capita and the percentage of registered voters who actually voted for a sample of U. S. counties in the 2004, 2008, and 2012 presidential elections.

a. Which variable is x? Which is y?
b. The researchers reported the slope of the least-squares line as 0.0205. The interpretation given in the article is, "an additional 10 impressions per capita [raises] voter turnout by almost 0.21 percentage points."[42] Comment on the author's interpretation.

10. Here again are the distances from Earth in megaparsecs (Mpc) and velocities (km/sec) for 24 galaxies observed by Edwin Hubble in the 1920s.

Galaxy	1	2	3	4	5	6	7	8
Distance (Mpc)	0.032	0.034	0.214	0.263	0.275	0.275	0.45	0.5
Velocity (km/sec)	170	290	-130	-70	-185	-220	200	290

Galaxy	9	10	11	12	13	14	15	16
Distance (Mpc)	0.5	0.63	0.8	0.9	0.9	0.9	0.9	1
Velocity (km/sec)	270	200	300	-30	650	150	500	920

Galaxy	17	18	19	20	21	22	23	24
Distance (Mpc)	1.1	1.1	1.4	1.7	2	2	2	2
Velocity (km/sec)	450	500	500	960	500	850	800	1090

a. Look at the scatterplot shown in section 2.4 and make a guess about what the correlation coefficient is.
b. Calculate the correlation coefficient. How close was your guess?
c. Find the equation of the least-squares line. Let x = distance and y = velocity.
d. Interpret the slope of the line in context.
e. The slope of the line in this context is known as the Hubble Constant and is related to the expansion of the universe. Find a current estimate of the Hubble Constant and compare it to the one you calculated. Explain the difference by doing some research.
f. Predict the velocity of a galaxy that is 1.5 Mpc from Earth.

11. Shown below are data for Western Hemisphere countries. The variables are the life expectancy for a newborn and average number of births per woman.[43]

Country	Life Expectancy	Births/woman	Country	Life Expectancy	Births/woman
Argentina	77	2.3	Guatemala	74	2.9
Bahamas	76	1.8	Honduras	74	2.4
Barbados	76	1.8	Jamaica	76	2.0
Belize	71	2.5	Mexico	77	2.2
Bolivia	69	2.8	Nicaragua	76	2.2
Brazil	76	1.7	Panama	78	2.5
Canada	82	1.5	Paraguay	73	2.5
Chile	80	1.8	Peru	75	2.4
Columbia	75	1.8	Puerto Rico	80	1.1
Costa Rica	80	1.8	U. S.	79	1.8
Dominican Republic	74	2.4	Uruguay	78	2.0
Ecuador	77	2.5	Venezuela	75	2.3
El Salvador	74	2.1	Virgin Islands	79	2.1

a. Make a scatterplot and describe the relationship between life expectancy and the average number of births per woman for these countries.
b. Find the equation of the least-squares line that will allow you to predict life expectancy from the average number of births per woman.
c. Interpret the slope of the line in context.
d. A newspaper publishes an article based on these data with the headline, "Birth Control Boosts Life Expectancy." Is this headline justified by the data? Explain.
e. The average number of births per woman in Iraq is 4.3. Predict life expectancy in Iraq. Give two reasons why your prediction may not be valid.

12. Shown here are 4 famous scatterplots summarizing relationships between 4 sets of x,y pairs. The data are often called "Anscombe's data" named after the statistician who created them.[44] For each relationship, the least-squares line is $\hat{y} = 3 + 0.5x$ and the correlation is $r = 0.82$.

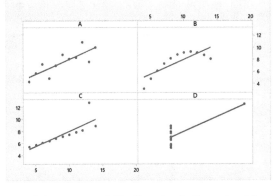

a. Comment on the use of the least-squares line and correlation for each plot.
b. What is the "moral of the story?"

13. Shown here are two scatterplots based on data collected on eruptions of the Old Faithful geyser in Yellowstone National Park, Wyoming.[45] The variables shown are: ATM = Actual Time in Minutes until the next eruption, DIM = Duration In Minutes of the eruption, DIS = Duration In Seconds of the eruption.

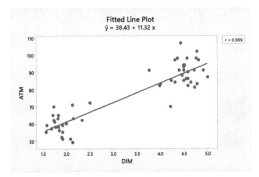

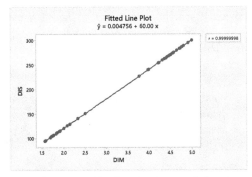

a. Suppose you're observing an eruption of Old Faithful that lasts 4.5 minutes. Predict the time until the next eruption.
b. Describe the relationship between DIM and DIS. Explain why you're seeing what you're seeing.
c. Why do you think the correlation between DIM and DIS isn't exactly 1?

14. You are employed in the marketing department of a large nationwide newspaper chain. The parent company is interested in investigating the feasibility of beginning a Sunday edition for some of its newspapers. However, before proceeding with a final decision, it needs to estimate the Sunday circulation that would be expected. In particular, it wants to predict the Sunday circulation that would be obtained by the newspaper in a metropolitan area that has a daily circulation of 200,000 papers. Circulation data from a random sample of 32 metro areas are obtained. The Sunday and Daily circulations (numbers of papers) for a particular newspaper for which the chain currently has both Sunday and daily editions are summarized here.

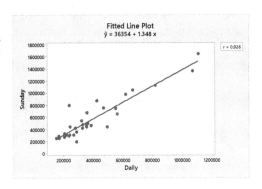

a. Describe the relationship between daily and Sunday circulation for these metro areas.
b. Interpret the slope of the least-squares line.
c. Predict the Sunday circulation for a metro area having a daily circulation of 200,000 papers.
d. The parent company asks you to use your data to predict the Sunday circulation for a metro area having a daily circulation of 2,000,000. What is the prediction? Is it valid? Explain.

15. The coach of a basketball team collected data on the 11 players across several games: how many points scored and how many personal fouls committed. The data are summarized here.

    a.   Interpret the slope of the least-squares line.
    b.   Do the data suggest that making more fouls is a way to increase point production for this team? Explain.

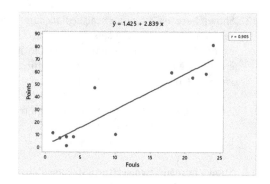

16. Researchers investigated whether the volume of chatter on Twitter could be used to forecast the box office revenues of movies. Opening weekend box office revenue (in millions of dollars) were collected for a random sample of five then-recent movies. Researchers also recorded the average number of tweets referring to the movie per hour for a time period of one week prior to the movie's release. The data (based on the study) are summarized here.[46]

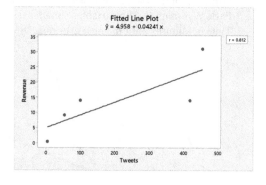

    a.   What issues would you need to consider before using these data as a basis for using Twitter activity to predict current new movie revenue?
    b.   Do these data suggest that trying to increase Twitter activity is a way to boost movie revenue? Explain.

17. Verify the equation of the least-squares line for the Ford Motor Company earnings per share data.

18. The Consumer Price Index (CPI) is a measure of the average cost of many consumer goods and services in the U.S.

Year	1915	1920	1925	1930	1935	1940	1945	1950
CPI	10.1	19.3	17.3	17.1	13.6	13.9	17.8	23.5

Year	1955	1960	1965	1970	1975	1980	1985	1990
CPI	26.7	29.3	31.2	37.8	52.1	77.8	105.5	127.4

Year	1995	2000	2005	2010	2015
CPI	150.3	168.8	190.7	216.7	233.7

    a.   Make a scatterplot and describe the behavior of consumer prices over time.
    b.   What do you think caused the dramatic change in the behavior of the CPI? (Making causal inferences is very difficult, but go ahead and speculate!)
    c.   Use the CPI data starting with 1980 to forecast the CPI for 2020.

19. Shown here are quarterly revenue data from Uber Technologies Inc.[47]

Year	Quarter	Revenue (billions of $)
2018	1	2.58
2018	2	2.77
2018	3	2.94
2018	4	2.97
2019	1	3.10

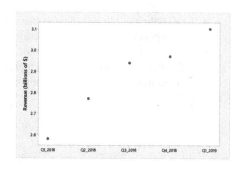

    a.   Why is the time series plot shown here misleading?

    b.   Make a more appropriate plot.

    c.   Forecast quarterly revenue for the second quarter of 2019.

20. Shown here are U.S. birth rates (number of births per 1,000 women aged 15 – 44) over time.[48]

Year	2010	2011	2012	2013	2014	2015	2016	2017
Birth Rate	13.0	12.7	12.6	12.4	12.5	12.4	12.2	11.8

    a.   Make a scatterplot with the y-axis scale starting at around 11.5 and going to around 13.5.

    b.   Make another scatterplot with the y-axis scale starting at 0 and going to around 13.5.

    c.   Which scatterplot would you use?

    d.   Forecast the birth rate for 2018.

    e.   As I'm writing this exercise, the birth rate for 2018 is not yet released but it likely will be by the time you're working this problem. If so, compare your forecast to the actual birth rate. How close was your forecast?

Notes

[1] U.S. Department of Education College Scorecard (as of 5/21/2019).

[2] Ibid.

[3] Heron M. (2018). "Deaths: Leading causes for 2016," *National Vital Statistics Reports*, 67(6), Hyattsville, MD: National Center for Health Statistics.

[4] Pew Research Center (2017). "Summer 2017 Political Landscape Re-interview Survey," conducted Aug. 15 – 17, 2017.

[5] Lazarus, S. C., Boushey, H. A., et al. (2001). "Long-Acting β_2-Agonist Monotherapy vs Continued Therapy with Inhaled Corticosteroids in Patients with Persistent Asthma: A Randomized Controlled Trial," *Journal of the American Medical Association*, 285 (20), 2583 – 2593.

[6] www.fidelity.com.

[7] www.transtats.bts.gov/OT_Delay.

[8] Schoen Consulting (2018). "Holocaust Knowledge and Awareness Study," commissioned by The Conference on Jewish Material Claims against Germany.

[9] www.statcan.gc.ca/tables-tableaux/sum-som/l01/cst01/demo02a-eng.htm.

[10] Azzalini, A. and Bowman, A. W. (1990), "A Look at Some Data on the Old Faithful Geyser," *Applied Statistics*, 39(3), 357 – 365.

[11] Hyndman, R. J. and Fan, Y. (1996), "Sample Quantiles in Statistical Packages," *The American Statistician*, 50(4), 361 – 365.

[12] U.S. Department of Education College Scorecard (as of 5/21/2019).

[13] Leonhardt, D. (July 28, 2000), "John Tukey, 85, Statistician; Coined the Word 'Software'," New York Times.

[14] www.wikipedia.com.

[15] www.espn.com/mlb.

[16] data.delaware.gov.

[17] O'Shea, A. (2020). "The Average 401(k) Balance by Age," www.nerdwallet.com.

[18] American Association of Individual Investors, www.aaii.com.

[19] census.gov/2010census/data.

[20] www.nhc.noaa.gov.

[21] Ibid.

[22] Lazarus, et al. (2001).

[23] ope.ed.gov/campussafety. U.S. Department of Education Campus Safety and Security website (2014).

[24] Based on a survey conducted June 10 – July 12, 2015, Pew Research Center.

[25] Lazarus, et al. (2001).

[26] www.trisomy18.org.

[27] data.delaware.gov. Data shown are for the 2017 school year.

[28] O'Shea, A. (2020).

[29] Adapted from Pew Research Center (2015). "Raising kids and running a household: how working parents share the load."

[30] ope.ed.gov/campussafety. U.S. Department of Education Campus Safety and Security website (2014).

[31] Kogan, M.D., Alexander, G. R., et al. (2000). "Trends in twin birth outcomes and prenatal care utilization in the United States, 1981 – 1997," *Journal of the American Medical Association*, 284 (3), 335 – 341, Table 2.

[32] Lindquist, T. M., Olsen L. M. (2007). "How much help is too much? An experimental investigation of the use of check figures and completed solutions in teaching intermediate accounting," *Journal of Accounting Education, 25 (3)*, 103 – 117.

[33] U.S. Department of Education College Scorecard (as of 5/21/2019).

[34] Hubble, E. (1929). "A Relation Between Distance and Radial Velocity Among Extra-Galactic Nebulae," *Proceedings of the National Academy of Sciences*, 15 (3), 168 – 173. Data are from Table 1.

[35] I used Zillow.com to perform the search and was able to obtain these data very easily.

[36] www.wikipedia.com.

[37] World Almanac & Book of Facts (1993), New York: Pharos Books.

[38] www.transtats.bts.gov/OT_Delay/OT_DelayCause1.asp

[39] www.marketwatch.com.

[40] Ibid.

[41] www.espn.com/mlb.

[42] Spenkuch, J. and Toniatti, D. (2018). "Political Advertising and Election Results," *Quarterly Journal of Economics*, 1981 – 2036, Table III.

[43] data.worldbank.org.

[44] Anscombe, F. J. (1973). "Graphs in Statistical Analysis," *American Statistician*, 27 (1), 17 – 21.

[45] Azzalini, A. and Bowman, A. W. (1990).

[46] Based on *IEEE International Conference on Web Intelligence and Intelligent Agent Technology*, 2010.

[47] www.marketwatch.com.

[48] Martin, J. A. et al. (2018). "Births: Final data for 2017," *National Vital Statistics Reports*, 67(8), Hyattsville, MD: National Center for Health Statistics.

Chapter 3: Producing Data

Introduction

Statistics is the science of extracting information from data in order to answer a question. It is, in a sense, data science. In chapter 2, we learned some of the basic methods used to describe distributions of data and that different tools are required depending on the type of data we have (numeric or categorical) and the data structure (one categorical variable, two numeric variables, numeric data collected over time, etc.). The science of data, however, goes beyond simply playing with data. It also involves careful consideration of *how* the data are collected. It turns out, the manner in which the data are collected can have a profound impact on the kinds of information we can and cannot extract from the data.

Suppose your friend tells you, "My Volvo has over 300,000 miles on it and it has never been in the shop for a serious repair." Would you therefore go out and buy a Volvo for your next car? I hope not. It could be that Volvos are generally very reliable cars or it could be that your friend is a very careful driver and takes good care of her car. It could be that the reason her car lasted so long is that she maintains it well and not necessarily that it's a Volvo. Do you see the difference? It is tempting to make the conclusion that the *reason* for the longevity of your friend's car is that it is a Volvo but the data supplied by your friend (the make of the car, the number of miles, and the repair history) do not allow conclusions about the *cause* of the longevity.

Also notice that these are data supplied by one person (your friend). The data set might look like this:

Make	Mileage	Serious Repair?
Volvo	300,000	no

There is a name for data sets having only one row. They are called *anecdotes*. Webster's dictionary defines an **anecdote** as "a short narrative concerning a particular incident or event of an interesting or amusing nature."[1] You can think of anecdotes as "show and tell for adults." The problem with anecdotal evidence like this is that it is good for little else except to generate more questions. We've already seen that it is impossible to draw any sort of cause/effect conclusion from these data. Further, there is no way for us to use the data to make a general statement about the longevity this particular model Volvo, regardless of the reason.

Here's another example of anecdotal evidence that must be taken with a grain of salt. A recent college graduate says, "Standardized tests don't mean anything. I got a 1040 on the SAT but I had a 3.6 GPA in college." His argument is that while his SAT score was mediocre, his performance in college was very good. The problem is that it's anecdotal evidence, a story about a particular incident:

SAT	GPA
1040	3.6

There is no way to determine whether this kind of thing is typical (mediocre SAT scores related to high GPAs) or an outlier. There is no way to assess the relationship between SAT scores and GPAs with one data point.

In this section we'll explore much better ways to obtain data to answer questions. While anecdotes can be interesting, they should only be used as the basis for generating questions, not for answering them. Do Volvos last longer on average than other cars? Is there a relationship between SAT scores and GPA? These are questions that require careful data collection to answer.

3.1 The Scientific Method

Statistics is a science and therefore uses the scientific method to answer questions. Here's a quick review of the scientific method:

1. Pose a Question
2. Make a Hypothesis
3. Collect Data
4. Analyze the Data
5. Make a Conclusion

The conclusion drawn may generate new questions, and the process iterates. That's how we've accumulated a large body of knowledge in the sciences.

Notice the crucial role data play in the scientific method. It's the data that drive the conclusions and allow us to answer the question. That's why it is so important that the data are of the highest quality possible. Getting good data to answer a question is like using good ingredients in a recipe. Using the finest available ingredients can make a dish delicious. Using low-quality ingredients can make the same dish taste nasty. It's the same way in data science. If you put garbage data into the scientific method process, you get garbage conclusions out of it. Using high-quality data can result in very reliable conclusions.

It is easy to focus on the "Analyze the Data" step of the scientific method and forget about the "Collect Data" step. Many folks think statistics is only about analyzing data. I hope by the time you finish reading this chapter, you won't be one of those folks.

There are many ways to collect data and many sources from which data can be collected. You've seen some of them in chapter 2. Data can also be extracted from images and sounds. Your phone might be able to recognize your fingerprint or your voice and it does that by converting the image of your fingerprint or the sound of your voice into data. You might wear a device that counts the number of steps you take. It does that by converting motion into data. New and interesting data sources are popping up all the time. Fortunately you don't have to be an expert in all of the data sources to understand some basic data collection principles.

The method of data collection goes back to the first step in the scientific method, the question. The way data are collected must be driven by the kind of question you ask. Remember those three common types of statistical inferences I introduced in chapter 1? Here they are again with some examples of the kinds of questions that statistics and data science can help us answer.

1. Predictive Inference: What is an object's estimated/predicted value of a particular variable? Where might a process be headed in the future?
    *   Is an email message spam or not? Should it be moved to the junk mail folder?
    *   What is the market value of a piece of real estate?
    *   What is the forecast for on-time percentage for Southwest Airlines next month?
    *   What is an estimate of a company's sales next quarter?

2. Generalization Inference: What are the characteristics of a large group of objects?
    *   What is the average starting salary for all entry-level accountants?
    *   By how much of a margin is a political candidate expected to win an election?
    *   What is the relationship between SAT score and GPA for all college students?

3. Causal Inference: Is an intervention effective?
    *   Would changing the layout of a webpage increase click-throughs on advertisements?
    *   Does one method of classroom instruction work better than another?

While these are three very common kinds of questions – the questions with which this book is primarily concerned – they are certainly not the only kinds. It would be impossible to try to classify them all because new kinds of questions are proposed every year. In section 2.5 we took a deep dive into predictive inference in the case of two numeric variables. In this chapter we'll take a look at the last two kinds of questions in detail. Section 3.2 deals with generalization inference: questions about the characteristics of a large groups of objects and random sampling as the best way to collect data to answer those questions. Section 3.3 deals with causal inference: questions about whether an intervention works and random assignment as the best way to collect data to answer those questions. While collecting high-quality data to answer questions about prediction/forecasting can be challenging (so that we avoid extrapolating), doing so for the last two kinds of questions requires extra special care.

Section 3.1 Exercises

1. List, in order, the steps in the scientific method.

2. Why are quality data so important in the scientific method?

3. List and define the three types of inferences.

4. Classify each question according to the type of statistical inference it requires: predictive, generalization, or causal.

    a. Does working more increase people's chances of stroke?
    b. How does the proportion of people in the U.S. who snore compare to that in Canada?
    c. What is the risk of default for Jill who wants to take out a loan?
    d. What percentage of all daily bank transactions are processed correctly?

    e.   What is the estimated college GPA for Jack who scored a 28 on the ACT?

    f.   Does room temperature have an effect on workers' performance?

5.   Classify each question according to the type of statistical inference it requires: predictive, generalization, or causal.

    a.   Will voter turnout in the next election be better or worse than in the last election?

    b.   How many parents are in favor of a new policy being considered by the school board?

    c.   What is the average weight of product in bags of chips?

    d.   Does eating chicken soup really help a cold get better faster?

    e.   How high is the average contaminant level in batches of raw milk collected from local farmers?

    f.   Which method of reading instruction (phonics or whole language) works better?

    g.   What is Google's revenue forecast for next year?

6.   Classify each question according to the type of statistical inference it requires: predictive, generalization, or causal.

    a.   Do a majority of people oppose abortion in the third trimester of pregnancy?

    b.   How much body fat is Juan, a 6'1" male weighing 180 pounds, estimated to have?

    c.   What effect does vitamin C have on cold prevention?

    d.   What fraction of the trees in a forest are ready for harvesting?

    e.   What kind of music, if any, enhances retention during studying?

    f.   Can you estimate the electricity usage of a house you're interested in buying?

    g.   How do different watering regimens affect tomato growth?

7.   Consider the following anecdotes and conclusions. For each one, write a related question that could be answered using one of the three types of statistical inference.

Example: "Jerry doesn't have very good social skills. It figures; he was homeschooled." Note that this implies that homeschooling causes poor social skills. A question that could be answered using causal inference might be, "Does homeschooling result in poorer social skills than traditional schooling?"

    a.   "There's another Maryland driver who just changed lanes without signaling. Maryland drivers just aren't as good as other drivers."

    b.   "See, now you have a cold. I told you to wear a coat when you go outside in cold weather."

    c.   "My broker told me to buy stock in XYZ company. They just had record-high earnings last quarter."

3.2 Generalizing to Populations from Random Samples

Here are some questions that involve a large group of objects.

- What is the average starting salary of all bachelor-level accountants? (Not knowing the answer could leave you susceptible to a lower-than-average offer from a potential employer.)

- Do a majority of U.S. adults favor a flat tax? (Not knowing the answer could end your political campaign.)

- Does the group of all potential customers like tables or booths in restaurants? (Not knowing the answer to this question could be disastrous for your new restaurant in a particular area.)

- What age group uses internet dating more? (Not knowing the answer could send your dating app startup company down the tubes.)

- How much time during the day do employees typically spend off-task? (Not knowing the answer could cost your company dearly in lost productivity.)
- What is the average price of gas in your area? (Not knowing the answer may cost you a few extra pennies at the gas pump.)

Each one of these questions is about a very large group of objects. It would be impractical at best and impossible at worst to collect data on each and every one of those objects. It is nearly impossible to ask every U.S. adult about their opinion on taxation. Even the U.S. Census, conducted once every 10 years, which hopes to collect data on every household, likely misses a few. More practical ways of answering questions like these involve generalizing information collected on a smaller (in many cases *much smaller*) subset of objects. We call such a subset a **sample**. The larger group of objects from which the sample is taken is called the **population**. It is easier to collect data on the smaller sample of objects than it is to collect data on every object in the population. That's the value in sampling.

The problem with sampling is that it leaves us with incomplete information about the population. The only data we'll get is that from the sample of objects; we won't see the entire population data set. Our job will be to make an educated guess about what the population looks like based on what we see in the sample data. A statement about a population based on data collected on a sample is called a **generalization inference**. That is, we must infer what the population looks like by generalizing the patterns we see in the sample.

Things can go wrong when dealing with only a sample from the population, not because of errors or mistakes in arithmetic (we'll always assume we've computed correctly), but because of the way the data are collected or summarized. If something about the inference process tends to systematically over- or under-estimate a population feature, we call that systematic mistake "bias." **Bias** results when the inference process systematically results in a generalization that is either too high or too low compared to the actual population feature. As we'll see, bias often results from the data collection process.

Consider the questions above. The following table shows the population, an example of a sample, and the necessary variables that would need to be observed on each object.

Question	Population	Sample	Variable(s)
What is the average starting salary of all bachelor-level accountants?	all bachelor-level accountants	10 bachelor-level accountants at a consulting firm in your city	starting salary
Do a majority of U.S. adults favor a flat tax?	all U.S. adults (over 300 million of them!)	100 adults leaving a grocery store on a Saturday	opinion on flat tax
How much time during the day do employees typically spend off-task?	all employees	20,000 employees of a large banking company	time per day spent off-task
What is the average price of gas in your area?	all gas stations in your area	7 gas stations along your street	price of regular unleaded gas

Question	Population	Sample	Variable(s)
What age group uses internet dating more?	all U.S. adults	1957 U.S. adults randomly selected	age and whether used internet dating

Which of those samples is the best one? You might say it's the 20,000 employees from the population of all employees. Or you might say it depends on the population size, that the largest sample size compared to the population size is the best. Perhaps there are only 20 gas stations in your area so that 7 of them would be a relatively large sample size. The answer is a resounding "No!" in each case.

The best sample is the one that *represents* the population best. It's the one that has the best chance of capturing all the features of the population that are relevant to the question of interest. If a sample over-represents one particular group and under-represents another, it's likely that the sample will give an incorrect impression of what the population looks like. While sampling has the advantage of reducing the number of objects we need to look at, it introduces the chance that the objects we look at are somehow different from the objects in the population in some way. If that's the case because of the way the sample is taken, then we say the sample does not represent the population well, and we say that there is *selection bias* in the sample data. **Selection bias** occurs when the sampling method systematically over-represents or under-represents a particular group of objects in the population.

Let's look at each of the samples above and comment on their ability to represent the population.

Question	Population	Sample	Variable(s)
What is the average starting salary of all bachelor-level accountants?	all bachelor-level accountants	10 bachelor-level accountants at a consulting firm in your city	starting salary

It could be that the cost of living in your city is higher than in most other places in the country and so starting salaries for most jobs are generally higher to compensate. Or perhaps this consulting firm is one of the best and generally pays its employees very well. If we only look at bachelor-level accountants at this one firm in your city, the 10 starting salaries we observe will give us an inflated impression of average starting salary of all bachelor-level accountants. This sample is very vulnerable to selection bias and *not representative* of the population.

Question	Population	Sample	Variable(s)
Do a majority of U.S. adults favor a flat tax?	all U.S. adults (over 300 million of them!)	100 adults leaving a grocery store on a Saturday	opinion on flat tax

What if Saturday shoppers (maybe those with steady weekday jobs) as a group generally have a different opinion of the flat tax than those that shop during the week? What if the store is in a predominantly Democratic area and hence those shoppers typically prefer the progressive tax (with different tax "brackets") to a flat tax? In either case, the 100 adults we ask will not give us an accurate picture of the U.S. adult population on this issue. This sample is very vulnerable to selection bias and *not representative* of the population.

Question	Population	Sample	Variable(s)
How much time during the day do employees typically spend off-task?	all employees	20,000 employees of a large banking company	time per day spent off-task

A sample size of 20,000 is quite large; no one would accuse you of not doing your work should you amass a data set on these employees! However, does the sample represent all employees with respect to time per day spent off-task? Are employees of this bank typical of the population of all employees? Suppose the working culture at this bank is not good, the employees generally do not enjoy their work, and tend to spend more time on social media and the internet during the day than the general population of all employees. Then the average time per day spent off-task would be higher in the sample than it is in the general population. The sample over-represents employees in the upper part of the off-task time distribution and under-represents those in the lower part. Even though the sample size is large, there is likely selection bias because of the way the sample was obtained; it is *not representative* of the population.

Question	Population	Sample	Variable(s)
What is the average price of gas in your area?	all gas stations in your area	7 gas stations along your street	price of regular unleaded gas

There can't be that many gas stations in your area, right? Seven should be a reasonable sample size, right? Perhaps, but again, the sample size takes a back seat to the method of data collection. What if the competition among those 7 gas stations on your street drives prices lower than elsewhere in your area where there are single gas stations scattered around? If so, then the mean price of gas in the sample would be lower than it would be in the population of all gas stations in your area. Then this sample would *not represent* the population because of selection bias.

Are you beginning to see the difficulty in collecting good data? It certainly is not an easy task. You might wonder if there is a good way to get a representative sample and avoid selection bias. Let's look at the last example.

Question	Population	Sample	Variable(s)
What age group uses internet dating more?	all U.S. adults	1957 U.S. adults randomly selected	age and whether used internet dating

In this sample, the adults were *randomly selected*. The phrase "randomly selected" or "chosen at random" in this context is a loaded phrase; it does not have one of the usual meanings! It does *not* mean "odd or unpredictable" as in "I have a random sense of humor." It does *not* mean "haphazard, without definite aim, purpose, or reason" as in "We just randomly drove around." It does *not* mean "unknown or suspiciously out of place" as in "A couple of random pelicans just flew by." The randomness we're referring to here means "structured uncertainty." A **random sample** is a sample obtained in a way such that the outcome of one random sample is uncertain but the collective outcomes of many random samples has a structure that can be predicted in advance. We'll look at randomness in much more detail in the next few chapters but for now the important point about a random sample is that it *eliminates any systematic selection bias*.

I know what some of you might be thinking: "What?! Leave it up to chance?! How can that be scientific?" If that's you, you're not alone.

Example: The American Community Survey is a survey of American households that provides annual information about the nation. Households are randomly selected from the population of all American households and the results are used to determine how billions of dollars of federal and state funds are distributed. As you might suspect, where there's money, there's controversy. In 2012, an article in the New York Times quoted a Florida congressman as saying, "We're spending $70 per person to fill this out. That's just not cost effective especially since in the end this is not a scientific survey. It's a random survey."[2] On the contrary; it is exactly the random selection process governed by the laws of probability that makes it a scientific survey. Using any other kind of method to select the households may plague the sample with selection bias.

One kind of random sample is a *simple random sample*. A **simple random sample (SRS)** is one in which each possible sample of size n from a population has the same probability of being the actual one chosen. Write the names of all the objects in the population on equal-sized slips of paper, put those slips of paper in a hat, mix well, and have a blindfolded person draw 4 slips from the hat. You've just taken a SRS of 4 from the population!

Unfortunately in practice, it's not so easy. In many populations, the hat would have to be the size of a swimming pool to accommodate all those slips of paper. In those cases, a computer can make the job easier.

Example: Shown below is a partial list of the 383 airports classified by the Federal Aviation Administration as primary airports: publically owned, receive scheduled passenger service, and had at least 10,000 passenger boardings in 2014.[3] The airports are listed in order of the number of boardings from most to least. The Hub variable indicates whether the airport is a large hub (L), medium hub (M), small hub (S), or nonhub (N) based on the number of passenger boardings.

State	Locid	City	Name	Hub	RandNum
GA	ATL	Atlanta	Hartsfield - Jackson Atlanta International	L	0.5987205
CA	LAX	Los Angeles	Los Angeles International	L	0.0003657
IL	ORD	Chicago	Chicago O'Hare International	L	0.9450574
TX	DFW	Fort Worth	Dallas/Fort Worth International	L	0.7920791
NY	JFK	New York	John F Kennedy International	L	0.1068305
CO	DEN	Denver	Denver International	L	0.6304081
CA	SFO	San Francisco	San Francisco International	L	0.1696685
NC	CLT	Charlotte	Charlotte/Douglas International	L	0.2100343
NV	LAS	Las Vegas	McCarran International	L	0.1149479
AZ	PHX	Phoenix	Phoenix Sky Harbor International	L	0.3047632
TX	IAH	Houston	George Bush Intercontinental/Houston	L	0.7879612
FL	MIA	Miami	Miami International	L	0.9144805
WA	SEA	Seattle	Seattle-Tacoma International	L	0.8025854
NJ	EWR	Newark	Newark Liberty International	L	0.8868462

State	Locid	City	Name	Hub	RandNum
...	...	...	...	...	...
MA	PVC	Provincetown	Provincetown Municipal	N	0.5044562
IL	UIN	Quincy	Quincy Regional-Baldwin Field	N	0.7947195

Suppose we have these data in an Excel spreadsheet. To take a SRS of 10 airports, add a random number column by using the RAND() function. This gives each airport a random number between 0 and 1. Then sort the entire spreadsheet by the RandNum column and take the first 10 airports as the SRS and collect data on only those 10 airports instead of all 383 airports.

State	Locid	City	Name	Hub	RandNum
CA	LAX	Los Angeles	Los Angeles International	L	0.0003657
MT	MSO	Missoula	Missoula International	N	0.001642
TN	MEM	Memphis	Memphis International	S	0.0024235
MN	BJI	Bemidji	Bemidji Regional	N	0.0025971
TX	AUS	Austin	Austin-Bergstrom International	M	0.0041356
TX	SJT	San Angelo	San Angelo Regional/Mathis Field	N	0.0072471
WY	LAR	Laramie	Laramie Regional	N	0.0075757
WI	GRB	Green Bay	Austin Straubel International	N	0.0103016
TX	ABI	Abilene	Abilene Regional	N	0.0182296
MA	MVY	Vineyard Haven	Martha's Vineyard	N	0.0186098
OH	LUK	Cincinnati	Cincinnati Municipal Airport Lunken Field	N	0.0186326
OH	LCK	Columbus	Rickenbacker International	N	0.0262652
NC	OAJ	Jacksonville	Albert J Ellis	N	0.0302729
AK	KSM	St Mary's	St Mary's	N	0.0306655
...	...	...	...	...	...
FL	VPS	Valparaiso	Eglin AFB	N	0.9958715
IN	IND	Indianapolis	Indianapolis International	M	0.9964839

In this way, every possible group of 10 airports has the same probability as being the sample you choose. You might notice that there are more nonhubs than small, medium, or large hubs in the sample. That's okay because there are many more airports in the population classified as nonhubs, so we would expect that they would be represented more heavily in a SRS. My hunch is that you recognize more airports in the first list than you do in the SRS. They are the largest airports; they get more air traffic (you might have flown through several of them); they are in the news much more frequently than smaller airports. If left to your own devices, you might select 10 airports with which you are familiar as your sample and thus over-represent the larger airports.

The SRS eliminates selection bias, a *systematic* over- or under-representation of certain groups. Over *many* SRSs, the *average* composition of the samples would mirror the population. However, there is no guarantee that any one *particular* SRS has exactly the same characteristics as the population. In the airport example above, it is possible that a SRS of 10 airports would contain no large hubs since there are only 30 large hubs in the population of all 383 primary airports. (There's a 43.8% chance of such a sample!) When you toss 10 fair coins, it is possible to get 8 heads and only 2 tails. A SRS could possibly result in a similar imbalance.

Here's an example of a population of 40 shapes: 20 dots and 20 squares. One possible SRS of 4 is shown below.

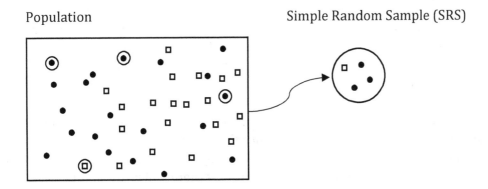

If we know ahead of time that there are certain groups we definitely want to accurately represent in a single random sample, we can conduct a fancier version of the SRS called a *stratified random sample*. A **stratified random sample (STRS)** is obtained by taking a SRS of objects in each of several subpopulations called *strata*.

What if that population of dots and squares was naturally divided into four subpopulations or strata as shown below? Notice that the dot/square ratio is very different depending on the strata. The SRS above failed to sample objects from the two strata having larger fractions of squares. If you could take a SRS of 1 from each strata, you have a better chance of the sample more accurately representing the population.

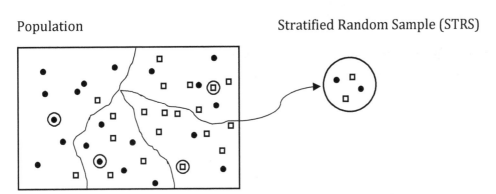

Example: The population of 383 primary airports is classified by the FAA according to size as large hub, medium hub, small hub, and nonhub. The frequency table is shown below.

Hub	Frequency	Percent
Large	30	7.8%
Medium	31	8.1%
Small	72	18.8%
Nonhub	250	65.3%
Total	383	100.0%

If you wanted to guarantee that a random sample of 100 airports has similar percentages of large hub, medium hub, small hub, and nonhub airports as the population, you could take a SRS of 8 large hub airports, a SRS of 8 medium hub airports, a SRS of 19 small hub airports, and a SRS of 65 nonhub airports. Putting the four SRSs together would constitute a STRS stratified by size.

Eliminating the chance of a "lopsided sample" is a big advantage of the STRS but it turns out there are other goodies that come free with a STRS. One is that it always allows a comparison of the strata based on the sample data. With a SRS there is a chance that objects from some strata/groups may not appear in the sample. In such a case there would be no way to extract information about those groups. A STRS eliminates this problem by guaranteeing that there will be data from each stratum/group.

The other huge statistical goodie is that the STRS is almost always *more efficient* than the SRS; you can usually get away with collecting data on fewer objects if you do a STRS compared to a SRS. That's because the STRS described here (by using sample sizes proportional to the strata sizes) results in estimates that are generally closer to the population figures than the SRS generates.

Example: Let's take both a SRS and a STRS (stratified by size) of 100 airports from the population of all 383 primary airports. For each airport we'll observe the number of passenger boardings in 2014 and use the sample mean as an estimate of the population mean. Here are the results:

SRS Sample Mean	STRS Sample Mean
$\bar{x} = 1,256,241$	$\bar{x} = 1,734,382$

But it's really hard to evaluate a statistical method based on only one application of the method. So let's repeat the sampling methods 1,000 times each and keep track of the sample means for all 1,000 samples:

	SRS Sample Mean	STRS Sample Mean
1	1,256,241	2,226,512
2	1,775,287	2,148,855
3	2,217,921	1,916,993
...	...	...
1,000	2,895,352	2,140,184

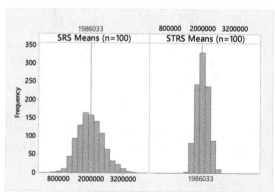

This simulation analysis gives us a really good idea of how these methods perform. The histograms show the distributions of the 1,000 sample means. The centers are both at the population mean number of boardings for all 383 airports

(1,986,033). Even though there was no one sample where $\bar{x} = 1,986,033$, both methods "get it right" on average.

But notice that the sample means for STRS don't vary nearly as much as the sample means for SRS! A STRS gives more consistent estimates and we wouldn't have to worry about getting an estimate as far away from 1,986,033 as we might from a SRS. The STRS eliminates the possible SRS's that come from only nonhub airports (which give very low estimates of the population mean) or only from large and medium hubs (which give very high estimates of the population mean). So the STRS method decreases the variability of the estimates compared to the SRS method! In order to get the same consistency of sample means from a SRS, we'd have to use a sample size of around 240 (see the figure to the right).

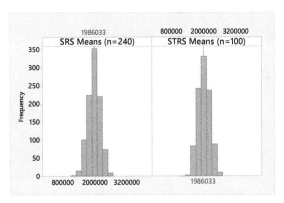

In this example, the strata were defined in terms of airport size which gave strata that were very different in terms of number of passenger boardings. It turns out that the more the strata differ with respect to the variable of interest, the more efficient the STRS is compared to the SRS. If the strata don't differ much, then the STRS will be about the same as the SRS in terms of efficiency. The moral? Choose a stratification method that creates strata that are as different as possible with respect to the variable of interest.

Both the SRS and the STRS require a complete list of the objects in the population of interest. In some circumstances, such a list may not exist or would be very difficult to compile. For example, consider the first question at the start of this section: "What is the average starting salary of all bachelor-level accountants?" A SRS or STRS of all bachelor-level accountants would require a list of all of them. A more practical approach might be to take a random sample of accounting firms and then collect data on every bachelor-level accountant in each of the selected firms. Such a random sample is called a *cluster random sample*. In a **cluster random sample (CRS)**, the population is (often naturally) divided into groups called clusters and a SRS of clusters is drawn and observed. The simplest cluster sampling method observes data on each object in the selected clusters. More complex cluster sampling methods might take a SRS of objects within each selected cluster or a SRS of subclusters in each selected cluster.

Here's our population of 40 objects again but this time suppose they naturally are found in groups of two. Taking a SRS of 2 clusters instead of 4 objects results in a cluster random sample.

Population   Cluster Random Sample (CRS)

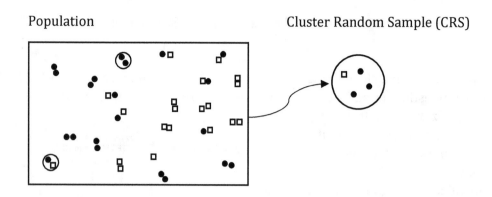

Example: In Uttar Pradesh (a state in India), it is estimated that only 43.5% of children in 5th grade can read at a basic level.[4] Incredibly, teachers in primary schools in India can be absent between 25% and 40% of the time![5] A study was conducted in the Juanpur district within Uttar Pradesh to compare several ways to get community members more involved in public services like education.[6] The idea is that conditions might improve if, for example, communities started holding teachers more accountable for their behavior. In order to collect some baseline data, the researchers needed a representative sample of 2800 households from all households in the Juanpur district. Finding a complete list of all households in the Juanpur district could be difficult (it may not even exist!), so taking a SRS of 2800 households doesn't work. Instead, the researchers used a variation of cluster sampling. Each district in India is divided into administrative blocks and each block contains about 100 villages, on average. So the researchers randomly selected 4 administrative blocks, then randomly selected about 70 villages within each administrative block, and then randomly selected 10 households in each village for a total of 4(70)(10)=2800 households. Once you get to the village level, obtaining a list of households in the village is not that hard, so cluster sampling is really the only way to go here.

The problem with the CRS is that it is usually *less efficient* than the SRS; you can usually get away with collecting data on fewer objects if you do a SRS compared to a CRS. That's because the CRS (collecting data on all objects in each of a SRS of clusters) usually results in estimates that are less consistent than the SRS generates. The more alike objects are within a cluster and the more differences there are between clusters, the less efficient the CRS will be compared to the SRS.

Example: Ethnicity is a variable that tends not to vary much within a household: individuals within a household tend to be all of the same ethnicity. Suppose we wanted to estimate the percentage of Hispanic individuals in a certain community. There are 5000 households in the community with 4 individuals per household for a total of 20,000 individuals. Let's say that there are 1000 purely Hispanic households giving 4000 (20%) Hispanic individuals in the population. Let's take both a SRS of 400 individuals and a CRS of 100 households (which will also give us 400 individuals). In each sample we'll compute the percentage of Hispanics. Here are the results:

SRS Sample % Hispanics	CRS Sample % Hispanics
19	14

But it's really hard to evaluate a statistical method based on only one application of the method. So let's repeat the sampling methods 1,000 times each and keep track of the sample percentages for all 1,000 samples:

	SRS Sample %	CRS Sample %
1	19.00	14
2	19.25	21
3	19.00	14
...	...	...
1,000	22.50	24

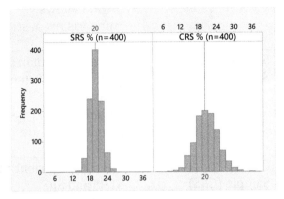

This simulation analysis gives us a really good idea of how these methods perform. The histograms show the distributions of the 1,000 sample percents. The centers are both at the population percentage of Hispanics (20%). But notice that the sample percents for SRS don't vary nearly as much as the sample percents for CRS! A SRS gives more consistent estimates and we wouldn't have to worry about getting an estimate as far away from 20% as we might from a CRS. The reason is that each household selected in the CRS provides 4 times as many objects as each individual selected in the SRS. Any chance imbalance will be magnified in the CRS. So the CRS method increases the variability of the estimates compared to the SRS method! In order to get the same consistency of sample percents from a CRS, we'd have to use a sample size of around 1600, or 400 households (see the figure to the right).

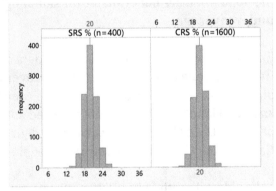

In this example, the clusters (households) could look very different from other clusters in terms of ethnicity and so the CRS is much less efficient than the SRS. It turns out that the more similar the clusters look with respect to the variable of interest, the more efficient the CRS. If each cluster looks exactly the same, then there would be no variability from one cluster random sample to another and the CRS would be more efficient; a single cluster would perfectly represent the population. That doesn't happen much in practice as there are usually cluster to cluster differences.

The silver lining with the CRS is that sampling clusters may be cheaper than sampling individual objects. For example, it might be much cheaper to travel to a few randomly selected hospitals and interview patients there than to travel all over the country to interview a SRS of patients nationwide. You may be able to offset the loss of efficiency in a CRS by being able to collect data on more objects.

The following table summarizes the three random sampling methods and includes recommendations for implementing each method.

Sampling Method	Notes	Recommendations
Stratified Random Sample (STRS)	• often more efficient than SRS and CRS • always allows comparisons of strata	• Define strata so that they are as *different as possible* with respect to the variable of interest. • Consult a statistician before collecting data.
Simple Random Sample (SRS)	• less efficient than STRS • often more efficient than CRS • default method assumed by most software packages	• Use a SRS if you will be doing the data analyses yourself.
Cluster Random Sample (CRS)	• often less efficient than STRS and SRS • may be more convenient/cheaper	• Use only when conducting a SRS or STRS is impractical or too expensive. • Consult a statistician before collecting data.

Just to be clear: when taking a random sample, we must sample from the entire population of interest. There must not be any part of the population unavailable for sampling. If any objects from the population are not available for sampling, then there is likely to be selection bias because of **undercoverage**.

Example: Suppose you want to collect data on people in your city and you take a SRS of 250 people from the phone book. Beware of undercoverage! Even though you've taken a random sample, the phone book very likely does not include everyone in your city. People who only have a cell phone and not a landline will likely not be listed and so your sample will not be representative of all people in the city. These days, random digit dialing is used in many good surveys so that both "landline only" and "cell phone only" people are reached.

Take a deep breath...SRS, STRS, CRS...starting to sound like alphabet soup? Just remember that the goal of these methods is to provide a sample that *represents* the population by avoiding selection bias. Random sampling is the best way to do that. Despite their differences, the SRS, STRS (with sample sizes proportional to the strata sizes), and CRS all provide estimates that are correct on average over many applications of the method. *Data collected from non-random samples may provide estimates that are incorrect on average because of selection bias.*

Example: In 1936 a magazine called The Literary Digest conducted a survey of over 10 million people asking who they planned to vote for in the 1936 presidential election. The Republican candidate was Alfred Landon, governor of Kansas, and Franklin D. Roosevelt was the Democratic incumbent. About 54% of the 2.4 million respondents said they planned to vote for Landon and only 41% said they planned to vote for Roosevelt.[7] The poll missed the mark badly. Roosevelt won the election with 61% of the popular vote. Landon won only two states: Maine and Vermont, not even his home state of Kansas.[8]

How did The Literary Digest's poll go so wrong with such a large sample size? For one, the poll was plagued by selection bias. The Digest took its sample of 10 million from readily-available sources: its own subscriber list and lists of registered automobile owners and telephone users. People on these lists tended to be wealthier and have more conservative political leanings than the general population who generally could not afford magazine subscriptions and cars during the Great Depression. The poll over-represented Republicans and under-represented Democrats because of selection bias.

This example is one of my favorites that illustrate the importance of *how* the data are collected over the *size* of the sample. People tend to get hung up on sample size and think that "larger samples are always better." Not true! In fact, I'd rather have a smaller, random sample than a much larger non-random sample.

Surveys and polls like election polling are one of the most common ways to collect a sample of data from a population of interest. Have you ever participated in a survey? Maybe you stayed on the line and answered a pollster's questions or perhaps you took the time to fill out a survey form and mail it back. If so, good for you! You did your part in the data collection process.

In addition to employing random sampling methods, good surveys have several other features that result in high-quality data appropriate for generalization inference. Good surveys have high response rates which help to minimize nonresponse bias and good surveys use carefully worded questions which help to minimize response bias. Let's look at these two features in more detail.

One of the other big problems with The Literary Digest's poll was that out of the 10 million or so surveys sent out, only 2.4 million came back, a dismal 24% response rate. Even if the 10 million would have been a representative sample of the population, data were observed on only a small fraction of them. It is likely that the other 76% who didn't bother to complete the survey form and send it back looked differently in terms of who they planned to vote for than the 24% who responded. Bias like this that is introduced by people who don't respond to surveys is called *non-response bias*. **Non-response bias** is a systematic over- or under-estimation of a population feature introduced by non-responders who are different than responders with respect to the variable of interest. Despite your efforts to get a good random sample, non-response can severely bias your results. Good surveys have follow-up protocols in place to try to reach people who don't initially respond. Be skeptical of results from surveys with low response rates.

The way in which a survey is administered can also impact the data values. For example, the way the questions are worded or how they are asked can affect the results. **Response bias** is a systematic over- or under-estimation of a population feature due to something about the way in which the survey is administered. A question that tends to elicit one response over another will result in response bias. Here are two questions about the death penalty:

Q1:   Don't you think the death penalty should be illegal since it is a cruel form of punishment that rarely deters criminals?

Q2:   Do you favor or oppose the death penalty for a person convicted of murder?

Q1 is worded to elicit a response of "oppose" much more than Q2. Also notice that Q2 puts a better context around the death penalty, restricting its use to murder cases.

Another source of response bias is a question about a sensitive topic. Consider a survey about illegal drug use. A question like, "Have you ever used illegal drugs?" may not get a truthful answer, especially if the survey is conducted by an in-person or telephone interview. Respondents who have used illegal drugs may be embarrassed to answer "yes" and so the data would underestimate the proportion of all individuals in a population who have used illegal drugs. One way around this problem is to use a *randomized response survey*. Here's an example of how it works.

If your mother's birthday is between January 1 and July 31, answer Q1. Otherwise, answer Q2.

Q1: Have you ever smoked marijuana?
Q2: Is your father's birthday between July 1 and December 31?

Answer "yes" or "no."

In a survey like this, someone's response doesn't tell you which question they're answering. A "yes" could mean either that the respondent has smoked marijuana or that the respondent's father was born between July 1 and December 31. You might wonder how any kind of useful information could be extracted from a survey like this. Here's how:

In a particular sample of college students at a large university there were 61 "yes" and 73 "no" responses. Using only these numbers and assuming that birthdays are uniformly spread out across the 12 months of the year, we can fill in the following contingency table.

	Yes	No	
Answered Q1	33	45	78
Answered Q2	28	28	56
	61	73	134

We'd expect about 7/12 of the 134 responses (about 78 students) to have answered Q1 about their marijuana use, leaving about 56 to answer the innocuous Q2 about dad's birthday. Of those 56, we'd expect 1/2 (28 students) to say "yes." That leaves 33 "yes" responses to have come from Q1 and so we'd estimate the proportion of those students who have smoked marijuana as 33/78 (42%). The trick is to build in enough uncertainty about which question people answer so that they feel anonymous and yet get more people to answer the question of interest.

Sometimes even the ordering of phrases within a question can affect responses.

Example: In a Pew Research Center survey of public opinion about the issue of mandatory health insurance[9], two versions of the same question were posed at random to respondents:

Question	Approve
Q1: As you may know, by 2014 nearly all Americans will be required to have health insurance. *People who do not buy insurance will pay a penalty* while <u>people who cannot afford it will receive financial help from the government</u>. Do you approve or disapprove of this policy?	47%

Question	Approve
Q2: As you may know, by 2014 nearly all Americans will be required to have health insurance. <u>People who cannot afford it will receive financial help from the government</u> while *people who do not buy insurance will pay a penalty*. Do you approve or disapprove of this policy?	34%

Notice how simply changing the order of the phrases dramatically affected the approval ratings! People were more likely to approve if they heard the phrase about getting government help last.

<

This example brings us to another use of randomness in producing data. The version of the question was posed to respondents *at random*. Once again, a decision about how data were collected was left up to chance. Why? The answers are in the next section.

Section 3.2 Exercises

1. For each scenario, identify the population, the sample, and the variable of interest.

    a. You want to determine the fraction of people in the U.S. who use alternative medical treatments that improve in their condition. You ask 10 of your friends who use alternative medical treatments about whether their condition has improved. Seventy percent said it has.

    b. In order to determine the percentage of all daily bank transactions that are processed correctly, the bank selects 100 transactions at the end of the day on Friday. It finds that 92% of those transactions were processed correctly.

    c. A school board wants to find out how many parents in the district are in favor of a new policy being considered. A poll is posted on the school district's website and only 10% of the 678 parents who responded were in favor of the new policy.

    d. A potato chip manufacturer wants to monitor the average weight of product in bags of chips. To do so it randomly selects 30 bags from the production line each hour. The average weight of the most recent sample was 12.05 ounces.

2. For each scenario, identify the population, the sample, and the variable of interest.

    a. To determine whether the average contaminant level in batches of raw milk collected from local farmers is at an acceptable level, a random sample of 10 local farmers is selected and the most recent batch of milk from each is tested.

    b. Do a majority of U.S. adults oppose abortion in the third trimester of pregnancy? To answer this question, a news agency mailed a questionnaire to 1000 randomly-selected U.S. adults. All 1000 respond and 25% indicate opposition to late-term abortion.

    c. What fraction of the trees in a forest are ready for harvesting? The landowner randomly selects 100 latitude/longitude coordinates in her forest and determines whether the nearest tree to each location is ready for harvest. 45% of the trees selected are ready.

3. To determine whether the average contaminant level in batches of raw milk collected from local farmers is at an acceptable level, a random sample of 10 local farmers is selected and the most recent batch of milk from each is tested. If there are 82 local farmers in the population, describe, in detail, how to select a SRS of 10 farmers.

4. Do a majority of U.S. adults oppose abortion in the third trimester of pregnancy? To answer this question, a news agency mailed a questionnaire to 1000 randomly-selected U.S. adults. There are about 210 million U.S. adults. Describe, in detail, how to select a SRS of 1000 U.S. adults.

5. A car dealership wants to estimate the proportion of students at a local university who own their own car. There are 10,000 freshman, 9,000 sophomores, 8,000 juniors, and 7,000 seniors. The dealership wants to take a STRS of 100 freshman, 90 sophomores, 80 juniors, and 70 seniors.

    a. Why would a STRS be preferable to a SRS in this case?
    b. Describe, in detail, how to select the STRS.

6. A business analyst wants to average debt-to-income ratio for all businesses in the state. There are 200 product-oriented businesses and 800 service-oriented businesses in the state. The analyst wants to take a STRS of 10 product-oriented businesses and 40 service-oriented businesses.

    a. Why would a STRS be preferable to a SRS in this case?
    b. Describe, in detail, how to select the STRS.

7. A wireless service provider wants to estimate the proportion of people (including children) in the country who have their own smartphone. The project manager for the study is considering taking a CRS of 1000 U.S. households.

    a. Why would a CRS be preferable to a SRS in this case?
    b. Describe, in detail, how to select the CRS.

8. The school board of a large school district wants to study the behavior of kindergarten students in the district. The study would require close observation of kindergarteners over several weeks and so it would not be possible to observe every student. The board is considering taking a CRS of 5 of the 30 kindergarten classes and placing observers in each of the 5 classes to observe each child and record the data. There are 20 children in each class.

    a. Why would a CRS be preferable to a SRS in this case?
    b. Describe, in detail, how to select the CRS.

9. For each of the following studies, identify the sampling method that was used: SRS, STRS, or CRS.

    a. A large auto manufacturer operates 9 plants worldwide. The company randomly selects 3 plants and has all employees at each plan complete a satisfaction survey.

    b. A large U.S. auto manufacturer employs 10 managers, 50 line supervisors, and 400 laborers at a particular plant. The company randomly selects 46 employees from the plant and has each one complete a satisfaction survey.

    c. A large U.S. auto manufacturer employs 10 managers, 50 line supervisors, and 400 laborers at a particular plant. The company randomly selects 1 manager, 5 supervisors, and 40 laborers from the plant and has each one complete a satisfaction survey.

10. For each of the following studies, identify the sampling method that was used: SRS, STRS, or CRS.

    a. To monitor ice cream production, Pheasant Valley Dairy randomly selects 100 cartons from the 5,000 scheduled for production per day.

    b. School health officials in a school district want to assess the summer reading habits of children in the district who will be entering 5th through 12th grade. The district randomly selects 100 households with children in those grades and sends a survey to the parents about each child's summer reading habits.

    c. A satellite internet provider wants to know the proportion of households in the country that have access to non-satellite high-speed internet service. A survey is designed where households are randomly-selected in each county in the U.S. The number of households selected in each county is proportional to the total number of households in the county.

11. Give one advantage and one disadvantage of using a STRS instead of a SRS.

12. Give one advantage and one disadvantage of using a CRS instead of a SRS.

13. Here are the annual salaries for everyone at a small company.

ID Number	Position	Salary
15938	Management	$100,000
94587	Management	$80,000
82400	Laborer	$30,000
28401	Laborer	$40,000
38427	Laborer	$35,000
58339	Laborer	$15,000
48927	Laborer	$25,000
60113	Laborer	$40,000

a. A SRS of 4 salaries is drawn and the sample mean salary is calculated. This process is repeated 20 times and the 20 sample means are shown here:

43750   50000   63750   35000   40000   51250   46250   42500   30000   57500
43750   55000   26250   46250   61250   60000   45000   30000   30000   47500

Construct a histogram of the 20 sample means and find their mean and standard deviation.

b. A STRS of 4 salaries is drawn (1 salary randomly selected from the managers and 3 salaries randomly selected from the laborers) and the sample mean salary is calculated. This process is repeated 20 times and the 20 sample means are shown here:

40000   46250   40000   45000   43750   47500   45000   46250   42500   53750
41250   46250   46250   45000   48750   43750   47500   47500   47500   45000

Construct a histogram of the 20 sample means and find their mean and standard deviation.

c. Compare the two distributions in a. and b.
d. How does the SRS compare to the STRS?

14. Consider the salary data in exercise 13. A STRS of 4 salaries is drawn (2 salaries randomly selected from the managers and 2 salaries randomly selected from the laborers) and the sample mean salary is calculated. This process is repeated 20 times and the 20 sample means are shown here:

57500   56250   58750   58750   60000   55000   62500   61250   62500   65000
58750   62500   56250   58750   61250   57500   61250   56250   58750   61250

a. Construct a histogram of the 20 sample means and find their mean and standard deviation.
b. Compare the distribution in part a. to the distribution in exercise 13b.
c. How does the STRS in part a. compare to the STRS in exercise 13b? What made the difference?
d. Make a general recommendation to anyone conducting a STRS.

15. A hospital conducts a patient satisfaction survey of its "critical units:" the emergency department (ED), the neo-natal intensive care unit (NICU), and the medical intensive care unit (MICU). The hospital selects a stratified random sample of 40 patients in each unit. It finds that 8 patients in the ED were satisfied overall with their care, 32 patients in the NICU were satisfied, 16 patients in the MICU were satisfied.

a. Describe, in detail, how to select the STRS. See part c. for the total number of patients in each unit.
b. Estimate of the proportion of all patients who were satisfied overall with their care by calculating $\frac{total \text{ \# } satisfied}{120}$.
c. During the same time period, the hospital decides to survey all patients in each of these three critical units. The results are shown below.

	ED	NICU	MICU
Number of patients	1000	300	700
Percent satisfied	20%	80%	40%

What proportion of all patients were satisfied overall with their level of care?

d. Compare your answers from parts b. and c. Why are they different? What could you do to correct the estimate from part b.?

16. Consider these two sampling scenarios:

Scenario I: A beverage company makes ginger beer and bottles it in 12-oz. bottles. The bottles are shipped in cases of 24 bottles. Recently customers have been reporting that some bottles have not been sealed properly and the company wants to investigate by randomly selecting bottles to check from the most recent 1000 cases.

Scenario II: In order to estimate the proportion of the 24,000 people in a certain town who attend weekly religious services, an organization randomly selects people and surveys them.

a. Identify the population, the sample, and the variable of interest in Scenario I.
b. Describe how to select both a SRS of 480 bottles and a CRS of 480 bottles for Scenario I.
c. Identify the population, the sample, and the variable of interest in Scenario II.
d. Describe how to select both a SRS of 480 people and a CRS of 480 people for Scenario II.
e. What is an appropriate summary of the sample data in scenario I? In scenario II? Why?
f. For one of the scenarios (I'm not telling you which one) the results of 1000 SRSs and 1000 CRSs are shown here. The relevant percentage was calculated for each sample and the distributions of the 1000 percentages are displayed for each sampling method. Compare the two distributions.

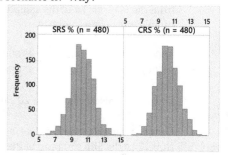

g. For the other scenario, the results of 1000 SRSs and 1000 CRSs are shown here. The relevant percentage was calculated for each sample and the distributions of the 1000 percentages are displayed for each sampling method. Compare the two distributions.

h. Which scenario goes with part f.? Which scenario goes with part g.? Why?

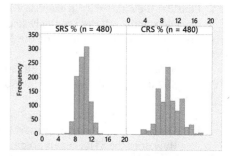

17. You want to determine the fraction of people in the U.S. who use alternative medical treatments that improve in their condition. You ask 10 of your friends who use alternative medical treatments about whether their condition has improved. Seventy percent said it has. What type of bias could be present? How would you have corrected it?

18. A school board wants to find out how many parents in the district are in favor of a new policy being considered. A poll is posted on the school district's website and only 10% of the 678 parents who responded were in favor of the new policy. What type of bias could be present? How would you have corrected it?

19. In order to determine the percentage of the daily transactions at a bank that are processed correctly, the bank selects 100 transactions at the end of the day on Friday. It finds that 92% of those transactions were processed correctly. What type of bias could be present? How would you have corrected it?

20. Do a majority of U.S. adults oppose abortion in the third trimester of pregnancy? To answer this question, a news agency mailed a questionnaire to 1000 randomly-selected U.S. adults. Of the 200 who responded, 25% indicated opposition to third trimester abortion. What type of bias could be present? How would you have corrected it?

21. A large university sent an exit survey to all of its 3,827 graduating seniors. One question was, "Would you recommend to others that they should attend the university?" The response choices were: 1 = not sure, 2 = not likely, 3 = likely, 4 = definitely, 5 = most definitely. Of the 780 graduates who responded, the mean response to this question was 4.33.

    a. What type of bias could be present? How would you have corrected it?
    b. Calculate the response rate for this survey.
    c. Is the mean an appropriate summary for the data? Why or why not?

22. Do a majority of U.S. adults oppose abortion in the third trimester of pregnancy? To answer this question, a news agency mailed a questionnaire to 1000 randomly-selected U.S. adults. The question was, "Do you think that a pregnant woman has the right to choose whether to abort her pregnancy in the third trimester?" All 1000 responded and 25% indicated opposition to third trimester abortion. What type of bias could be present? How would you correct it?

23. A survey by a conservation organization asked this question: "Do you think we should do everything we can to save the Earth's rain forests?" The answer choices were: Yes, definitely; Yes, probably; Don't know; Probably not; Definitely not. The survey was sent to 500 randomly-selected households in a certain state. Of the 500 responses, 450 answered either "Yes, definitely" or "Yes, probably." What type of bias could be present? Rewrite the question so that it elicits more "No" responses.

3.3 Determining Cause/Effect from Random Assignment

Suppose someone conducts a study of a random sample of high school students and finds that there is a dramatically higher center in the distribution of GPAs for those who play an instrument compared to those who don't. What do you make of that finding? Does that mean playing an instrument is a way to improve grades? Many people would jump to that conclusion. It is true that we may be able to generalize the difference to all high school students (a generalization inference) but it is not a good idea to say that playing an instrument is the reason for the difference. The way the data were collected does not permit a conclusion about the effect of playing an instrument on the GPAs of high school student.

Here are more questions that involve determining whether changing something causes a change in something else.

-   Does a red or green "Buy Now" button or putting it at the top or bottom work better on a certain webpage? (Not knowing the answer could cost a company millions in lost revenue.)
-   Is medication A better than medication B for treating asthma? (Not knowing the answer may leave thousands of asthma patients without a safe and effective treatment.)
-   Is it better to include review problems with every homework assignment or not? (Not knowing the answer might mean students retain less throughout a course.)
-   If pregnant women got free bed nets, would that decrease the incidence of malaria during their pregnancy? (Not knowing the answer might result in poor birth outcomes that could have been prevented.)

In order to answer these questions, the researcher has to intervene or change something and then observe the effects on a variable of interest. The process of collecting data with the goal of isolating the effect of an intervention is called an **experiment**. The process of using the data to determine whether the change worked, that is, whether the intervention was the actual cause of any effects on the data, is called **causal inference** and is the topic of this section. Seems pretty simple, right? Just change something and see what happens. How hard can that be? In practice, conducting a well-designed experiment that results in high-quality data that are appropriate for causal inference is not easy. As we'll see, it is far too easy to jump to conclusions about the cause of an effect that are not justified by the data.

In an experiment, the **factor** is the variable the researcher changes in the hopes of observing the effect of the change on another variable of interest, called the **response**. The researcher implements the change by intentionally assigning values of the factor to objects/subjects. The settings or values of the factor used by the researcher are called **levels** which determine the **treatments**. If there is only one factor used, the levels of the factor are the treatments. If the researcher uses two or more factors, the treatments are the combinations of the levels of each factor. The following table shows the factors (with the levels of each factor in parentheses), the treatments, and a possible response for each of the questions above.

Question	Factors (Levels)	Treatments	Response
Does a red or green "Buy Now" button or putting it at the top or bottom work better on a certain webpage?	color (red, green) location (top, bottom)	red button at top red button at bottom green button at top green button at bottom	sale? (yes, no)
Is medication A better than medication B for treating asthma?	medication type (A, B)	medication A medication B	volume of air expelled from lungs in 1 second
Is it better to include review problems with every homework assignment or not?	homework structure (new+review problems, new problems only)	new+review problems new problems only	final exam score
If pregnant women got free bed nets, would that decrease the incidence of malaria during their pregnancy?	bed net status (bed net, no bed net)	bed net no bed net	got malaria? (yes, no)

In an **observational study**, the researcher does not intervene by intentionally assigning the levels of the factor to the objects; the levels are simply observed for each object. For example, in the music study, the factor is instrument status and the levels are "play" and "not play," but they were not really "treatments" that the students were forced to receive. Their status was simply observed. The response was GPA. As I hope to convince you by the end of this section, causal inferences are not justified by observational studies.

The goal of causal inference is to answer the question, "What would have happened if objects would have gotten a different treatment?" If their response would have been the same, then we'd know the factor has no effect on the response. If their response would have been different, then we'd know the factor caused the change in the response. This **comparison** aspect of an experiment is very important; we can't just apply one treatment without *comparing* its effects to another treatment.

Example: Suppose we gave a group of people free bed nets and observed that 5% got malaria. Is that good? Does that mean the free bed nets worked? I hope you can see that it is impossible to answer those questions without having comparison data. It would help a little to know what percent of the general population got malaria during the same time period. Suppose it's 10%. But what if these people lived in an area less susceptible to malaria-carrying mosquitos than the general population. Even if the bed nets didn't work at all, these folk would have less exposure to malaria and therefore contract the disease less frequently.

The fundamental problem with experiments and causal inference is that we can't go back in time and repeat the experiment using a different treatment. It would be ideal if we could because then we'd be able to observe exactly the same objects in exactly the same circumstances with the *treatment as the only difference*. Unfortunately we can never know what would have happened to the *same* objects at the *same* time with a different treatment. Instead we must use *different* objects at the *same* time or the *same* objects at *different* times as the comparison. Either way is like comparing apples to oranges. Because different objects will differ on variables other than the factor, it can be difficult (or even impossible) to know whether the intervention was the cause of the effect on the response or whether it was another variable that was the cause. If we use the same objects in different treatments at different times, it can be difficult (or even impossible) to make sure nothing else changed between the two time periods. Other variables that are related to the factor that may be responsible for causing changes in the response are called **confounding variables**. Confounding variables can create imbalance among the treatment groups so that they differ in more ways that just the treatment.

Now you might be saying, "What?! Factors, responses, confounding variables... My head is spinning." You're not alone. We're not used to thinking about more than two variables at a time. In fact, we've been trained not to! Remember your algebra days when you dealt with equations like $y = 3x - 5$? In an equation like this there are only two variables, x and y, in the universe; we are not trained to think about more than two variables at a time until a multivariable calculus course and many of us don't get that far in our study of mathematics. To make matters worse, we're used to thinking like this: increasing x by 1 *causes* y to increase by 3. That's fine as long as x and y are the only variables in the universe. However, our universe is incredibly complex with many variables changing at once. Much of our experience cannot be reduced to a simple causal connection between only two variables. The following examples illustrate this complexity.

Example: Let's consider how we might answer the question, "Does a red or green "Buy Now" button or putting it at the top or bottom work better on a certain webpage?" Suppose we design four different versions of the webpage, one for each treatment. Starting at 12am, we show each version for an hour and at 4am we start over with the first version and repeat this process for several weeks. During this time we keep track of whether or not visitors to the page actually buy the product and find that the "red button at top" version got the most buys by far. What do we conclude? Many would be tempted to say that putting a red button at the top *caused* more sales. It's so easy to focus on only the factors and response variables but if you stop and think for a minute, you can probably come up with confounding variables that may have actually caused more sales. Notice that the "red button at top" was shown during the 12pm-1pm and 8pm-9pm time slots. If many people do online shopping during their lunch breaks or in the evening, time of day is a confounding variable and there would be no way to determine whether it was the treatment "red button at top" or time of day that caused more sales. Designing an experiment like this is not a good idea because of the potential for the treatments to be imbalanced with respect to time of day.

Example: How about the question, "Is it better to include review problems with every homework assignment or not?" Suppose we use the "new+review problems" treatment in the 12pm section of a course and the "new problems only" treatment in the 8am section of a course taught by the same instructor. We keep many other variables constant across the two sections: lectures, projects, and exams. Suppose that the mean final exam score for students in the 12pm section was 5 points higher than that for students in the 8am section. Was the "new+review problems" treatment effective? Many would be tempted to say so. It's easy to focus on only the homework structure and final exam score variables and conclude that, since there was a 5 point difference, changing the homework structure variable *caused* the change in mean final exam score. Resist the temptation! Think about possible confounding variables. Could there have been other differences between the 8am section and the 12pm section other than the homework structure? Absolutely. Perhaps there were more upperclassmen in the 12pm class because they could register earlier and chose the more desirable time. As upperclassmen with more experience, they might have done better on the final exam even with the same homework structure as the 8am section. Perhaps the instructor was able to deliver a more polished and effective lesson at 12pm because she got all the kinks worked out at 8am. Either way, homework structure would be confounded with class time and there would be no way to determine whether it was the treatment "new+review problems" or class standing or lesson effectiveness that caused better final exam scores. Designing an experiment like this is not a good idea because of the potential for the treatments to be imbalanced with respect to confounding variables.

Using the same objects can take care of some of these difficulties but can introduce others.

Example: "If people got free bed nets, would that decrease the incidence of malaria?" Suppose we tried to answer this question by giving a group of people bed nets on March 1 and observe which ones get malaria by April 30. Then we take the bed nets away on May 1 and observe which ones get malaria by June 30. If fewer people get malaria between March 1 and April 30, does that mean the bed nets were effective? I hope you're beginning to see that such a simplistic "two variable universe" conclusion is not valid. What if the rainy season begins in May and there are more mosquitos around during the rainy season? If so, then there would be more cases of malaria in May and June even without the use of the bed nets in March and April. Because bed net status would be confounded with the season of the year, there would be no way to determine whether people got malaria because of the "no bed net" treatment or because of the rainy season. In spite of using the same people, designing an experiment like this is not a good idea because of the potential for the treatments to be imbalanced with respect to season of the year.

There is a way to get the treatment groups balanced with respect to confounding variables. It involves (surprise) the use of chance!

Example: Consider this question: "Is medication A better than medication B for treating asthma?" Suppose we *randomly assign* a group of 300 volunteers to one of two treatments: medication A or medication B. We administer the treatments to the patients, making each treatment look, taste, and smell the same. At the end of the study period, a trained technician measures the volume of air that each patient can force out of their lungs in one second. The trained technician doesn't know who is on which treatment. The results show that the patients on medication A generally had higher volumes on average than those on medication B. Does that mean medication A is better? It's possible because of the random assignment.

Random assignment to treatment should *balance the treatment groups with respect to <u>any</u> potential confounding variable.* In addition to comparison, it is the second important characteristic of a well-designed experiment. Because of random assignment, the subjects in the treatment groups in the asthma study should be similar with respect to gender, ethnicity, asthma severity, age, cardiovascular health, and any other variable you might think of that could affect lung function other than the treatment. The beauty of random assignment is that you don't have to think of all of the potential confounding variables; random assignment takes care of them all!

Here's an illustration of how the random assignment process works for the asthma example. The sample of 300 volunteers is randomly divided into two groups. Those two groups of 150 should look similar with respect to any potential confounding variable. The average age, gender split, ethnic makeup, asthma severity, and overall health should be similar for both groups before the treatments are applied. So the effect of the treatments should be the only difference between the two groups.

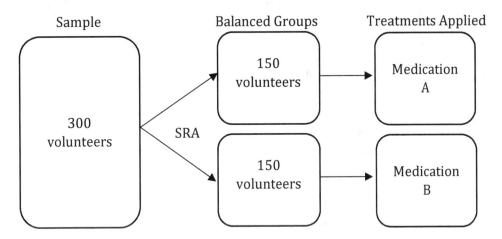

Perhaps the simplest way to carry out a random assignment to two treatment groups is to use a fair coin. In the homework example, we could toss a coin for each student in the course. If the coin comes up heads, we assign that student to the "new+review problems" treatment. If the coin comes up tails, we assign that student to the "new problems only" treatment. There would be no systematic favoring of stronger students over weaker students, females over males, seniors over freshmen. The groups should be balanced so that any differences we observe in final exam scores should be due to only the differences in the homework structure.

Let's go back to the asthma study where we found that the patients on medication A generally had better lung function. You might be thinking, "Isn't there a chance that the healthier people were randomly assigned to medication A?" You're right, there is. It is possible that we ended up with an unfortunate random assignment that unfairly favored medication A. The good news is that we can compute how likely that would have been. Much more on that in a later chapter. Random assignment takes care of balancing the effects of confounding variables and we can deal with the issue of unfortunate assignments later.

Random assignment eliminates the effect of confounding variables in a long-run sense: over *many* random assignments, the *average* composition of the treatment groups would be the same. However, there is no guarantee that any one *particular* random assignment results in balanced treatment groups. You might also notice that with the coin method, there is a chance that everyone gets assigned to the same treatment! Just as with random sampling, there are ways to improve the random assignment.

We can place a restriction on the random assignment that ensures there are at least some objects in each treatment group. A **simple random assignment (SRA)** is one where all possible ways to assign objects to treatment groups *of fixed group sizes* are equally-likely. The researcher can determine what those group sizes are.

Example: In the homework study, suppose there are 40 students in the course. For small studies like this, you could use the hat method. Write the names of all 40 students on equal-sized slips of paper, put those slips of paper in a hat, mix well, and have a blindfolded person draw 20 slips from the hat. Those students are assigned to the "new+review problems" treatment and the rest are assigned to the "new problems only" treatment. You've just conducted a SRA of the 40 students to the two treatment groups!

In some cases, the hat method doesn't work well.

Example: Consider the webpage study on the color and location of the "Buy Now" button. Suppose we wanted to use data on 1000 visitors to this webpage. People are continuously visiting the webpage so we don't have a list of all of them at the start of the study. The hat method won't work. Instead create a list of visitor numbers 1 – 1000 in an Excel spreadsheet. List each treatment 250 times in another column. To conduct a SRA of the 1000 visitors to the four treatments, add a random number column by using the RAND() function. Then sort only the Treatment column by the RandNum column. You now have a way to determine which visitor gets which treatment that will give you a SRA of 250 visitors to each treatment. For example, the first visitor will see a green button at the bottom, the second visitor will see a green button at the top, and the 1000th visitor will see a red button at the top.

List of visitors, treatments, and random numbers.		
Visitor	Treatment	RandNum
1	red button at top	0.243522
2	red button at top	0.432201
...	...	...
250	red button at top	0.298031
251	red button at bottom	0.799107
252	red button at bottom	0.630926
...	...	...
500	red button at bottom	0.418659
501	green button at top	0.203969
502	green button at top	0.416651
...	...	...
750	green button at top	0.456285
751	green button at bottom	0.922316
752	green button at bottom	0.736581
...	...	...
1000	green button at bottom	0.205148

Sort only the treatments by the random numbers.		
Visitor	Treatment	RandNum
1	green button at bottom	0.000486
2	green button at top	0.000573
...	...	...
250	red button at top	0.244933
251	red button at bottom	0.245735
252	red button at top	0.246362
...	...	...
500	red button at top	0.496706
501	red button at top	0.497812
502	red button at top	0.498362
...	...	...
750	red button at top	0.745664
751	red button at bottom	0.750231
752	red button at bottom	0.751145
...	...	...
1000	red button at top	0.999980

While a SRA guarantees that there will be no empty treatment groups, it does not guarantee that a *particular* SRA will result in balanced treatment groups on a confounding variable. In the homework study, a SRA of 20 students to each treatment could result in a GPA imbalance where more students with higher GPAs are assigned to one of the treatments. The good news is that we can calculate the probability that this happened by chance. The bad news is that the probability is not zero. If we know ahead of time that we want to balance the effects of a certain confounding variable across the treatments, we can conduct a fancier version of the SRA called a *randomized block assignment*. A **randomized block assignment (RBA)** is obtained by conducting a SRA of objects to treatments in each of several subgroups called *blocks*. The blocks are defined by the confounding variable we want to control and we refer to the confounding variable in that case as the *blocking variable*.

Example: Suppose that in the homework example the GPA letter grade distribution for the 40 students in the course is shown below.

GPA	Frequency	Percent
A	4	10%
B	12	30%
C	18	45%
D/F	6	15%
Total	40	100%

If you wanted to guarantee that a random assignment of 20 students to each treatment results in the same numbers of A, B, C, and D/F students per treatment, use a RBA with GPA as the blocking variable. Conduct a SRA of the A students (2 to each group), a SRA of the B students (6 to each group), a SRA of the C students (9 to each group) and a SRA of the D/F students (3 to each group). Putting the four SRAs together would constitute a RBA blocked by GPA. The figure below illustrates the design.

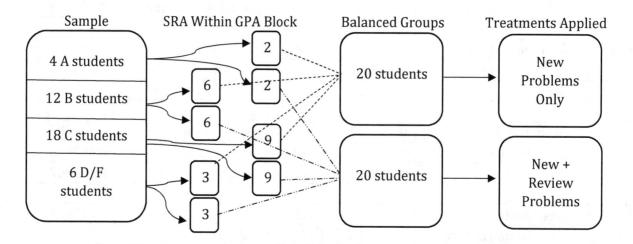

Eliminating the chance of unbalanced treatment groups with respect to the blocking variable is a big advantage of the RBA but it turns out there is another statistical advantage of a RBA; it can be *more efficient* than the SRA. If so, you can get away with using fewer objects in the experiment by conducting a RBA compared to a SRA. That's because in those cases the RBA results in treatment comparisons that are more consistent.

In order to see how this works, let's isolate the random assignment aspect of an experiment by assuming (you could say "hypothesizing") there are no differences among the treatments. *That means if we were able to go back in time and assign each object to a different treatment, the responses would be exactly the same. Under the "no difference" hypothesis, it wouldn't matter to which treatments objects are assigned; their responses would remain unchanged by the treatment.* Those last two sentences are so important that I've italicized them so go back and read them again. The idea is to hypothesize what would happen if we could go back in time and then compare the actual results of the experiment to the results we'd expect under that hypothesis.

Example: Consider the homework experiment and suppose there is no difference between the "new+review problems" and "new problems only" treatments. *The final exam scores of students would be the same regardless of the treatment to which each is assigned.* The only differences we will see between the treatment groups are due to the random assignment; we'll assign some higher scores to one group and lower scores to the other group by chance.

Shown below are the final exam scores for each student and their GPA. Let's conduct both a SRA and a RBA (blocked by GPA) of the 40 students to the two treatments (20 per treatment) and assume their final exam score would be unaffected by the treatment. Since these students have already taken the final exam and we won't actually be implementing these treatments, we'll see only the effects of the random assignment.

Student	GPA	Final	Treatment (SRA)	Treatment (RBA)
1	A	95	new+review	new only
2	A	86	new+review	new only
3	A	92	new only	new+review
4	A	90	new+review	new+review
5	B	92	new only	new only
6	B	84	new only	new only
7	B	88	new only	new only
8	B	69	new only	new only
9	B	76	new+review	new+review
10	B	79	new+review	new+review
11	B	86	new only	new only
12	B	80	new only	new+review
13	B	80	new only	new+review
14	B	81	new+review	new+review
15	B	96	new+review	new+review
16	B	77	new only	new only
17	C	80	new+review	new only
18	C	85	new+review	new+review
19	C	88	new only	new only
20	C	65	new only	new+review
21	C	70	new only	new only
22	C	72	new only	new only
23	C	79	new only	new only
24	C	76	new only	new+review
25	C	73	new only	new only
26	C	84	new only	new only
27	C	80	new only	new only
28	C	79	new+review	new+review
29	C	79	new+review	new only
30	C	82	new only	new+review
31	C	50	new+review	new+review
32	C	93	new+review	new+review
33	C	66	new+review	new+review
34	C	64	new+review	new+review
35	D/F	80	new+review	new only
36	D/F	63	new+review	new+review
37	D/F	60	new only	new only
38	D/F	69	new+review	new only
39	D/F	70	new+review	new+review
40	D/F	70	new+review	new+review

Notice that in the SRA, 3 of the 4 A-students were assigned to the new+review group but in the RBA, two A-students were assigned to each group. The RBA forces the random assignment to give us the same number of students in each treatment from each GPA group.

Now let's calculate the difference in the mean final exam score ($\bar{x}_{nr} - \bar{x}_n$); compute $\bar{x}_{nr}$, the mean score for the 20 students in the "new+review" group, compute $\bar{x}_n$, the mean score for the 20 students in the "new only" group, and take the difference. Remember, we're expecting no difference between the treatments so a random assignment of these final exam scores to the two treatments is really only a random dividing of the 40 scores into two groups. Here are the results:

SRA Difference	RBA Difference
$\bar{x}_{nr} - \bar{x}_n = -1.3$	$\bar{x}_{nr} - \bar{x}_n = -2.7$

For this one random assignment, both methods gave assignments where the final exam scores were a bit higher on average in the "new problems only" treatment (by 1.3 for the SRA and 2.7 for the RBA).

On the surface, it looks like the RBA gave a more imbalanced assignment than the SRA because more of the higher scores were assigned to the "new problems only" treatment. However, it's really hard to evaluate a statistical method based on only one application of the method. So let's repeat the random assignment methods 1,000 times each and keep track of the difference in the mean scores for all 1,000 assignments:

	SRA Difference $\bar{x}_{nr} - \bar{x}_n$	RBA Difference $\bar{x}_{nr} - \bar{x}_n$
1	-1.3	-2.7
2	-1.8	6.2
3	3.0	2.0
⋮	⋮	⋮
1,000	1.9	1.9

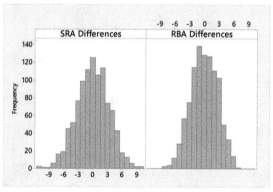

This simulation analysis gives us a really good idea of how these methods perform. The histograms show the distributions of the difference $\bar{x}_{nr} - \bar{x}_n$ for the 1,000 randomizations. The centers are both at 0, right where we'd expect them to be if there is no difference between the "new+review" and "new problems only" treatments. Both SRA and RBA reflect the "no difference" hypothesis on average.

But notice that the differences for RBA don't vary quite as much as the differences for SRA. The standard deviations of the 1,000 differences are 3.3 and 2.7 for SRA and RBA, respectively. A RBA gives more consistent differences that are closer to 0, on average. The RBA eliminates the random assignments that are unbalanced with respect to GPA. For example, the RBA eliminates random assignments that put more students with higher GPAs in the "new+review" treatment (which give higher, positive values of $\bar{x}_{nr} - \bar{x}_n$) and random assignments that put more students with lower

GPAs in the "new+review" treatment (which give lower, negative values of $\bar{x}_{nr} - \bar{x}_n$). The RBA method decreases the variability of the differences compared to the SRA method.

In this example, the blocks were defined by GPA which gave blocks that were quite different in terms of a response variable like final exam score. It turns out that the more the blocks differ with respect to the response, the more efficient the RBA is compared to the SRA. If the blocks don't differ much, then the RBA could be about the same as the SRA in terms of efficiency. The moral? Choose a blocking method that creates blocks that are as different as possible with respect to the response.

One special case of the RBA is where each object serves as a block. Each object gets each treatment in a random order. This design can be very efficient, especially if there is a lot of variability in the response from object to object.

Example: A company that makes a waterproofing product for boots wants to compare its product to its competitor's product. A group of volunteers are recruited and each volunteer is provided with a pair of boots to wear every day. One boot is treated with the company's product and the other with the competitor's product. Which boot (left or right) is treated with which product is determined at random. Each volunteer wears the pair of boots for several months and at the end of the study period a damage score is determined for each boot with higher numbers indicating more damage.

Wear-and-tear on boots will likely be quite variable among individual people; some people might work outside and therefore the boots will become much more damaged than for people who work indoors. Some people might wear the boots more than others. However, since each person wears one boot treated with each product, the comparison of the products can be made within the same person, therefore cancelling out all of the effects of variables that differ across people. Randomly deciding which boot gets which product avoids the situation where the company's product gets applied to only the right boots and the competitor's product gets applied to only the left boots.

In some circumstances, conducting a SRA of objects to treatments may be impractical or counterproductive. For example, consider the last question at the start of this section: "If pregnant women got free bed nets, would that decrease the incidence of malaria during their pregnancy?" A SRA of individual women to get a free bed net or not would result in women in both treatment groups in the same village. That may lead to those who didn't get the free bed net resenting those who did. It might be better to randomly assign entire villages to the same treatment using a *cluster random assignment*. In a **cluster random assignment (CRA)**, the objects under study are divided (often naturally) into groups called clusters and a SRA of clusters to treatments is conducted. The important difference is that clusters of objects are randomly assigned to treatments instead of individual objects.

Example: Getting malaria is bad enough; getting it while a woman is pregnant can result in poor outcomes for both the mother and the baby, including anemia and low birth weight. In a study of pregnant women in Ghana from 1993 to 1995, 1961 pregnant women participated.[10] The women were divided into 96 clusters according to the communities in which they lived. Forty-eight clusters were randomly assigned to receive bed nets treated with insecticide to prevent mosquito bites and the other 48 got nothing. The CRA ensured that women in the same community were in the same treatment group and helped to ease any tensions that might have arisen had some women gotten nets and others had not. The design is illustrated below.

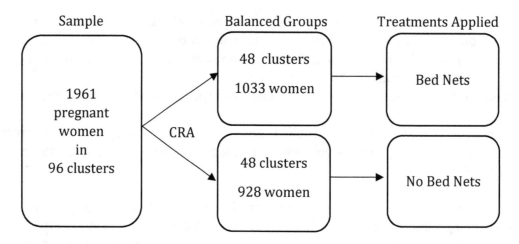

Just like cluster random sampling, the problem with a CRA is that it is often *less efficient* than the SRA. That's because the CRA usually results in treatment comparisons that are less consistent than the SRA generates. The more alike the objects are within a cluster and the more differences there are between clusters, the less efficient the CRA will be compared to the SRA.

In order to see how this works, let's isolate the random assignment aspect of an experiment by hypothesizing that there are no differences among the treatments. *That means if we were able to go back in time and assign each object to a different treatment, the responses would be exactly the same. Under the "no difference" hypothesis, it wouldn't matter to which treatments objects are assigned; their responses would remain unchanged by the treatment.*

Example: Suppose there are 2000 pregnant women we want to study in a certain region prone to malaria. The region is naturally divided into 100 villages (clusters) of 20 pregnant women each. Let's suppose that some villages are more prone to malaria than others so that the malaria rate ranges from 40% to 50% to 60% to 70% and that there are 25 villages with each rate. Let's conduct both a SRA of 1000 women to each of two treatments (bed nets and nothing) and a CRA of 50 villages to each treatment (which will also give us 1000 women on each treatment).

Now let's calculate the difference in the percentage who get malaria between the two treatments (bed nets – nothing). Remember, we're expecting no difference between the treatments so a random assignment of these women to the two treatments is really only a random dividing of the 2000 women into two groups. Here are the results:

SRA Difference	CRA Difference
- 1.3%	- 3.6%

Here we found the malaria rate was higher in the group that got nothing by 1.3% for the SRA and by 3.6% in the CRA. However it's really hard to evaluate a statistical method based on only one application of the method. So let's repeat the random assignment methods 1,000 times each and keep track of the difference in malaria rates for all 1,000 assignments:

	SRA Difference	CRA Difference
1	- 1.30%	- 3.6%
2	0.40%	- 1.2%
3	2.55%	2.4%
...	...	...
1,000	- 0.60%	- 1.2%

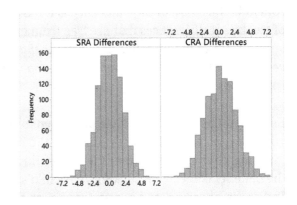

This simulation analysis gives us a really good idea of how these methods perform. The histograms show the distributions of the difference in malaria rate for the 1,000 randomizations. The centers are both at 0, right where we'd expect them to be if there is no difference between the "bed nets" and "nothing" treatments. Both SRA and CRA reflect the "no difference" hypothesis on average.

But notice that the differences for SRA don't vary quite as much as the differences for CRA. The standard deviations of the 1,000 differences are 1.9 and 2.3 for SRA and CRA, respectively. A SRA gives more consistent differences that are closer to 0, on average. The reason is that each village randomly assigned in the CRA provides 20 times as many women as each woman assigned in the SRA. Any chance imbalance will be magnified in the CRA. So the CRA method increases the variability of the treatment differences compared to the SRA method!

The following table summarizes the three random assignment methods and includes recommendations for implementing each method.

	Notes	Recommendations
Randomized Block Assignment (RBA)	• often more efficient than SRA and CRA • permits a more sophisticated analysis	• Define blocks so that they are as *different as possible* with respect to the response. • Consult a statistician before conducting the experiment.
Simple Random Assignment (SRA)	• often less efficient than RBA • often more efficient than CRA • data are relatively simple to analyze	• Use a SRA if you will be doing the data analyses yourself.
Cluster Random Assignment (CRA)	• often less efficient than RBA and SRA • may be more convenient/ethical	• Use only when conducting a SRA or RBA is impractical or unethical. • Consult a statistician before conducting the experiment.

More alphabet soup...SRA, RBA, CRA... Remember that the goal of these methods is to balance the treatment groups with respect to confounding variables so that their effects cancel out when we measure treatment differences. Random assignment is the best way to do that so that we can conclude the cause of the difference is the treatment. The SRA, RBA, and CRA all provide measures of treatment differences that are free from effects of confounding variables on average over many applications of the method. *Experiments that do not employ random assignment to treatments are very vulnerable to the effects of confounding variables which may prohibit making cause/effect conclusions.*

Example: The title of an article published in the New York Times should be attention-catching to you once you learn that determining the cause of an effect is difficult without random assignment: "Sucking Your Child's Pacifier Clean May Have Benefits."[11] Notice the implication that the cause of the benefits is the act of sucking your child's pacifier clean. The study, published in the journal *Pediatrics*, observed 184 infants from birth to age 3, symptoms of allergies, and their parents' methods of cleaning their pacifiers.[12] The data showed that there were fewer cases of asthma, eczema, and airborne and food sensitivities among the children whose parents sucked their pacifiers clean, hence the title. To its credit, the article does point out that it might not be the act of sucking a pacifier clean that caused the difference. Perhaps something else about the lifestyles of the parents who weren't as "germ conscious" actually caused the lower rates of these conditions. Perhaps they (and therefore their children) had better diets or were less likely to smoke.

This is not to say that passing saliva to your children in this way couldn't possibly be helpful in preventing asthma or eczema or food allergies. The point is that the design of this study and the way the data were collected (without random assignment to suck or not to suck) do not allow such a conclusion.

Human Subjects

When experiments involve people as the objects (we then call them "subjects"), designing an experiment with a comparison group and random assignment may not be enough. Human beings are intelligent and self-aware; we can sense when something about our environment changes and we may adjust our behavior accordingly. Have you ever been in a situation where you sensed that "something just wasn't right?" We have an uncanny ability to perceive the imperceptible.

To make matters worse, our thought processes can be easily swayed. Remember the survey question about the health care law from section 3.2? Our opinions can be changed just by re-ordering the words in a question. Do you think a doctor evaluating a subject in an experiment could make fair evaluation if she knew the subject was assigned to a treatment hypothesized to be less effective? These issues make performing experiments on human subjects all the more challenging. We may need more than a comparison group and random assignment.

When human subjects are involved in an experiment, it is often beneficial to hide information from the people involved. This practice is called **blinding**. One common approach is to blind the subjects from knowing the treatment they get so that their behavior is not influenced by knowing. In medical studies, this is done by making all the treatments look, feel, and taste the same. Any labels are removed. Pills or injections are made to look the same. Inhalers are all the same size and shape and either don't have any labels or have a common label. If all of the subjects are blinded to treatment, the experiment is called **single-blind**. But because the doctors or other study personnel who interact with the subjects may also be influenced in their ratings or diagnoses or use of equipment by knowing the treatment a subject is getting, they are often blinded as well. If so, the experiment is called **double-blind** because both groups of people are blinded to treatment. My first job out of college was as a data analyst in a university's college of medicine. One of the projects I worked on was an asthma experiment with three treatments. Those of us who were analyzing the data were blinded to treatment. The treatment groups shown in the data sets were simply A, B, or C; we didn't know what treatments those represented. I suppose you could call that experiment triple-blind! It was an extra layer of blinding so that even those analyzing the data would not be influenced by knowing which subject was getting which treatment. Only a few people in our department who interacted with neither the subjects, the physicians, nor the data knew which treatments A, B, and C represented.

Sometimes, the simple fact of giving a subject something that they think will make them better actually makes them better! It's a psychological phenomenon called the **placebo effect**, named after a placebo treatment. A **placebo** is an inactive treatment that looks, feels, and tastes just like the active treatments. If a new drug is in pill form, a placebo pill is made to look and taste like the real drug but is made of inactive ingredients. You may have heard the term "sugar pill" used to refer to placebos. But using actual sugar would only work if the real drug tasted the same way! Besides, sugar actually has effects on our bodies so actual placebos are not really "sugar pills." Good drug experiments have a placebo group as one of the treatments to be used as a comparison with the active drug treatment. (In the asthma study, we had two active asthma drug groups and one placebo group.) A placebo group allows us to measure the placebo effect so that a drug is deemed effective only if subjects' responses are better than those on placebo. Without a placebo treatment, if subjects improved while taking the drug, it would be impossible to determine whether the improvement was due to the drug itself or due to the person knowing they were taking something that might make them better. The placebo effect is another reason for using a comparison group in an experiment.

The placebo effect can work outside the medical context where subjects can react (sometimes for the better) to any change that they perceive. If it is not possible to blind subjects to treatment, this can make things really sticky. The tendency for people to improve just because they think they are being treated differently from others (in a good way) has been referred to as the **Hawthorne effect**, named for a series of experiments in the early 1900s on workers at the Hawthorne plant of the Western Electric Company located in Chicago.[13] Experiments that varied the lighting conditions, the length of the workday, and the length of breaks showed that workers' productivity improved after merely changing these conditions, whether or not the change was thought to be for the better. For example, a group of workers' average hourly productivity improved even after lengthening the workday when it was thought that lengthening the workday would decrease productivity because of fatigue. Data collected on the workers helped to shed some light on a potential reason: the group of workers in the experiment experienced a camaraderie because they worked with each other every day and were separated from the rest of the workers. They enjoyed being asked how they were doing. Other workers were envious of their special treatment. Could it be that those workers performed better just because someone was paying attention to them? If so, then the Hawthorne effect could be the true cause of the increased productivity and not the physical changes in the workplace. The experiment itself could introduce confounding variables!

It is for this reason that subjects are sometimes blinded to the true nature of the experiment by intentionally deceiving them. Suppose you wanted to study the effect of having people work in cubicles compared to working at desks without physical barriers. You randomly assign each person who volunteers for the study to one of two treatments: work in a room with cubicles or work in an open room with flat desks. Because you don't want them to know the true purpose of the experiment, you tell them that the two groups differ on what kind of job they'll be doing: transcribing a document set in one of two different fonts. They don't really know that one font group will be working in a room with cubicles and the other font group will be working in a room with desks. The researchers really don't care about how well the subjects transcribe the document; they'll be observing responses related to how the subjects interact with each other.

The Hawthorne effect has implications for the long-term efficacy of any intervention. If an intervention is effective because subjects know they're being treated in a new way, what will happen when the experiment ends? Might the effect disappear? If so, then the intervention may not have been all that effective.

The artificial environments of experiments, while designed to limit the effects of confounding variables, can limit the extent to which the results can be generalized. Unless the objects in the experiment were randomly selected from a larger population, it is difficult to generalize any cause/effect conclusions to that larger population. Any efforts to treat each treatment group the same that impose unnatural conditions on the objects will limit conclusions about the efficacy of the intervention to only those conditions. For example, in the cubicles vs. desks study, any conclusions about how the desks group differed from the cubicles group could not be easily generalized to other kinds of group interactions. Since volunteers were used, conclusions would also be limited to only those people, or people similar to them. If the volunteers were all college students, the study results could not be easily generalized to older or younger people. The study would need to be replicated with people in different age groups in order to have some validity in those populations. We say that experiments that do not use representative samples of objects suffer from the lack of *external validity*. A study has **external validity** if the results of the experiment can be generalized to a group larger than just the study objects.

Section 3.3 Exercises

1. A study reported that in a random sample of over 60,000 U.S. adults, those who rated themselves as "a lot less active" than their peers had about twice the mortality risk compared to those who rated themselves as "a lot more active."[14] Perceived activity was recorded as "a lot more active", "a little more active," "about as active," a little less active," or "a lot less active." Subjects were followed for up to 21 years and time until death was recorded.

    a. Identify the factor(s), the levels of each factor, and the response.
    b. Is this study an observational study or an experiment? Explain your answer.
    c. Which of these headlines would be an appropriate way to report the results of the study?
        i.   "How Our Beliefs Can Affect Our Longevity"
        ii. "Perceived Activity Found to be Related to Length of Life"
        iii. "How Long You Live Associated with How You Compare Activity to Others"
        iv. "Thinking of Yourself as Less Active Can Reduce Lifespan"
        v.   "Prolong Your Life by Thinking of Yourself as More Active"

2. In one study, 160 college student volunteers were randomly assigned to either an extra-large sheet of paper or a half-sheet of paper. Each student was given a pair of scissors and randomly assigned to either cut their sheet of paper into four pieces or not cut the paper at all. Whether each student disposed of their paper in the recycle can or the garbage can on the way out of the room was observed. About 80% of those who cut up their extra-large sheet recycled the pieces compared to about 40% who cut up their half-sheet. About 80% of those who did not cut up their sheets recycled them.[15]

    a. Identify the factor(s), the levels of each factor, and the response.
    b. Is this study an observational study or an experiment? Explain your answer.
    c. Identify the treatments.
    d. Can you conclude that cutting the half-sheet into 4 pieces decreases the chances of those pieces getting recycled? Why or why not?

3. Transporting animals can induce stress which can have negative effects on outcomes like weight. One study investigated the use of ascorbic acid (AA) as a way to reduce stress in goats during transport. Twenty-four goats were divided into four groups at random: one group got a dose of AA prior to transport, one group got the same dose after transport, one group got a dose of saline solution instead of AA, and one group got a dose of saline solution but was not transported at all. Immediately after transport, each goat was weighed. Only the goats who got the AA dose prior to transport had statistically similar mean weights to those who weren't transported at all. The other two groups showed statistically lower average weights.[16]

    a. Identify the factor(s), the levels of each factor, and the response.
    b. Is this study an observational study or an experiment? Explain your answer.
    c. Identify the treatments.
    d. Can you conclude that the dose of AA prior to transport improved weight outcomes for goats like these? Why or why not?

4. A rural ice cream shop wants to see if changing the name of one of their flavors from "Chocolate Mud" to "Chocolate Dream" would change sales of that flavor. They keep track of sales (in dollars) of Chocolate Mud for each day during May. The name is then changed to Chocolate Dream and sales are observed each day during June. The data show that sales were higher on average for June.

    a. Identify the factor(s), the levels of each factor, and the response.
    b. Is this study an observational study or an experiment? Explain your answer.
    c. Identify the treatments.
    d. Would you advise the shop to permanently change the name to Chocolate Dream? Why or why not?

5. Jack and Jill want to see if eating candy makes you run faster. They decide who gets to eat candy by tossing a fair coin; Jill wins the coin toss. They both start at the bottom of a hill and race to the top to see who gets to the pail of water first. Jill beats Jack by 0.5 seconds. Jill thinks it was the candy that won her the race. How would you respond?

6. All winter long, Gunhilde's pet dragon's scales are dull and colorless. Gunhilde thinks that lack of exercise is the culprit so she decides to take it outside for walks every day. It's spring and the weather is starting to get nice. After a month of going for walks outside every day, the dragon's scales have tuned a deep shade of green and are getting nice and shiny. Was it the exercise that did it? Explain.

7. A farmer records data on a factor: rooster status (crows, doesn't crow) and a response: sun status (rises, doesn't rise) every morning for a month. Every single morning the rooster crows and every single morning, a short while later, the sun rises. The farmer concludes that the rooster crowing is what makes the sun rise. Comment on the farmer's conclusion.

8. In exercise 1, give one potential confounding variable. Does the study control the effect of that variable on the response? Explain.

9. In exercise 2, give one potential confounding variable. Does the study control the effect of that variable on the response? Explain.

10. In exercise 3, give one potential confounding variable. Does the study control the effect of that variable on the response? Explain.

11. In exercise 4, give one potential confounding variable. Does the study control the effect of that variable on the response? Explain.

12. Consider the asthma study comparing medications A and B. Let's say 60 volunteers sign up to participate in the study.

    a. Describe, in detail, how to use the hat method to conduct a SRA of the two medications to the volunteers so that each treatment group has the same number of people.
    b. Describe, in detail, how to use a computer to conduct a SRA of the two medications to the volunteers so that each treatment group has the same number of people.

13. Consider the homework study comparing the "new problems only" and "new+review" problems treatments. In a course with 40 students, describe, in detail, how to use a computer to conduct a SRA of the two treatments to the students so that each treatment group has the same number of students.

14. Refer to the recycling study in exercise 2. Describe, in detail, how to conduct a SRA of the treatments to the students so that each treatment group has the same number of students.

15. Refer to the ascorbic acid study in exercise 3. Describe, in detail, how to conduct a SRA of the treatments to the goats so that each treatment group as the same number of goats.

16. Give one advantage and one disadvantage of using a RBA instead of a SRA.

17. Give one advantage and one disadvantage of using a CRA instead of a SRA.

18. Elizabeth has two garden plots each having space for 10 tomato plants. She plants a tomato plant in each of the 20 spots. One plot is near a large maple tree and the other is far from tall trees and buildings. She wants to compare the yields of plants treated with two brands (A and B) of organic fertilizer.

    a. Describe, in detail, how to assign fertilizer treatments to the 20 tomato plants using SRA.
    b. Describe, in detail, how to assign fertilizer treatments to the 20 tomato plants using a RBA. (What is a natural choice for the block?)
    c. Which would you recommend: a SRA or a RBA with garden plot as the block? Explain your choice.

19. Thirty-six college students participated in a study where each student received each of three treatments: 1) listening to 10 minutes of a Mozart piano sonata, 2) listening to 10 minutes of relaxation instructions, and 3) sitting in silence for 10 minutes. After each treatment, the student took a variation of a spatial reasoning test. Higher scores indicated greater spatial reasoning ability.[17]

    a. Identify the factor(s), the levels of each factor, and the response.
    b. Identify the blocking variable.
    c. Describe, in detail, how to assign treatments in a RBA.
    d. An alternative design is a SRA where 108 students are randomly assigned to only one of the three treatments. This design would also result in 36 responses per treatment. What is one advantage of using a RBA instead of a SRA?

20. One common measure of lung function in asthma studies is the volume of air a subject can expel from their lungs in 1 second. This is often referred to as "forced expiratory volume in 1 second" or FEV1 for short. Since men tend to have larger lungs than women, they may also have higher FEV1 values. Suppose a study comparing medications A and B recruits 40 male and 60 female asthma patients. The males are randomly assigned to a medication (20 to A and 20 to B) and the females are randomly assigned to a medication (30 to A and 30 to B).

a. The variables in this study are: FEV1, medication, sex. Label each variable with its role in the study (e.g. "factor," "response," etc.).
b. Which method was used to assign subjects to treatments?
c. What potential problem with the SRA does the RBA solve?

21. Consider the asthma study described in exercise 20. A simulation was conducted to see how the RBA and SRA perform. The simulation *assumed that there is no difference in the effects of the two medications on FEV1.* One thousand SRAs were performed and the difference between the two treatment means $(\bar{x}_A - \bar{x}_B)$ in liters was calculated for each SRA. Similarly, 1000 RBAs were performed and the difference in the FEV1 means was calculated for each RBA. Some summaries of the two sets of differences are shown here but they are not labeled as to which is from the SRA and which is from the RBA.

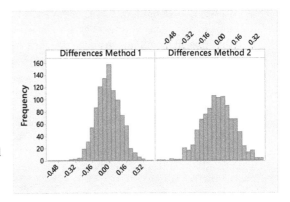

	N	Mean	StDev	Median	IQR
Differences Method 1	1000	-0.00731	0.11096	-0.00938	0.14857
Differences Method 2	1000	-0.00432	0.15156	-0.00347	0.20577

a. Compare the two distributions.
b. How do the two distributions reflect the assumption of no difference in the effects of medications A and B?
c. Which distribution came from the SRA simulation? from the RBA simulation? Explain.
d. Let's say that, in the actual study, the group assigned to medication A had FEV1s that were 0.25 liters higher on average than those assigned to medication B, i.e. $\bar{x}_A - \bar{x}_B = 0.25$. Find the z-score for 0.25 in the SRA distribution and in the RBA distribution. Under which method would a difference in treatment means of 0.25 liters be more unusual?

22. Consider Elizabeth's tomato experiment in exercise 18.

a. In a simulation, 15 SRAs were performed *under the assumption that there is no difference in the effects of the two fertilizers on yield.* The difference between the two fertilizer means $(\bar{x}_A - \bar{x}_B)$ was calculated for each SRA. The differences (in pounds) are shown here:

-2.7	-0.7	0.3	-2.1	-0.7	-3.2	2.8	2.2	-2.0	-1.0
2.2	2.1	0.8	3.2	-0.7					

Construct a boxplot of these differences.

b. In another simulation, 15 RBAs were performed *under the assumption that there is no difference in the effects of the two fertilizers on yield.* The difference between the two fertilizer means $(\bar{x}_A - \bar{x}_B)$ was calculated for each RBA. The differences (in pounds) are shown here:

-1.6	0.4	0.2	-0.4	-1.1	-0.2	0.5	0.7	-0.7	1.5
-1.5	-0.8	0.5	0.7	0.8					

Construct a boxplot of these differences on the same scale as your boxplot in part a.

c. Compare the two distributions in a. and b.
d. Why are the centers of both distributions near 0?
e. Why is the IQR smaller in the RBA distribution of differences?
f. Let's say that, in the actual study, the plants assigned to fertilizer A had yields that were 2 pounds higher on average than those assigned to fertilizer B, i.e. $\bar{x}_A - \bar{x}_B = 2$. Find the z-score for 2 in the SRA distribution and in

the RBA distribution. Under which method would a difference in treatment means of 2 pounds be more unusual?

23. Consider the study in Ghana designed to assess the efficacy of bed nets in preventing malaria.

    a. Describe, in detail, how to assign the 96 community clusters to the two treatments (bed net, no bed net) using CRA so that each treatment has the same number of community clusters.
    b. Describe, in detail, how to assign the 1961 individual women to the two treatments (bed net, no bed net) using SRA so that each treatment has (almost) the same number of women.

24. You want to compare two programs (A and B) designed to prepare high school students for the Advanced Placement (AP) Statistics Exam. You recruit 200 high schools nationwide to participate in the study. You list the high schools in a spreadsheet containing the names of the 200 schools in one column. You use a random number generator to put 200 random numbers between 0 and 1 in another column. After you sort both columns by the random number column, you assign the first 100 schools to program A and the rest to program B. You record the scores on the AP Statistics Exam for each student in each participating high school.

    a. Which method was used to assign students to programs?
    b. Name two potential confounding variables. Does the study control the effects of these confounding variables on exam score? Explain.
    c. What is a difficulty in using SRA to assign individual students in the 200 schools to a program?
    d. Someone suggests you should use a placebo group in your study. How would you respond?

25. Consider two programs (A and B) designed to prepare high school students for the Advanced Placement (AP) Statistics Exam, similar to exercise 24, but let's say that only 2 schools participate and there are only 2 students taking AP Statistics at each school. Kirk High School is known for its stellar AP Statistics program so that students are expected to do well regardless of any additional intervention. Students at Clayton High School have traditionally struggled on the AP Statistics exam. A CRA is used to assign the programs to the students. Shown below is a table showing the results of the experiment. Also shown is another column with the other possible treatment assignment, *assuming that the programs do not differ in their effect on AP Score.*

Student ID	School	Program Assigned	AP Score	Other CRA
183	Kirk HS	A	4	B
274	Kirk HS	A	5	B
099	Clayton HS	B	1	A
156	Clayton HS	B	0	A

Same scenario, but now a SRA is used to assign the programs to the students. Shown below is a table showing the results of the experiment. Also shown are 5 other columns with the other possible treatment assignments, *assuming that the programs do not differ in their effect on AP Score.*

Student ID	School	Program Assigned	AP Score	Other SRA 1	Other SRA 2	Other SRA 3	Other SRA 4	Other SRA 5
183	Kirk HS	A	4	A	A	B	B	B
274	Kirk HS	B	5	A	B	B	A	A
099	Clayton HS	A	1	B	B	A	B	A
156	Clayton HS	B	0	B	A	A	A	B

a. Complete the tables below, calculating the treatment mean scores and the differences for each possible random assignment.

	$\bar{x}_A$	$\bar{x}_B$	$\bar{x}_A - \bar{x}_B$
Original CRA Assignment	4.5	0.5	4.0
Other CRA Assignment			
Original SRA Assignment	2.5	2.5	0.0
Other SRA Assignment 1			

	$\bar{x}_A$	$\bar{x}_B$	$\bar{x}_A - \bar{x}_B$
Other SRA Assignment 2			
Other SRA Assignment 3			
Other SRA Assignment 4			
Other SRA Assignment 5			

    b.   Make a dotplot for each distribution of differences, one for the two CRA assignments and one for the 6 SRA assignments.

    c.   Compare the two distributions.

    d.   Find the mean and standard deviation in each distribution.

    e.   Why are the means both 0? Why is the CRA distribution of differences more variable?

26. Here's a rare example of when the CRA can outperform the SRA. Consider two programs (A and B) designed to prepare high school students for the Advanced Placement (AP) Statistics Exam, similar to exercise 24, but let's say that only 2 schools participate and there are only 2 students taking AP Statistics at each school. Students at both Miller High School and Parks High School tend to perform similarly on the AP exam. A CRA is used to assign the programs to the students. Shown below is a table showing the results of the experiment. Also shown is another column with the other possible treatment assignment, *assuming that the programs have no effect on AP Score.*

Student ID	School	Program Assigned	AP Score	Other CRA
294	Miller HS	A	3	B
385	Miller HS	A	2	B
100	Parks HS	B	2	A
267	Parks HS	B	3	A

Same scenario, but now a SRA is used to assign the programs to the students. Shown below is a table showing the results of the experiment. Also shown are 5 other columns with the other possible treatment assignments, *assuming that the programs have no effect on AP Score.*

Student ID	School	Program Assigned	AP Score	Other SRA 1	Other SRA 2	Other SRA 3	Other SRA 4	Other SRA 5
294	Miller HS	A	3	A	A	B	B	B
385	Miller HS	B	2	A	B	B	A	A
100	Parks HS	A	2	B	B	A	B	A
267	Parks HS	B	3	B	A	A	A	B

    a.   Complete the tables below, calculating the treatment mean scores and the differences for each possible random assignment.

	$\bar{x}_A$	$\bar{x}_B$	$\bar{x}_A - \bar{x}_B$
Original CRA Assignment	2.5	2.5	0.0
Other CRA Assignment			
Original SRA Assignment	2.5	2.5	0.0
Other SRA Assignment 1			
Other SRA Assignment 2			
Other SRA Assignment 3			
Other SRA Assignment 4			
Other SRA Assignment 5			

    b.   Make a dotplot for each distribution of differences, one for the two CRA assignments and one for the 6 SRA assignments.

    c.   Compare the two distributions.

    d.   Find the mean and standard deviation in each distribution.

    e.   Why are the means both 0? Why is the SRA distribution of differences more variable?

    f.   Make a general statement about when the CRA will result in less variable treatment mean differences compared to the SRA.

27. Suppose that in the Ghana bed net study, the women who received bed nets had a lower incidence of malaria compared to those who did not receive bed nets. Suppose that a statistician calculates that the probability of that happening by chance (by the random assignment alone) is only 0.01. We might then have convincing evidence that the bed nets were effective in reducing the incidence of malaria for women like the ones in the study.

    a.   Give one reason why we still should be cautious in concluding that the bed nets were effective.
    b.   Give one reason why we should be cautious in concluding that the bed nets would work the same way for people in Burma.

28. Consider the study of the effect of ascorbic acid (AA) on the stress of transported goats. The dose of AA prior to transport was found to be effective in preventing weight loss. A local farmer wants to try the same treatment on Angus steers that are transported by truck. How would you advise the farmer?

Notes

[1] Webster's Encyclopedic Unabridged Dictionary of the English Language (1989), dilithium Press, Ltd., New York.

[2] Rampell, C. (May 19, 2012), "The Beginning of the End of the Census?" New York Times.

[3] Source: http://www.faa.gov/airports/planning_capacity/passenger_allcargo_stats/categories/

[4] Pratham Organization (2009), "Annual Status of Education Report (Rural) 2008," Mumbai, India. Pratham Resource Center, http://asercentre.org/asersurvey/aser08/pdfdata/aser08.pdf.

[5] Chaudhury, N., et al. (2006), "Missing in Action: Teacher and Health Worker Absence in Developing Countries," *Journal of Economic Perspectives*, 20(1), 91 – 116.

[6] Banerjee, A. V., et al. (2010), "Pitfalls of Participatory Programs: Evidence from a Randomized Evaluation in Education in India," *American Economic Journal: Economic Policy*, 2(1), 1 – 30.

[7] "Landon, 1,293,699; Roosevelt, 972,897: Final Returns in The Digest's Poll of Ten Million Voters," (October 31, 1936), *The Literary Digest*, 122(18), Funk & Wagnalls Co., New York. For an example of a news report at the time, see "Straw Vote Fight Arouses Interest," (November 2, 1936), The Pittsburgh Press.

[8] Diamond, R. A., editor (1975), *Congressional Quarterly's Guide to U.S. Elections*, Congressional Quarterly, Inc., Washington, D.C.

[9] Pew Research Center, March 7 – 11, 2012.

[10] Browne, E. N. L., et al. (2001), "The Impact of Insecticide-Treated Bednets on Malaria and Anaemia in Pregnancy in Kassena-Nakana District Ghana: A Randomized Controlled Trial," *Tropical Medicine and International Health*, 6(9), 667 – 676.

[11] O'Connor, A. (May 6, 2013), "Sucking Your Child's Pacifier Clean May Have Benefits," New York Times.

[12] Hesselmar, B., et al. (2013), "Pacifier Cleaning Practices and Risk of Allergy Development," *Pediatrics*, 131(6), 1829 – 1837.

[13] Roethlisberger, F. J. (1940), *Management and the Worker*, Harvard University Press, Cambridge and Landsberger, H. (1958), *Hawthorne Revisited*, Cornell University, New York.

[14] Zahrt, O. H. and Crum A. J. (2017), "Perceived Physical Activity and Mortality: Evidence from Three Nationally Representative U.S. Samples," *Health Psychology*, 36(11), 1017 – 1025.

[15] Trudel, R. and Argo, J. J. (2013), "The Effect of Product Size and Form Distortion on Consumer Recycling Behavior," *Journal of Consumer Research*, 40(4), 632 – 643.

[16] Nwunuji, T. P., et al. (2014), "The ameliorative effect of ascorbic acid on the oxidative status, live weight and recovery rate in road transport stressed goats in a hot humid tropical environment," *Animal Science Journal*, 85, 611 – 616.

[17] Rauscher, F. H., et al. (Oct. 14, 1993), "Music and spatial task performance," *Nature*, 365, 611.

Chapter 4: Probability

Introduction

"But Bryan," you say, "I thought this was a book about data. Why do we need probability?" I feel your pain. Maybe you've heard about this part of the course…and it wasn't good news. Maybe you've struggled with probability before. There is a very good reason why we study probability in a course about data science and statistics. It has everything to do with random mechanisms.

In chapter 3 you studied two random mechanisms (random sampling and random assignment) that are very useful in producing high-quality data to answer two kinds of questions. Random sampling helps to ensure that samples represent populations so that generalization inferences are justified. Random assignment helps to ensure that treatment groups are balanced so that causal inferences are justified. You also saw that the word *random* in this sense does not have the usual meaning from casual conversation. *Random* does not mean "odd," "haphazard," "without definite aim," or "suspiciously out of place." Instead, you know that a *random* sample, for instance, is a sample obtained in a way such that the outcome of one random sample is *uncertain* but the collective outcomes of many random samples has a *structure* that can be predicted in advance. Drawing names out of a hat or using a computer to randomly order a list of objects are two examples of the randomness we're talking about. Randomness for the statistician means *structured uncertainty*. We know what will happen if we repeat the random process many times.

Fortunately we don't have to actually repeat the random process many times! We can use models to describe the structure of the uncertainty for random mechanisms. These models are called probability models.

Before we get to the details of probability models, let's think about what exactly we mean by "probability." For example, if I tell you that the probability that a thumb tack will land point down is 0.4 – denoted $P(point\ down) = 0.4$ – what do I mean by that? Does it mean that there's a 40% chance that the tack will land point down? That doesn't help much; "40% chance" is just a synonym for "probability of 0.4." Does it mean the tack will land point down 4 times in 10 drops? That's closer but not exactly true either. For the purposes of this course, we'll take probability to mean *long-run proportion*. So $P(point\ down) = 0.4$ really means that if I drop the tack many, many times, it will land point down in about 40% of those drops. The long-run proportion of "point downs" will approach 0.40, although it may never exactly get there. If you've taken a calculus course, this idea can be expressed using limits:

$$P(point\ down) = \lim_{B \to \infty} \frac{\#\ point\ downs}{B\ drops}$$

If you've not taken calculus, never mind! The important thing to remember is that probability is the long-run proportion of times some event occurs. This concept is appealing because you can evaluate my proposed probability of 0.4 by dropping the tack a bunch of times. For example, if you get 600 point downs in 1000 drops, you might question my claim that $P(point\ down) = 0.4$. How many times do we need to drop the tack to get a good test of my claim? How far would the proportion of "point downs" have to be from 0.4 before we start to question the claim? The answers to those questions come from probability models. Let's get to it.

4.1 Elements of Probability

Our ultimate goal in learning to use probability tools is so that we can apply them to the random mechanisms of random sampling and random assignment. There are four elements of probability theory that are important for our application to data and statistics. They are:

- Experiment
- Sample Space
- Assignment of Probability
- Random Variable

An **experiment** is an action whose outcome is not known with certainty ahead of time. We are not talking about the kind of experiment you may have done in a physics class where you studied projectile motion. The outcomes of those experiments are determined by the laws of motion and how objects behave in a gravitational field. The experiments we'll study are those with uncertain outcomes but where that uncertainty has a structure; our job is to discover the structure.

Example: Suppose Aaron is a 60% free-throw shooter from the foul line. Aaron attempts 3 free-throws and observes whether or not he makes each shot.

The experiment here is Aaron's 3 free-throw attempts. The outcome (which ones he makes and which ones he misses) is not known with certainty before he actually attempts the three shots. One possible outcome is XOX (letting X = made shot and O = missed shot). Even though we don't know the outcome of Aaron's experiment, we do know what the possible outcomes look like. Knowing what the possible outcomes are is part of the structure of the uncertainty and that brings us to the second element of probability, the sample space. The **sample space** is the set of all possible outcomes of an experiment. We often denote the sample space as S.

Example: In Aaron's free-throw experiment, the sample space is the set of all possible sequences of X's and O's:

$$S = \{OOO, XOO, OXO, OOX, XXO, XOX, OXX, XXX\}$$

A tree diagram is one way to identify all the outcomes in the sample space in experiments like this that can naturally be divided into several stages. A tree diagram for the three-stage free-throw experiment is shown here.

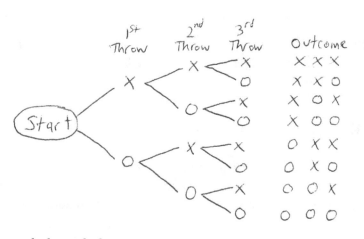

At each stage, you only need to think about the possibilities at the next stage (here there are only two: X = make and O = miss). To get any outcome of the entire experiment, follow a particular path through the tree.

Once we have the sample space defined, we assign each one a probability (a long-run proportion). This **assignment of probability** needs to follow two rules:

1. A probability must be a number between 0 (impossible) and 1 (certain). While we sometimes talk about probabilities as percentages (e.g. "...a 20% chance of rain today..."), they are numbers in the interval [0, 1].

2. The sum of the probabilities for all outcomes in S must be exactly 1. It's like we get a bucket of probability of size 1 and must *distribute* all of it across the outcomes in S.

Consider Aaron's free-throw experiment with eight possible outcomes. Here is one possible assignment of probability to the eight outcomes:

Outcome	*000*	*X00*	*OXO*	*00X*	*XXO*	*XOX*	*OXX*	*XXX*
Probability	0.125	0.125	0.125	0.125	0.125	0.125	0.125	0.125

Each probability is between 0 and 1 and they add to 1. At least we've followed the rules. But what about the assumption that Aaron is a 60% shooter? Wouldn't that mean that he should make more shots than he misses? An assignment of probability should take that into account. We want the *XXX* outcome to be more likely than the *000* outcome. The tree diagram is a handy tool to compute the probabilities of the outcomes by labeling the branches with the probabilities of made and missed shots.

The probability of an outcome is found by multiplying the probabilities along that path through the tree.

Outcome	000	XOO	OXO	OOX	XXO	XOX	OXX	XXX
Probability	0.064	0.096	0.096	0.096	0.144	0.144	0.144	0.216

This is a better assignment of probabilities because, in addition to following the rules, it matches the 60% assumption. Since probabilities are long-run proportions, this table describes the behavior of Aaron's experiment over many, many repetitions. Imagine if Aaron were able to repeat his set of 3 free-throws thousands and thousands of times. If he is really a 60% shooter (and that characteristic doesn't change over time), he'll make all three shots in about 21.6% of those sets of 3 free-throws.

You may notice that some of the sample space outcomes share a common characteristic; some of them have the same number of made and missed shots. A numeric characteristic of a sample space outcome is called a **random variable**. A random variable is the probability analog to a variable in a data context, but random variables in probability can only be numeric; there is no such thing as a categorical random variable. We call the set of possible values of a random variable the **range** of the random variable. You can think of the range of a random variable as a set of values on the number line. In statistics, our goal in any experiment is to determine the *probability distribution* of a random variable. The **probability distribution** of a random variable describes how probability is distributed (spread out) across its range.

Example: Consider Aaron's free-throw experiment. Let X be the number of shots he makes in a set of 3 free-throws. This is a very natural way to describe the outcome of the experiment, especially if you don't care about the ordering of the made/missed shots. The range of X is $\{0, 1, 2, 3\}$. To get the probability distribution of X, we simply figure out which sample space outcomes correspond to which values of X. If there are multiple sample space outcomes that correspond to the same value of X, add those probabilities together.

x	Outcomes in S	$p(x)$
0	000	0.064
1	XOO, OXO, OOX	$0.096 + 0.096 + 0.096 = 0.288$

x	Outcomes in S	$p(x)$
2	XXO, XOX, OXX	$0.144 + 0.144 + 0.144 = 0.432$
3	XXX	0.216

The table shown here is one way to describe the probability distribution of X. In this example, I used the notation $p(x)$ to denote the probability of a value of X. In practice, we usually omit the middle column. Imagine if Aaron were able to repeat his set of 3 free-throws thousands and thousands of times. If he is really a 60% shooter (and that characteristic doesn't change over time), he'll make 2 out of the 3 shots in about 43.2% of those sets of 3 free-throws.

Example: Maria purchases a car insurance policy with a $1300 annual premium. (The premium of an insurance policy is what you pay the insurance company in return for their agreement to cover costs related to damages to your car, the property of other people, or personal injuries from an accident.) The table below gives the possible annual costs that the insurance company may have to pay on Maria's behalf along with her risk profile.

Cost	$0 (accident-free!)	$1000 (fender-bender)	$15,000 (wrecks her car)	$250,000 (multi-car accident with hospital bills)
Risk	85%	10%	4.9%	0.1%

Let X be the net gain for the insurance company per year on Maria's policy. Find the probability distribution of X.

Let's look at the elements of this problem.

- Experiment: Maria purchases the car insurance policy and observes her accident status for the year.
- Sample Space: {accident-free, fender-bender, wrecks her car, multi-car accident with hospital bills}
- Assignment of Probability: The table shown above (almost) does this based on Maria's risk profile. (Determining Maria's "risk profile" is a very important job for the insurance company and is often done by actuaries using accident data collected on many people like Maria.) Remember to use decimals as probabilities instead of the percentages; probabilities must be numbers between 0 and 1.
- Random Variable: The net gain for the insurance company is the premium Maria pays them minus any costs they pay on Maria's behalf. The table below shows how we get the range of X and the probability distribution.

Cost	x	$p(x)$
$250,000	$1300 - $250,000 = -$248,700	0.001
$15,000	$1300 - $15,000 = -$13,700	0.049
$1000	$1300 - $1000 = $300	0.1
$0	$1300 - $0 = $1300	0.85

Notice that in some cases, the insurance company will take a loss on Maria's policy because they're paying out more to cover her costs than she is paying in premium. However, those cases are fairly low probability outcomes.

Aside: Beware of statements about probability or risk that compare the probability of two events. Suppose I tell you that Maria is almost 50 times as likely to wreck her car than she is to have a multi-car accident with hospital bills. While that is technically a true statement, it may give you the false impression that a wreck for Maria is right around the corner. However, fifty times a very small probability (like 0.001) is still a small probability! Whenever you hear statements like this, always try to determine the actual probabilities. You may be surprised by how small they really are.

I'll close this section by distinguishing between two types of random variables. A **discrete random variable** is a random variable having a countable range. For our purposes, that means the possible values in the range have gaps between them on the number line. The two random variables we've seen so far are discrete.

Random Variable	Range
X = number of shots made out of 3	$\{0, 1, 2, 3\}$
X = net gain of insurance company	$\{-\$248700, -\$13700, \$300, \$1300\}$

Each range would graph as a set of points on the number line having gaps between them. It is impossible for Aaron to make 1.5 shots out of 3. The insurance company cannot lose $500 on Maria's policy.

You might notice that I simplified the insurance example to four cases. In reality, the insurance company's cost can be any dollar amount in some range and is modeled by using a continuous random variable instead of a discrete one. A **continuous random variable** is a random variable where the range forms an interval on the number line with no gaps between possible values. Here are two examples:

Random Variable	Range
X = time to finish an exam (minutes)	$0 < x \le 120$
X = weight gain in two weeks (pounds)	$-40 < x < 40$

While we often report variables like time and weight to the nearest minute, hour, ounce, pound, or kilogram, that does not mean the actual values are limited. Time passes on a continuum. Weight is measured on a continuum. Each range above would graph, not as a discrete set of points, but as an interval on the number line with no gaps between possible values.

The difference between discrete and continuous random variables is important because we play with probability differently for discrete random variables than we do for continuous ones. More about continuous random variables in a later section.

Section 4.1 Exercises

1.   Consider the experiment of tossing a fair coin 3 times and recording the sequence of heads and tails you get.

     a.   List all the possible outcomes in the sample space.
     b.   Assign probabilities to the outcomes.
     c.   What is the probability of getting exactly 2 tails?
     d.   What is the probability of getting at least 2 tails?
     e.   Let X be the number of tails you get. What are the possible values of X?
     f.   Find $P(X \leq 2)$.

2.   In Monopoly, you move around the board by rolling two fair dice. Let's say that one is green and one is blue. Consider a single Monopoly roll.

     a.   List all the possible outcomes in the sample space.
     b.   Assign probabilities to the outcomes.
     c.   If you roll doubles, you get to take an extra turn. What is the probability of rolling doubles?
     d.   You move your game token around the board according to the sum of the two numbers on the dice. If you're sitting at Virginia Avenue, what's the probability your roll lands you at Free Parking?

3.   Darryl is a baseball player who has a 0.250 batting average. That means that over some period of time, Darryl had a hit in 1 out of 4 at-bats. Consider the experiment where Darryl has 4 at-bats in a game and observes whether or not he gets a hit each time.

     a.   List all the possible outcomes in the sample space.
     b.   Assign probabilities to the outcomes.
     c.   What is the probability that Darryl gets 2 hits in that game (i.e. "goes 2 for 4")?
     d.   What is the probability that Darryl gets at most 1 hit in that game?
     e.   Let X be the number of hits Darryl gets in the game. What are the possible values of X?
     f.   Find $P(X \geq 3)$.

4.   A department store runs a promotion where customers spin a wheel twice at the checkout to determine their discount. The wheel has 10 equal-sized slices: six are labeled "2%," three are labeled "10%," and one is labeled "20%."

     a.   List all the possible outcomes in the sample space.
     b.   Assign probabilities to the outcomes.
     c.   What is the probability that a customer spins 10% twice?
     d.   What is the probability that a customer spins at least one 20%?
     e.   A customer's actual discount is the larger of the two spins so let X be the larger discount. What are the possible values of X?
     f.   Give the probability distribution of X in table form.
     g.   Find $P(X = 2\%)$.

5.   There are two traffic lights on Maria's way to work. If both lights are green, she can get to work in 20 minutes. Stopping at the first light adds 1 minute to her commute time and stopping at the second light adds 2 minutes. The probabilities that she stops at the first and second lights are 0.3 and 0.2, respectively. Consider the experiment where Maria records what happens at each light one morning on her way to work.

     a.   List all the possible outcomes in the sample space.
     b.   Assign probabilities to the outcomes.
     c.   What is the probability that Maria stops at at least one light?
     d.   What is the probability that Maria stops at at most one light?
     e.   What is the probability that Maria stops at exactly one light?
     f.   Let X be Maria's commute time. What are the possible values of X?
     g.   Give the probability distribution of X in table form.

6. One drawer in our kitchen is the "junk drawer." Among other things in there are batteries. Let's say there are 10 batteries in the drawer but 3 of them are totally dead. I select 2 batteries at random for my son's train.

    a. What is the probability that both of them are dead?
    b. What is the probability that at least one of them is dead?
    c. What is the probability that both of them are good?
    d. Let X be the number of good batteries I select. Give the probability distribution of X in table form.

7. Which of the following are legitimate probability distributions? For those that are not, explain why not.

    a.
x	0	1	2
$p(x)$	0.4	0.5	0.6

    b.
x	-1	0	1
$p(x)$	0.1	0.5	0.4

    c.
x	100	200	300
$p(x)$	0.9	-0.4	0.5

    d.
x	-0.5	0	0.5
$p(x)$	0.2	0.7	0.1

8. Consider these probability distributions.

    I.
x	0	1	2
$p(x)$	0.3	0.2	

    II.
x	0	1	2
$p(x)$	0.2		

    III.
x	0	1	2
$p(x)$			0.1

    IV.
x	0	1	2
$p(x)$			

    a. Find the missing value in distribution I.
    b. If $P(X \le 1) = 0.7$, find the missing values in distribution II.
    c. If $P(X \ge 1) = 0.8$, find the missing values in distribution III.
    d. If $P(X \le 1) = 0.15$ and $P(X \ge 1) = 0.95$, find the missing values in distribution IV.

9. Brian, Art, Sam, and Phil audition for America's Got Luck. The audition process consists of tossing a fair coin for each person. Heads they get on; tails they don't.

    a. What's the probability that all four guys get a spot on the show?
    b. What's the probability that at least one of them gets a spot on the show?
    c. Let X be how many of these guys get a spot on the show. Find $P(X = 2)$.

10. Brian, Art, Sam, and Phil audition for America's Got More Luck. The audition process consists counting the number of folks in line to audition and putting half that number of red marbles in a bucket and half that number of green marbles in the bucket. As people come in for their "audition," a marble is selected from the bucket without replacement. Red they get on; green they don't. Suppose that these four guys are the only ones in line.

    a. What's the probability that all four guys get a spot on the show?
    b. What's the probability that at least one of them gets a spot on the show?
    c. Let X be how many of these guys get a spot on the show. Find $P(X = 2)$.

11. Micah purchases a car insurance policy with a $3000 annual premium. The table below gives the possible annual costs that the insurance company may have to pay on Micah's behalf along with the probabilities.

Cost	$0	$1500	$20,000	$300,000
Probability	0.80	0.15	0.045	0.005

    Let X be the net gain for the insurance company per year on Micah's policy. Find the probability distribution of X.

12. Hannah purchases a homeowner's insurance policy with a $2000 annual premium. The table below gives the possible annual costs that the insurance company may have to pay on Hannah's behalf along with the probabilities.

Cost	$0	$1000	$50,000	$450,000
Probability	0.90	0.08	0.019	0.001

    Let X be the net gain for the insurance company per year on Hannah's policy. Find the probability distribution of X.

13. Consider the game of spinning the spinner shown here twice and observing the random variable X = sum of the numbers on both spins.

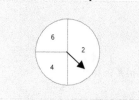

   a.   What is the probability of getting a sum of 6?
   b.   Find the probability distribution of X.
   c.   What is the probability that the sum is more than 4?

14. You want to dunk your friend in the dunk tank at the carnival. Your friend sits on a ledge above a tank of water and you throw baseballs at a target. If you hit the target, the ledge collapses and your friend gets dunked. You figure you have a 20% chance of hitting the target with each throw. It costs $2 per throw and you only have $8.

   a.   Let X be the amount you spend until you dunk your friend or you run out of money. Find the probability distribution of X.
   b.   What's the probability that you dunk your friend before you run out of money?

15. An experiment is designed to see whether 3rd graders can write faster with a pen or pencil. Four third graders participate, 2 boys and 2 girls. For each child, a fair coin is tossed; if heads, the child gets a pen and if tails, the child gets a pencil. Once two children are assigned to one group, the others are automatically assigned to the other group. The children are assigned in this order: boy 1, girl 1, boy 2, girl 2.

   a.   List the possible outcomes in the sample space.
   b.   What is the probability that both boys or both girls get a pen?

16. An experiment is designed to see whether 3rd graders can write faster with a pen or pencil. Four third graders participate, 2 boys and 2 girls. For each child, a marble is drawn without replacement from a bucket containing 2 red marbles and 2 blue marbles. If a red marble is selected, the child gets a pen and if blue, the child gets a pencil. The children are assigned in this order: boy 1, girl 1, boy 2, girl 2.

   a.   List the possible outcomes in the sample space.
   b.   What is the probability that both boys or both girls get a pen?

17. Four hospital patients are waiting to be discharged; two from labor and delivery with their newborns and two from the emergency department. You want to select a SRS of two patients for a patient satisfaction survey. What is the probability that both patients are from the same department?

18. Consider exercise 17. If you select a STRS of two patients stratified by department, what is the probability that both patients are from the same department?

19. For each of the random variables below, say whether it is discrete or continuous.

   a. X = number of seeds that do not germinate in a package of 20 seeds
   b. X = hand span of a college student
   c. X = time you have to wait for the walk sign at an intersection
   d. X = number of billboards along I95 in Delaware that are electronic

20. For each of the random variables below, say whether it is discrete or continuous.

   a. X = distance to a star
   b. X = annual return on a mutual fund
   c. X = number of employees at a company

d. X = number of views of a webpage in one hour

4.2 Features of Probability Distributions

Probability distributions have features that are analogous to distributions of data. We can refer to the center, shape, and variability of a probability distribution just like we can do with a distribution of numeric data values. The difference is that a probability distribution is not an actual data distribution. Rather it describes the behavior of data values over many, many repetitions of the experiment. Likewise, the features of a probability distribution describe the long-run behavior of the experiment. That's why we can often use a random variable and its probability distribution as a model for a distribution of numeric data.

Let's look again at the probability distribution of X = number of shots made out of 3 for Aaron, a 60% free-throw shooter.

x	$p(x)$
0	0.064
1	0.288
2	0.432
3	0.216

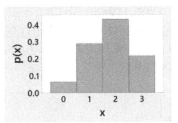

Constructing a *probability histogram*, like the one shown here, is a nice way to visualize the distribution. Note that this histogram is not like the histograms we've seen before because it does not represent an actual set of data values. Rather, it is a model for what would happen over many, many sets of 3 free-throws. Notice that the histogram bars are centered at the possible values of x and have width 1. The height of each bar is the probability of that value of x. The areas of the bars, therefore, also represent probabilities. This idea of probability being represented geometrically by area is an important one that we'll use in the next section.

Notice how the important features of the distribution are clear in the probability histogram. The distribution is a bit left-skewed with a center somewhere between 1 and 2. The range is 3 but other measures of variability may be smaller than 3 since probability is largely concentrated around the center with lower probabilities at 3 and (especially) at 0. Let's learn how to get precise values for the center and variability.

Center: Expected Value

A common measure of center in a distribution of numeric data is the mean, the balance point in the distribution. In a probability distribution, we call this measure of center the **expected value of X**, denoted $E(X)$. $E(X)$ is the average value of X over many, many repetitions of the experiment. Run the experiment, write down the value of X, compute the average of all the x's so far, and repeat many (say B) times. You would find that the averages start to converge to some value and we call that value the long-run average value or expected value of X:

$$E(X) = \lim_{B \to \infty} \frac{x_1 + x_2 + \cdots + x_B}{B}$$

This is very consistent with our notion of probability as long-run proportion.

Here's how it would work with Aaron and his set of 3 free-throws experiment:
- Aaron attempts 3 free-throws.
- We record the value of X, the number he makes.
- We calculate the average of all the values of X we've recorded so far.

Here are the results for 1,000 repetitions of this process. By the time Aaron finishes all 1,000 experiments (3,000 free-throws), the average of all the values of X is about 1.8. That is, average number of shots made out of 3 across is about 1.8, at least in this set of 1,000 experiments. Notice how the running average converges to the 1.8 figure. If you're satisfied that the running average isn't converging to anything other than 1.8 after 1,000 repetitions, you could say that $E(X) = 1.8$.

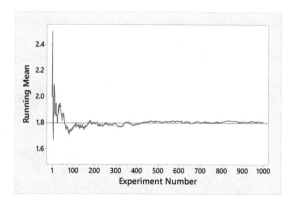

Experiment Number	x	Running Average
1	2	2.000
2	3	$(2+3)/2 = 2.500$
3	1	$(2+3+1)/3 = 2.000$
4	2	$(2+3+1+2)/4 = 2.000$
5	1	$(2+3+1+2+1)/5 = 1.800$
6	1	$(2+3+1+2+1+1)/6 = 1.667$
$\vdots$	$\vdots$	$\vdots$
998	3	$(2+3+1+2+1+1+\cdots+3)/998 = 1.809$
999	3	$(2+3+1+2+1+1+\cdots+3+3)/999 = 1.810$
1,000	1	$(2+3+1+2+1+1+\cdots+3+3+1)/1000 = 1.810$

Of course, by this time, Aaron is so tired and grumpy that he will never help you again. While the idea of repeating the experiment many times helps us to understand the meaning of expected value, it is not helpful at all for actually computing $E(X)$. To do that, we use the table of x's and $p(x)$'s. Let's suppose there are k values of X in its range. Using the probabilities as weights, we can calculate the expected value of a discrete random variable X this way:

x	$p(x)$
x_1	p_1
x_2	p_2
$\vdots$	$\vdots$
x_k	p_k

$$E(X) = x_1 p_1 + x_2 p_2 + \cdots + x_k p_k$$

Multiply each value of X by its probability and add up the products. $E(X)$ is a weighted average of the values of X; the weights are the probabilities. Just remember not to divide by anything at the end!

Example: Let's calculate $E(X)$ for Aaron's free-throw experiment without his help. The probability distribution of X = number of shots made out of 3 and the expected value of X is:

x	$p(x)$
0	0.064
1	0.288
2	0.432
3	0.216

$$E(X) = 0(0.064) + 1(0.288) + 2(0.432) + 3(0.216) = 1.8$$

Interpretation: Over many repetitions of his 3-shot experiment, the average number of shots he will make is 1.8 out of 3. The 1.8 figure doesn't mean anything in the context of one set of 3 shots. It's impossible to make 1.8 shots out of 3!

Example: Let's calculate the expected net gain in one year for the insurance company on Maria's car insurance policy. The random variable X represented the net gain and so we use the probability distribution to do the calculation:

x	$p(x)$
−$248,700	0.001
−$13,700	0.049
$300	0.1
$1300	0.85

$$E(X) = -\$248{,}700(0.001) - \$13{,}700(0.049) + \$300(0.1) + \$1300(0.85) = \$215$$

Interpretation: For many policies like Maria's, the average net gain for the insurance company is $215 per policy. The $215 figure doesn't mean anything in the context of Maria's single policy; the insurance company cannot realize a net gain of $215 on her policy in one year.

Note that the expected net gain is positive (good news for the insurance company) but does not account for operating costs so the insurance company's profit margin will be less than $215.

A few notes about interpreting expected value:
- Never try to round $E(X)$ to a whole number. Since it is a long-run average, $E(X)$ does not need to be an integer.
- Never try to round $E(X)$ to one of the possible values of X. Since it is a long-run average, $E(X)$ does not need to be one of X's possible values.
- Always interpret $E(X)$ with respect to the "many repetitions" concept. It is meaningless in the context of a single run of the experiment.
- Just like the mean is the balance point in a data histogram, $E(X)$ is the balance point in a probability histogram.

Variability: Standard Deviation and Variance

In addition to $E(X)$, it is helpful to have measures of the variability of the values of X over many, many repetitions of the experiment. Again we'll use data concepts like variance and standard deviation but apply them to a probability context. In a probability distribution, the **standard deviation of X**, denoted $SD(X)$ is the standard deviation of the many, many values of X obtained over many, many repetitions of the experiment, a long-run standard deviation. Similarly, the **variance of X**, denoted $Var(X)$, is the long-run variance. Run the experiment, write down the value of X, compute the standard deviation of all the x's so far, and repeat many times. You would find that the standard deviations start to converge to some value and we call that value the long-run standard deviation of X. This is also consistent with our notion of probability as long-run proportion.

Here's how it would work with Aaron and his set of 3 free-throws experiment:
- Aaron attempts 3 free-throws.
- We record the value of X, the number he makes.
- We calculate the standard deviation of all the values of X we've recorded so far.

Here are the results for 1,000 repetitions of this process. By the time Aaron finishes all 1,000 experiments (3,000 free-throws), the standard deviation of all the values of X is about 0.856. That is, the number of shots made out of 3 differs from the mean of 1.810 by about 0.856, on average, in this set of 1,000 experiments. Notice how the running standard deviation converges to about 0.85. If you're satisfied that the running average isn't converging to anything other than about 0.85 after 1,000 repetitions, you could say that $SD(X) = 0.85$.

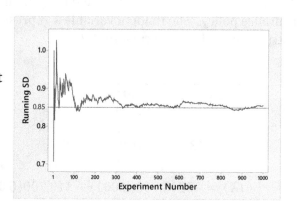

Experiment Number	x	Running Average	Running SD
1	2	2.000	$\sqrt{\frac{(2-2)^2}{1-1}}$ is undefined.
2	3	2.500	$\sqrt{\frac{[(2-2.5)^2+(3-2.5)^2]}{2-1}} = 0.707$
3	1	2.000	$\sqrt{\frac{[(2-2)^2+(3-2)^2+(1-2)^2]}{3-1}} = 1.000$
⋮	⋮	⋮	⋮
1,000	1	1.810	$\sqrt{\frac{[(2-1.810)^2+(3-1.810)^2+(1-1.810)^2+\cdots+(1-1.810)^2]}{1000-1}} = 0.856$

While the idea of repeating the experiment many times helps us to understand the meaning of standard deviation, it is not helpful at all for actually computing $SD(X)$. To do that, we again use the table of x's and $p(x)$'s. Let's suppose there are k values of X in its range. Using the probabilities as weights, we can calculate the standard deviation and variance of a discrete random variable X this way:

x	$p(x)$
x_1	p_1
x_2	p_2
$\vdots$	$\vdots$
x_k	p_k

$$SD(X) = \sqrt{[x_1 - E(X)]^2 p_1 + [x_2 - E(X)]^2 p_2 + \cdots + [x_k - E(X)]^2 p_k}$$

$$Var(X) = [SD(X)]^2$$

First calculate $E(X)$. Then square the differences between each value of X and the expected value, multiply each squared difference the corresponding probability, add up the products, and then take the square root. Just remember not to divide by anything at the end!

Example: Let's calculate $SD(X)$ and $Var(X)$ for Aaron's free-throw experiment without his help. The probability distribution of X = number of shots made out of 3 and the expected value of X is:

x	$p(x)$
0	0.064
1	0.288
2	0.432
3	0.216

$$E(X) = 0(0.064) + 1(0.288) + 2(0.432) + 3(0.216) = 1.8$$

$$SD(X) = \sqrt{(0 - 1.8)^2(0.064) + (1 - 1.8)^2(0.288) + (2 - 1.8)^2(0.432) + (3 - 1.8)^2(0.216)}$$
$$\approx 0.8485$$

$$Var(X) = [SD(X)]^2 = 0.72$$

Interpretation: Over many repetitions of his 3-shot experiment, the average difference between the number of shots made out of 3 and 1.8 is about 0.85. Just like expected value, the 0.85 figure doesn't mean anything in the context of one set of 3 shots.

Example: Let's calculate $SD(X)$ and $Var(X)$ in the car insurance example. The random variable X represented the net gain in one year for the insurance company on Maria's policy and so we use the probability distribution to do the calculation:

x	$p(x)$
−$248,700	0.001
−$13,700	0.049
$300	0.1
$1300	0.85

$$E(X) = -\$248{,}700(0.001) - \$13{,}700(0.049) + \$300(0.1) + \$1300(0.85) = \$215$$

$$Var(X) = (-248{,}700 - 215)^2(0.001) + (-13{,}700 - 215)^2(0.049)$$
$$+ (300 - 215)^2(0.1) + (1300 - 215)^2(0.85) = 72{,}447{,}775$$

$$SD(X) = \sqrt{Var(X)} \approx \$8511.63$$

Interpretation: For many policies like Maria's, the average difference between the net gains for the insurance company and $215 is about $8511.63 per policy. The $8511.63 figure doesn't mean anything in the context of Maria's single policy.

Note that while the expected net gain is positive, the variability in net gains from policy to policy is quite large. While most policies generate positive net gains for the insurance company, the less frequent serious accidents will generate net losses very far below the expected net gain. We'll see in a later section how the insurance company can take advantage of the large number of policyholders like Maria to deal with the large variability from policy to policy.

Changing Units

Suppose someone asks you to report the expected net gain and the standard deviation on Maria's policy in British pounds. One approach would be to convert the four possible net gain values from U.S. dollars to British pounds and then recompute $E(X)$ and $SD(X)$. As I'm writing this section, one U.S. dollar is equivalent to about 0.76 British pounds. The calculation would look like this:

x (dollars)	x (pounds)	$p(x)$
−$248,700	−£189,012	0.001
−$13,700	−£10,412	0.049
$300	£228	0.1
$1300	£988	0.85

$$E(X) = -£189,012(0.001) - \$£10,412(0.049) + £228(0.1) + £988(0.85) = £163.4$$

$$Var(X) = (-189,012 - 163.4)^2(0.001) + (-10,412 - 163.4)^2(0.049)$$
$$+ (228 - 163.4)^2(0.1) + (988 - 163.4)^2(0.85) = 55,060,309$$

$$SD(X) = \sqrt{Var(X)} \approx £7,420.26$$

Fortunately, there is an easier way to change the units of expected values and standard deviations without having to start over. Here are the general rules.

Let X be a random variable and let Y be another random variable representing X in different units. Suppose the units change can be expressed as a linear function of X:

$$Y = aX + b$$

where a and b are constants. Then the expected value, variance, and standard deviation of Y are:

$$E(Y) = aE(X) + b$$

$$Var(Y) = a^2 Var(X)$$

$$SD(Y) = |a|SD(X)$$

To get the expected value of X in the new units, just use the same units-conversion formula on $E(X)$. To get the standard deviation of X in the new units, multiply $SD(X)$ by a and forget about b. Why does b disappear from the variance and standard deviation? Aside from the mathematical derivation (which I'm guessing you aren't that interested in anyway), it's because constants by themselves don't vary; they have zero variability and so they contribute nothing to measures of variation.

Many units changes can be expressed as $aX + b$ by choosing the correct a and b, so these tools are really handy.

Example: Let X be the net gain (in dollars) in one year for the insurance company on Maria's policy. To convert to pounds, let $Y = 0.76X$. This is a linear change of units where $a = 0.76$ and $b = 0$ so

$$E(Y) = aE(X) + b = 0.76(215) = £163.4$$

$$Var(Y) = a^2Var(X) = 0.76^2(72{,}447{,}775) = 55{,}060{,}309$$

$$SD(Y) = |a|SD(X) = |0.76|(8511.63) \approx £7{,}420.26$$

Isn't that much simpler than starting over?

Example: Suppose X is a continuous random variable that measures temperature in degrees Fahrenheit and that $E(X) = 60°F$ and $SD(X) = 20°F$. To convert these values to degrees Celsius, let $Y = \frac{5}{9}(X - 32)$. First put the formula for Y into the form $aX + b$:

$$Y = \frac{5}{9}(X - 32) = \frac{5}{9}X - \frac{5}{9}(32) = \frac{5}{9}X - \frac{160}{9}$$

This is a linear change of units where $a = \frac{5}{9}$ and $b = -\frac{160}{9}$ so

$$E(Y) = aE(X) + b = \frac{5}{9}(60) - \frac{160}{9} = \frac{140}{9} \approx 15.6°C$$

$$SD(Y) = |a|SD(X) = \left|\frac{5}{9}\right|(20) = \frac{100}{9} \approx 11.1°C$$

Section 4.2 Exercises

1. Consider these probability distributions.

I.

x	0	1	2
$p(x)$	1/3	1/3	1/3

III.

x	0	1	2
$p(x)$	0.2	0.6	0.2

II.

x	0	1	2
$p(x)$	0.1	0.8	0.1

IV.

x	0	1	2
$p(x)$	0.5	0	0.5

a. Without calculating anything, compare these four probability distributions with respect to their centers, $E(X)$, and variabilities, $SD(X)$.

b. Calculate $E(X)$, $Var(X)$, and $SD(X)$ for each distribution.

2. Consider this probability distribution.

x	-0.5	0	0.5
$p(x)$	0.2	0.7	0.1

   a. Find $E(X)$.
   b. Find $Var(X)$.
   c. Find $SD(X)$.

3. Consider this probability distribution. If $E(X) = 215$, find a and b.

x	100	200	300
$p(x)$	a	b	1/5

4. Toss a fair coin 3 times and let X be the number of tails you get.

   a. Find and interpret $E(X)$.
   b. Find and interpret $SD(X)$.

5. Darryl is a baseball player who has a 0.250 batting average. That means that over some period of time, Darryl had a hit in 1 out of 4 at-bats. Consider the experiment where Darryl has 4 at-bats in a game and observes whether or not he gets a hit each time. Let X be the number of hits Darryl gets in the game.

   a. Find and interpret $E(X)$.
   b. Find and interpret $SD(X)$.

6. A department store runs a promotion where customers spin a wheel twice at the checkout to determine their discount. The wheel has 10 equal-sized slices: six are labeled "2%," three are labeled "10%," and one is labeled "20%." A customer's actual discount is the larger of the two spins so let X be the larger discount.

   a. Find and interpret $E(X)$.
   b. Find and interpret $SD(X)$.

7. There are two traffic lights on Maria's way to work. If both lights are green, she can get to work in 20 minutes. Stopping at the first light adds 1 minute to her commute time and stopping at the second light adds 2 minutes. The probabilities that she stops at the first and second lights are 0.3 and 0.2, respectively. Consider the experiment where Maria records what happens at each light one morning on her way to work. Let X be Maria's commute time.

   a. Find and interpret $E(X)$.
   b. Find and interpret $SD(X)$.

8. One drawer in our kitchen is the "junk drawer." Among other things in there are batteries. Let's say there are 10 batteries in the drawer but 3 of them are totally dead. I select 2 batteries at random for my son's train.

   a. Find the expected number of good batteries I select.
   b. Interpret your answer in context.

9. Brian, Art, Sam, and Phil audition for America's Got Luck. The audition process consists of tossing a fair coin for each person. Heads they get on; tails they don't.

   a. Find the expected number of these guys who get a spot on the show.
   b. Interpret your answer in context.

10. Micah purchases a car insurance policy with a $3000 annual premium. The table below gives the possible annual costs that the insurance company may have to pay on Micah's behalf along with the probabilities.

Cost	$0	$1500	$20,000	$300,000
Probability	0.80	0.15	0.045	0.005

Find the expected net gain for the insurance company per year on Micah's policy.

11. Hannah purchases a homeowner's insurance policy with a $2000 annual premium. The table below gives the possible annual costs that the insurance company may have to pay on Hannah's behalf along with the probabilities.

Cost	$0	$1000	$50,000	$450,000
Probability	0.90	0.08	0.019	0.001

Let X be the net gain for the insurance company per year. $E(X) = 520$. Which of these interpretations of $E(X)$ is correct?

a. Hannah expects to pay $520 per year for this insurance.
b. The insurance company will net $520 next year on Hannah's policy.
c. The average net gain for the insurance company next year on Hannah's policy will be about $520.
d. Over many policies like Hannah's, the average net gain for the insurance company will be about $520 per policy.

12. Consider the game of spinning the spinner shown here twice.

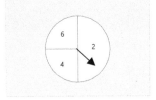

a. Find the expected sum of the numbers on both spins.
b. Fill in the blanks below with your answer to part a. Which of these then is the best interpretation?
    i.   For many sets of two spins, the average sum will be close to _____.
    ii. Over many spins, the average spin value will be close to _____.
    iii. Someone playing this game can expect to get a sum of _____.
    iv. The average sum for someone playing this game is about _____.

13. You want to dunk your friend in the dunk tank at the carnival. Your friend sits on a ledge above a tank of water and you throw baseballs at a target. If you hit the target, the ledge collapses and your friend gets dunked. You figure you have a 20% chance of hitting the target with each throw. It costs $2 per throw and you only have $8. Find the expected amount you spend at the dunk tank.

14. An experiment is designed to see whether 3rd graders can write faster with a pen or pencil. Four third graders participate, 2 boys and 2 girls. For each child, a fair coin is tossed; if heads, the child gets a pen and if tails, the child gets a pencil. Once two children are assigned to one group, the others are automatically assigned to the other group. The children are assigned in this order: boy 1, girl 1, boy 2, girl 2. Let X be the number of boys assigned to a pen.

a. Find $E(X)$.
b. Find $SD(X)$.

15. An experiment is designed to see whether 3rd graders can write faster with a pen or pencil. Four third graders participate, 2 boys and 2 girls. For each child, a marble is drawn without replacement from a bucket containing 2 red marbles and 2 blue marbles. If a red marble is selected, the child gets a pen and if blue, the child gets a pencil. The children are assigned in this order: boy 1, girl 1, boy 2, girl 2. Let X be the number of boys assigned to a pen.

a. Find $E(X)$.
b. Find $SD(X)$.
c. Compare your answers to exercise 14.

16. Six hospital patients are waiting to be discharged; three from labor and delivery with their newborns and three from the emergency department. You want to select a SRS of two patients for a patient satisfaction survey. Let X be the number of patients that are from labor and delivery. Find $E(X)$ and $SD(X)$.

17. Consider exercise 16 but instead select a STRS of two patients stratified by department. Let X be the number of patients that are from labor and delivery.

   a.   Find $E(X)$

   b.   Find $SD(X)$.

   c.   Compare your answers to exercise 16.

18. In Monopoly, you move around the board by rolling two fair dice. Let's say that one is green and one is blue. Consider a single Monopoly roll.

   a.   Find the expected roll value.

   b.   Find the standard deviation of roll values.

   c.   Find and interpret the z-score for a roll which takes you from Ventnor Avenue to Boardwalk.

19. Let X be a random variable having $E(X) = 0.4$ and $Var(X) = 0.0024$. Let $Y = \frac{12}{7}X - \frac{5}{14}$.

   a.   Find $E(Y)$.

   b.   Find $Var(Y)$ and $SD(Y)$.

20. Let X be a random variable having $E(X) = 100$ and $SD(X) = 15$. Let $Y = 10X + 30$.

   a.   Find $E(Y)$.

   b.   Find $Var(Y)$ and $SD(Y)$.

21. Let X be a random variable having $E(X) = 68$ and $SD(X) = 5$. Let $Y = \frac{X}{2.54}$.

   a.   Find $E(Y)$.

   b.   Find $Var(Y)$ and $SD(Y)$.

22. The heights in inches of a certain group of adults can be adequately modeled using a random variable having an expected value of 68 and a standard deviation of 5. Find the expected value and standard deviation of the related random variable: height in meters.

23. You spin a roulette wheel (18 red spaces, 18 black spaces, 2 green spaces) three times, each time betting $2 U.S. on red.

   a.   Find the expected amount won in U.S. dollars.

   b.   Find the expected amount won in Canadian dollars. Use the current exchange rate.

24. Refer to exercise 11. Find the expected net gain for the insurance company per year in British pounds. Use the current exchange rate.

25. A meteorologist forecasts the high temperature for tomorrow by using a probability distribution instead of a single number forecast. The distribution for X, the high temperature tomorrow in °F, is shown here. Let Y be the high temperature tomorrow in °C.

x	$p(x)$
54	0.05
55	0.20
56	0.40
57	0.25
58	0.10

   a.   Convert each x to °C first and then find $E(Y)$.

   b.   Find $E(X)$ first and then use it to find $E(Y)$.

26. Let X be a random variable having $E(X) = p$ where p is a number between 0 and 1. Let $Y = \frac{12}{7}X - \frac{5}{14}$.

   a.   Find $E(Y)$.

   b.   Note that both $E(X)$ and $E(Y)$ are lines (linear functions of p). Plot both lines on the same graph. Use p as the horizontal axis and expected value as the vertical axis.

   c.   For what values of p is $E(Y)$ negative? greater than 1?

   d.   Where do the lines intersect?

   e.   For what values of p will $E(Y)$ be less than $E(X)$? greater than?

4.3 Continuous Distributions and the Normal Distribution

The random variables and their probability distributions we studied in detail in the last section related to Aaron's free-throw experiment and Maria's car insurance policy were examples of discrete random variables. Remember that a discrete random variable has a countable range; its possible values have gaps between them on the number line:

Random Variable	Range
X = number of shots made out of 3	$\{0, 1, 2, 3\}$
X = net gain of insurance company	$\{-\$248700, -\$13700, \$300, \$1300\}$

Continuous random variables differ fundamentally from discrete ones in that their ranges are uncountable; the possible values form an interval on the number line. There are infinitely many possible values between any two. Here are three examples:

Random Variable	Range
X = time to finish an exam (minutes)	$0 < x \le 120$
X = weight gain in two weeks (pounds)	$-40 < x < 40$
X = birth time of the day (hours)	$0 \le x < 24$

While we might report time to the nearest minute, it is possible to measure time to any fraction of a minute or second we like, as long as we have a precise stopwatch. While we often report weight gain of a person to the nearest pound, it is possible to measure weight to any fraction of a pound we like, as long as we have a sensitive scale.

For a continuous random variable, we cannot create a table of x's and $p(x)$'s like we did in the last section; that would be impossible since there are infinitely-many possible x's in any interval. For continuous random variables, probabilities are assigned to *intervals, not individual x's*. To do that, we use a probability density function, denoted $f(x)$, and draw a sketch of $f(x)$ vs. x. Here are examples of realistic probability density functions for the three examples.

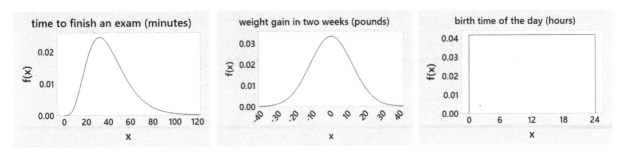

We may think that it is very likely that a student will finish a two-hour exam within the first hour, so a somewhat right-skewed distribution of probability would make sense. If we think that it is equally likely that a person will gain weight or lose weight in two weeks and that many people will gain or lose relatively little weight, a bell-shaped distribution might make sense. To show that we don't think babies are any more likely to be born at one time of day compared to another time of day, we'd use a uniform distribution of probability across the 24 hours of the day.

How do we assign probability to *intervals* and *not to individual values*? We use areas under curves. The *area* under the probability density function in any interval is the *probability* that the random variable will be in that interval (see the picture below). Isn't that wild?! Areas are probabilities.

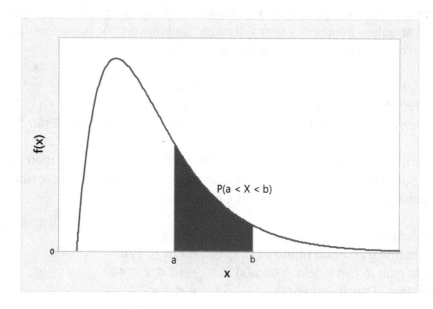

That shouldn't come as too much of a surprise because the areas of the probability histogram bars are probabilities for discrete random variables.

There are some interesting consequences of using areas as probabilities.

1. $P(a < X < b) = \int_a^b f(x)\,dx$
   For those of you who have studied integration, this is one very important application of integration. In general, to compute probabilities for continuous random variables, we evaluate a definite integral. For those of you who have no idea what that elongated S symbol means, don't worry. We won't need to actually do integration.

2. $P(X = a) = 0$
   Because probabilities are assigned to *intervals* and *not individual values*, there is no probability assigned to a single value of X. There is no area under the curve at a single value.

3. $P(a < X < b) = P(a \leq X \leq b)$
   Because there is no probability assigned to individual values, it doesn't matter whether or not we include single values in any interval. It matters for discrete random variables but not for continuous ones.

4. The area under the curve over the entire range of X is 1. We often say that the total area under the curve is 1. This is analogous to the fact that the sum of all the $p(x)$'s in a discrete distribution is 1.

Example: Suppose we select an expectant mother at random from the records of a certain clinical site and observe at what time during the day she gives birth. Let $X =$ birth time of the day (hours) and let X have a uniform probability distribution on the interval $0 \leq x < 24$. Find the probability that the expectant mother gives birth between 6am and 3pm.

Solution: For uniform distributions, probabilities are simply areas of rectangles. The probability density function for a uniform random variable is a horizontal line. The height of the line must be such that the total area under the curve is 1. Here, $f(x) = 1/24$. To find the probability between 6am and 3pm (between hours 6 and 15 of the day), we find the area under the horizontal line between 6 and 15. In other words, find the area of the rectangle with a width of $15 - 6 = 9$ and a height of $1/24$:

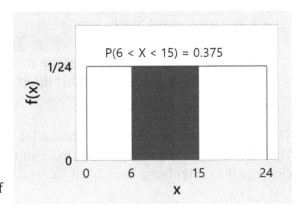

$P(6 < X < 15) = 9(1/24) = 3/8 = 0.375.$

Notes:

- $P(X = 6) = 0$
  You might protest and say, "but what if the baby actually *was* born at 6:00am?" The answer is that, although we usually report times like these to the nearest minute, the baby was not born at exactly 6:00am on the nose. The actual time was either a few seconds (or nanoseconds) earlier or later.

- Both $P(6 < X < 15)$ and $P(6 \leq X \leq 15)$ are equal to 0.375. It doesn't matter whether we include the times of 6:00am and 3:00pm or not because neither of those individual times have any probability/area associated with them.

Normal Distribution

One of the more common probability models for continuous distributions is the normal distribution. Here is a picture along with some important properties.

- Shape: perfectly symmetric, bell-shaped
- Center: the expected value of X, $E(X)$.
- Variability: governed by the standard deviation of X, $SD(X)$.
- Probabilities are areas under the curve.
- The total probability/area under the curve is 1.
- The curve never quite touches the x – axis; it extends infinitely far in either direction.

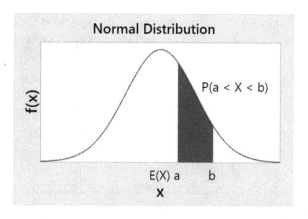

The normal model (or normal distribution) is appropriate for certain continuous random variables. We might expect the possible values to be spread out symmetrically around the center with values closer to the center being more likely than values farther from the center.

Be careful though: the normal model is not appropriate for every continuous random variable. That's where statistics and data science come in. We can use *data* collected on a random variable to evaluate the appropriateness of the *probability model* we're considering for an experiment. Putting data and models together is one of the most fascinating (and difficult) aspects of statistics. More on that in the next section.

In order to compute probabilities, we need to find areas under the normal curve. Unfortunately, that is not an easy problem; it involves approximating areas with a bunch of very narrow rectangles. The good news is that the problem has already been solved for one particular normal curve, the standard normal curve. The **standard normal distribution** is the normal distribution that results when converting any normal random variable to a z-score using the usual z-score formula:

$$Z = \frac{X - E(X)}{SD(X)}$$

In other words, the z-score changes the units of a number from whatever the original units are (inches, kilograms, etc.) to the number of standard deviations the number is from the expected value. This process is called *standardization* and results in a z-score where the *standard* deviation is the unit, hence the name *standard* normal distribution.

Because standardization is just a linear change of units, we can apply the rules from the last section to get the expected value and standard deviation of the new z-score. Both $E(X)$ and $SD(X)$ are constants:

$$Z = \frac{X - E(X)}{SD(X)} = \frac{1}{SD(X)}X - \frac{E(X)}{SD(X)} = aX + b$$

where

$$a = \frac{1}{SD(X)}, \quad b = -\frac{E(X)}{SD(X)}$$

and now use the rules for changing units:

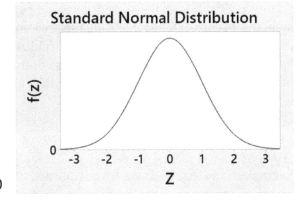

Standard Normal Distribution

$$E(Z) = aE(X) + b = \frac{1}{SD(X)}E(X) - \frac{E(X)}{SD(X)} = 0$$

$$Var(Z) = a^2 Var(X) = \left(\frac{1}{SD(X)}\right)^2 Var(X) = \frac{1}{Var(X)}Var(X) = 1$$

$$SD(Z) = |a|SD(X) = \left|\frac{1}{SD(X)}\right| SD(X) = 1$$

So the standard normal distribution is the normal distribution centered at 0 with a standard deviation of 1. Z-scores as large as 2 or 3 or as small as -2 or -3 are fairly unusual. That means values of a normal random variable that are 2 or 3 standard deviations (or more) from the center occur fairly rarely. (You may be wondering whether the units change to z-scores changes the "normalness" of the distribution. The answer is "no" but showing why not is way beyond the scope of this book. If that question didn't cross your mind, forget I brought it up.)

Areas under the standard normal curve to the left of a particular z-score are tabulated in Table I. The smallest z-score shown is -3.99 with very little area to its left. Z-scores smaller than -3.99 will have even less area to the left under the curve. The largest z-score shown is 3.99 with almost all the area to its left. Z-scores larger than 3.99 will have even more area to the left under the curve (but still less than 1).

The goal of this section is to become comfortable working with z-scores and areas under the normal curve. When working normal curve problems, ALWAYS sketch the curve and properly label your picture.

Example: Let Z have a standard normal distribution. Find $P(Z < 0.84)$.

Solution: First sketch the standard normal curve centered at 0. Then label the z-score of 0.84 and shade the area under the curve to its left. That's the probability you want. Table I gives this area directly:

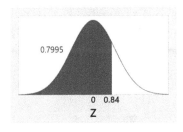

$$P(Z < 0.84) = 0.7995$$

Note that we don't care whether 0.84 is included or not.

Example: Let Z have a standard normal distribution. Find $P(Z > 0.84)$.

Solution: First sketch the standard normal curve centered at 0. Then label the z-score of 0.84 and shade the area under the curve to its right. That's the probability you want. Table I gives the area to the *left* of 0.84, not to the right, so 0.7995 is not the answer. However, you also know that the total area under the curve is 1, so the area you want is what's left over:

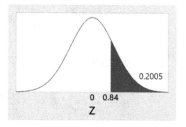

$$P(Z > 0.84) = 1 - 0.7995 = 0.2005$$

Example: Let Z have a standard normal distribution. Find $P(-1.14 \leq Z < 0.92)$.

Solution: First sketch the standard normal curve centered at 0. Then label the z-scores of -1.14 and 0.92 and shade the area under the curve between them. That's the probability you want but Table I doesn't give areas like this. However, you can get the area you want by subtracting the area to the left of -1.14 from the area to the left of 0.92:

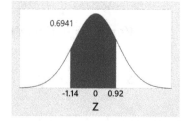

$$P(-1.14 \leq Z < 0.92) = 0.8212 - 0.1271 = 0.6941$$

The pictures below show this process graphically.

Start with this area...   ...then take away this area...   ...to get this area.

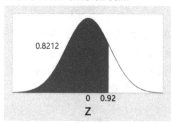

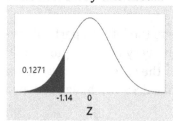

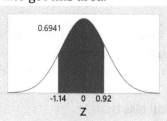

Example: Let Z have a standard normal distribution. Find b such that $P(Z \leq b) = 0.6064$.

Solution: Note that this problem is giving you the area and asking you to find the z-score that makes the probability statement true. These kind of problems are a bit more difficult but you should always start by sketching the standard normal distribution centered at 0. Next you should label what you're given and what you're trying to find. That requires you to think about two things simultaneously: 1) how much area you're given and 2) what the interval looks like that captures that area. Here you know that an area of 0.6064 lies to the left of b under the curve. That's over half the total area under the curve so you know to get that much area to the left, the z-score, b, must be above 0. Now you can mark where b is on the horizontal axis and label the area to its left. Finally use Table I to do a simple lookup:

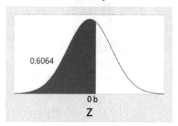

$$b = 0.27$$

Example: Let Z have a standard normal distribution. Find b such that $P(Z > b) = 0.9292$.

Solution: First sketch the standard normal curve centered at 0. Since you know that an area of 0.9292 (over half the total area) lies to the right of b under the curve, b must be below 0. Now you can mark where b is on the horizontal axis and label the area to its right. The problem is that Table I gives areas to the left, not to the right. What to do? Use the property that the total area under the curve is 1 and subtract to get an area of $1 - 0.9292 = 0.0708$ to the left of b. Then look up an area of 0.0708 in Table I and read off the corresponding z-score:

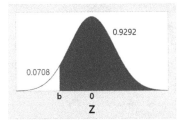

$$b = -1.47$$

So far we've only dealt with the standard normal distribution but the advantage of the standard normal distribution is that it allows us to deal with *any* normal distribution via z-scores.

Example: Let X have a normal distribution with $E(X) = 100$ and $SD(X) = 16$.

    a.   Find $P(X > 128)$.
    b.   Find $P(80 < X \le 92)$.
    c.   Find b such that $P(X < b) = 0.05$.

Solution:
    a.   You know the drill: first sketch the normal curve, but this time mark the center at 100. Then label 128 on the horizontal axis and shade the area to its right. That's the probability you want. Now convert 128 to a z-score to take advantage of the standard normal curve:

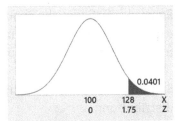

$$z = \frac{X - E(X)}{SD(X)} = \frac{128 - 100}{16} = 1.75$$

(I like to show a second horizontal scale for the z-scores below the x-axis.) Now the problem is to find $P(Z > 1.75)$ which you can do using Table I:

$$P(X > 128) = P(Z > 1.75) = 1 - 0.9599 = 0.0401$$

    b.   Sketch the normal curve centered at 100. Label both 80 and 92 on the horizontal axis and shade the area between them. That's the probability you want. Convert both 80 and 92 to z-scores:

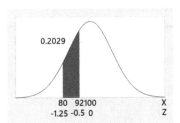

$$z = \frac{X - E(X)}{SD(X)} = \frac{80 - 100}{16} = -1.25$$

       and

$$z = \frac{X - E(X)}{SD(X)} = \frac{92 - 100}{16} = -0.50$$

Then use Table I to find the area between them:

$$P(80 < X \le 92) = P(-1.25 < Z \le -0.50) = 0.3085 - 0.1056 = 0.2029$$

c. Sketch the normal curve centered at 100. Since the area to the left of b is only 0.05 (less than half the area under the curve), b must be below 100. Use Table I to find the z-score that has an area of 0.05 to its left; this is the z-score corresponding to b. Finally, "unstandardize" to convert the z-score to b, the x-value you want to find:

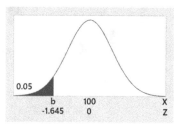

$$-1.645 = \frac{b - 100}{16} \quad \Rightarrow \quad b = 100 - 1.645(16) = 73.68$$

Note that since the area of 0.05 is exactly halfway between two values in Table I (0.0495 and 0.0505) we'll use the z-score that's the average of the corresponding z-scores (-1.65 and -1.64).

Now that you know how to work with the normal distribution, you can use it in the context of practical experiments where a normal distribution may be appropriate.

Example: Suppose the annual percent return on an investment has a normal distribution with an expected return of 18.5% and a standard deviation of 27.2%. What is the probability of making money on this investment in a year?

Solution: You can set this up as an experiment having a random variable with a normal distribution:

        Experiment: Observe the performance of this investment for one year.
        Random variable: $X =$ annual percent return on the investment
        Distribution: normal, $E(X) = 18.5, SD(X) = 27.2$
        You want: $P(X > 0)$

Sketch the normal curve centered at 18.5 and shade the area under the curve to the right of 0. That's the probability you want. Convert 0 to a z-score:

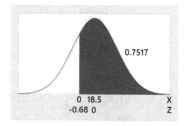

$$z = \frac{X - E(X)}{SD(X)} = \frac{0 - 18.5}{27.2} = -0.68$$

Then use Table I to find the area to the right:

$$P(X > 0) = P(Z > -0.68) = 1 - 0.2483 = 0.7517$$

There's about a 75% chance of making money on this investment in one year.

The normal distribution can also be used as a model for a set of data values. If the histogram shows a fairly symmetric, bell shape, the normal distribution could be a good approximation to the distribution.

Here's a histogram of 1000 IQ scores. Notice that a normal curve does a good job of showing the primary features of the data: where the center is, what the shape is, and how much variability the IQ scores have.

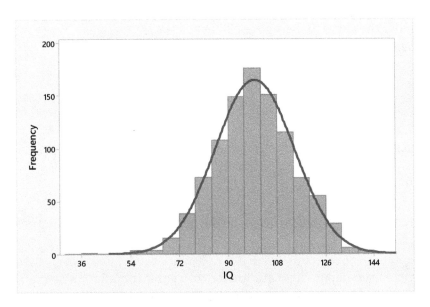

There are some subtle differences between using a normal curve as a probability model and using it as a model for an actual set of data:

- In a probability context, we use $E(X)$ and $SD(X)$ to denote the center and variability (standard deviation) of the distribution. In a data context, we'd use μ and σ to denote the center (mean) and variability (standard deviation), respectively, of a population data set and $\bar{x}$ and s to denote the center (mean) and variability (standard deviation) of a sample data set.
- In a probability context, an area under the curve is a probability. In a data context, an area under the curve is an approximation of histogram bar areas and can be interpreted as the predicted proportion of values in a certain interval.

The following table summarizes these differences.

	Normal Curve as a Probability Model	Normal Curve as a Population Data Model	Normal Curve as a Sample Data Model
Center	$E(X)$	μ	$\bar{x}$
Standard Deviation	$SD(X)$	σ	s
Meaning of Area under Curve	probability	predicted proportion	predicted proportion

Example: Suppose the daily milk production of cows in a large herd can be closely approximated with a normal model having a mean of 6.5 gallons and a standard deviation of 1.5 gallons.

   a. Use the model to predict the percentage of cows in the herd that produce more than 9 gallons per day.
   b. Estimate the 20th percentile of milk productions.

Solution:

   a. Sketch the normal curve centered at 6.5 and shade the area under the curve to the right of 9. That area represents the percentage you want. Convert 9 to a z-score:

   $$z = \frac{x-\mu}{\sigma} = \frac{9-6.5}{1.5} = 1.67$$

   Then use Table I to find the area to the right:

   $$1 - 0.9525 = 0.0475$$

   You predict that 4.75% of the cows in the herd produce more than 9 gallons of milk per day.

   Note that I used population notation for the center (population mean) and standard deviation since I used the normal model for a population data set. For the same reason, I didn't use the probability notation $P(X > 9)$ since the area under the curve represents a proportion of cows and not a probability.

   b. Sketch the normal curve centered at 6.5. Let b represent the 20th percentile. Since only 20% of the cows produce less than b gallons per day, b must be below 6.5. The area below b is 0.20. Use Table I to find the z-score that has an area of 0.20 to its left; this is the z-score corresponding to b. Finally, convert the z-score to a value in gallons:

   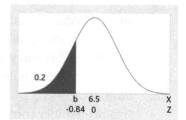

   $$-0.84 = \frac{b-6.5}{1.5} \quad \Rightarrow \quad b = 6.5 - 0.84(1.5) = 5.24$$

   The estimated 20th percentile of milk productions in this herd is 5.24 gallons per day.

"But Bryan," you ask, "you told me that the normal curve is not always an appropriate model. How can I tell when to use it and when not to use it?" That's a great question! Go back and look at all of the examples in this section. They all start out by telling us to use the normal distribution. In practice though, you have to make the call. The answer to the question about whether the normal curve is a good model or not involves putting the concepts of models and data together which we'll explore in the next section.

Section 4.3 Exercises

1. Why do we use a probability density function instead of a table to describe a continuous probability distribution?

2. What is the fundamental difference between a continuous random variable and a discrete one?

3. How are probabilities calculated for continuous random variables?

4. Let X be a continuous random variable. Indicate whether each statement is true or false. If false, explain why.

    a. $P(X = 7) = 0$
    b. $P(90 \leq X \leq 110) > P(90 < X < 110)$
    c. $P(0 \leq X \leq 10) \leq P(0 \leq X \leq 20)$
    d. The probability that $X = a$ is $f(a)$.
    e. The function $f(x) = \sin(x)$ can be used as a probability density function for x's between 0 and 2π.

5. Let X be the time in seconds you wait for the walk sign at an intersection. If X has a uniform distribution on the interval $0 \leq x \leq 60$, find

    a. $P(X < 10)$.
    b. $P(X > 45)$.
    c. $P(X \geq 45)$.
    d. $P(0 < X \leq 30)$.
    e. the probability that you wait fewer than 5 seconds or more than 50 seconds.

6. The actual amount of soft-drink that a machine dispenses into bottles varies. Suppose the machine fills 10 oz. bottles and the actual amount dispensed varies between 9.7 oz. and 10.5 oz. according to a uniform distribution. Find the probability that a bottle

    a. will be under-filled.
    b. will be over-filled.
    c. will contain an amount that's within 0.5 oz. of the bottle label.
    d. will contain an amount that's not within 0.5 oz. of the bottle label.

7. Let Z be a standard normal random variable. Indicate whether each statement is true or false. Correct the false ones.

    a. $E(Z) = 1$
    b. $Var(Z) = 0$
    c. $SD(Z) = 1$
    d. $SD(Z) = Var(Z)$
    e. $P(Z < -1) = P(Z \geq 1)$
    f. $P(Z < 1.4) < P(Z > -1.4)$

8. Let Z have a standard normal distribution. Find the following probabilities.

    a. $P(Z < 1.56)$
    b. $P(Z \leq -0.77)$
    c. $P(Z > -2.04)$
    d. $P(Z \geq 1.96)$
    e. $P(-1.15 < Z < 0.99)$
    f. $P(0.50 \leq Z < 1.63)$
    g. $P(|Z| < 1.00)$
    h. $P(|Z| > 2.00)$

9. Let Z have a standard normal distribution. Find the following probabilities.

    a. $P(Z > 1.65)$
    b. $P(Z \geq -2.57)$
    c. $P(Z \leq -0.84)$
    d. $P(Z < 1.28)$
    e. $P(-0.68 < Z \leq 1.49)$
    f. $P(-2.15 \leq Z < -0.33)$
    g. $P(|Z| \geq 0.77)$
    h. $P(|Z| \leq 3.00)$

10. Let Z have a standard normal distribution. Find b such that

    a. $P(Z \le b) = 0.1357$
    b. $P(Z < b) = 0.9591$
    c. $P(Z \ge b) = 0.1112$
    d. $P(Z > b) = 0.6562$
    e. $P(-b < Z < b) = 0.2960$
    f. $P(|Z| > b) = 0.1236$
    g. $P(|Z| < b) = 0.95$
    h. $P(0.50 < Z \le b) = 0.30823$

11. Let Z have a standard normal distribution. Find b such that

    a. $P(Z < b) = 0.8997$
    b. $P(Z \le b) = 0.0401$
    c. $P(Z > b) = 0.1003$
    d. $P(Z \ge b) = 0.9505$
    e. $P(-b < Z < b) = 0.9836$
    f. $P(|Z| \le b) = 0.7994$
    g. $P(|Z| \ge b) = 0.2460$
    h. $P(-1.25 \le Z \le b) = 0.2918$

12. Let Z have a standard normal distribution. Approximate

    a. $P(Z \ge 4.56)$
    b. $P(Z < -5.99)$
    c. $P(Z > -4.01)$
    d. $P(Z \le 8.51)$

13. Let X have a normal distribution with $E(X) = 100$ and $SD(X) = 15$. Find the following probabilities.

    a. $P(X = 100)$
    b. $P(X > 100)$
    c. $P(X < 100)$
    d. $P(X \le 104)$
    e. $P(X > 110)$
    f. $P(X \ge 92)$
    g. $P(87 < X \le 120)$
    h. $P(105 \le X < 138)$

14. Let X have a normal distribution with $E(X) = 20$ and $SD(X) = 4$. Find the following probabilities.

    a. $P(X < 25)$
    b. $P(X > 20)$
    c. $P(X < 19)$
    d. $P(X > 17)$
    e. $P(X < 22.8)$
    f. $P(X > 15.5)$
    g. $P(13.2 < X < 18.2)$
    h. $P(21.9 \le X \le 30.5)$

15. Let X have a normal distribution with $E(X) = 0.45$ and $SD(X) = 0.022$.

    a. Find $P(X > 0.50)$
    b. Find $P(0.40 < X < 0.50)$
    c. Find $P(X \ge 0.38)$
    d. Find $P(X \le 0.47)$
    e. Find b such that $P(X > b) = 0.01$.
    f. Find b such that $P(X < b) = 0.05$.
    g. Find b such that $P(X \ge b) = 0.90$.
    h. Find b such that $P(X \le b) = 0.80$.

16. Let X have a normal distribution with $E(X) = 0.2$ and $SD(X) = 0.08$.

    a. Find $P(X < 0.36)$
    b. Find $P(X > 0.16)$
    c. Find $P(X \ge 0.60)$
    d. Find $0.28 < X \le 0.40)$
    e. Find b such that $P(X \le b) = 0.05$.
    f. Find b such that $P(X \ge b) = 0.10$.
    g. Find b such that $P(X > b) = 0.80$.
    h. Find b such that $P(X < b) = 0.95$.

17. Let X have a normal distribution with $SD(X) = 100$ and $P(X \le 1500) = 0.6554$. Find $E(X)$.

18. Let X have a normal distribution with $SD(X) = 0.05$ and $P(X > 0.39) = 0.9681$. Find $E(X)$.

19. Let X have a normal distribution with $E(X) = 9.9$ and $P(X > 5.0) = 0.8708$. Find $SD(X)$.

20. Let X have a normal distribution with $E(X) = 1/5$ and $P(X \ge 3/5) = 0.0071$. Find $SD(X)$.

21. The annual return on an investment has a normal distribution with an expected return of 7% and a standard deviation of 5%.

    a. What is the probability that this investment returns more than 10% in one year?
    b. What is the probability of losing money on this investment in one year?
    c. Fill in the blank: There is a 90% chance that the annual return will be more than _____.
    d. Fill in the blank: The probability that the annual return will be at most _____ is 0.75.

22. The tensile strength of a bolt is the pulling force necessary to elongate the bolt. Suppose the tensile strength of a bolt has a normal distribution with an expected strength of 8000 psi and a standard deviation of 300 psi.

    a. What is the probability that the bolt's tensile strength will exceed 8500 psi?
    b. What is the probability that the bolt's tensile strength will be at most 7600 psi?
    c. What is the probability that the bolt's tensile strength will be between 7900 psi and 8200 psi?
    d. A mechanic mistakenly uses a bolt like this in an application requiring a bolt having a tensile strength of at least 8700 psi. If this bolt experiences a load of 8700 psi, what is the probability that it will elongate?

23. A student is selected at random and asked how fast they've ever driven a car. Suppose that a normal model with an expected value of 103 mph and a standard deviation of 19 mph is a good model for the student's response.

    a. Find the probability that the student will report having driven a car in excess of 120 mph.
    b. Find the probability that the student's answer will be less than 70 mph.
    c. Fill in the blank: The probability that the student's answer will be at least _____ mph is 0.10.

24. The tread life of a certain tire has a normal distribution having expected value of 40,000 miles and a standard deviation of 2700 miles.

    a. Find the probability that a tire's tread life will be less than 35,000 miles.
    b. Find the probability that a tire will have a tread life between 42,000 and 46,000 miles.
    c. Fill in the blank: The probability that the tire's tread life will exceed _____ miles is 0.90.
    d. A retailer wants to offer a refund to customers whose tires do not last above a certain mileage figure. However, the retailers doesn't want to pay refunds to too many customers. What mileage figure should the retailer use so that the chances of a refund to any one customers are only 1 in 20?

25. Consider the casino game of Roulette where players place $1 bets on color (either red or black). The probability of winning a single bet is 18/38, slightly in the casino's favor.

    a. For 1000 such bets, the expected gain for the casino is $52.63, the standard deviation is $31.58, and the distribution of possible gains is approximately normal. Find the probability that the casino makes money overall after 1000 bets.
    b. For 5000 such bets, the expected gain for the casino is $263.16, the standard deviation is $70.61, and the distribution of possible gains is approximately normal. Find the probability that the casino makes money overall after 5000 bets.

26. The normal distribution with a mean of 1020 and a standard deviation of 160 is a reasonable model for SAT scores. Answer these questions based on this model.

    a. Predict the proportion of SAT scores above 1300.
    b. Predict the percentage of SAT scores that are at most 1200.
    c. Estimate the 30th percentile.
    d. Estimate the 85th percentile.

27. Suppose the fuel efficiencies in miles per gallon (mpg) for passenger cars have a normal distribution with $\mu = 38$ and $\sigma = 10$. Answer these questions based on this model.

    a. Predict the proportion of cars that get more than 40 mpg.

b.   Predict the proportion of cars that get between 20 and 30 mpg.
c.   Estimate the 3rd quartile.
d.   Predict the percent of cars that would be classified as outliers according to the z-score criterion.
e.   Fill in the blank: We predict that eighty-two percent of cars get at least _____ mpg.

28. A logger estimates that the mean diameter of the trees in a plot of land is 18 inches and that the standard deviation is 5 inches. Assume that a normal distribution is a good model for the tree diameters. Answer these questions according to this model.

a.   Predict the proportion of trees that are over 2 feet in diameter.
b.   Predict the percentage of trees that are between 12 and 20 inches in diameter.
c.   Fill in the blank: We predict that sixty percent of the trees have diameters greater than _____ inches.
d.   Fill in the blank: We predict that ninety percent of the trees have diameters less than _____ inches.

29. A potato chip manufacturer fills 12 oz. bags with potato chips. The weight of the chips in the bag varies slightly from bag to bag according to a normal model. Suppose the mean weight is 12.1 oz. and the standard deviation is 0.08 oz.

a.   Predict the proportion of bags that are underweight.
b.   The manufacturer can adjust the process to change the mean weight. What should the mean weight setting be so that only an estimated 2% of bags would be underweight?

30. The normal model is a good model for SAT math scores. If the median score is 515, find the standard deviation.

31. The normal model is a good model for SAT verbal scores. If the 90th percentile is 650 and 82% score above 400, find μ and σ.

32. According to the National Center for Health Statistics, $Q1$ and $Q3$ for the weights of 10-year-old boys are 63 pounds and 80 pounds, respectively. If weights of 10-year-old boys are normally distributed, find the mean and standard deviation.

4.4 The Normal Probability Plot

In this section, we'll use data collected on a variable (or random variable in an experiment) to assess whether the normal curve is a good model or not. This tool is so handy that we'll use it again and again throughout the book. It's called a **normal probability plot** and it works by plotting ordered pairs on an x-y coordinate system. The idea is to compare the normal curve model to an actual set of data by comparing percentiles. Here's how it works:

- Order the data from low to high. These data values are the **actual values**.

- Determine the **percentile rank** (p) for each data value. To do this, use the method from chapter 2 to find the address (k) of a percentile. Then solve for p:

$$k = \frac{p}{100}(n+1) \qquad \Rightarrow \qquad p = 100\left(\frac{k}{n+1}\right)$$

- Find the mean and standard deviation of the data and use these as the $E(X)$ and $SD(X)$, respectively, of a normal curve model.

- Using the normal curve model, find the corresponding percentile for each data value's percentile rank. Call these the **predicted values**; they are predictions of the actual values if the data set were exactly normal.

- Construct a plot of the n ordered pairs $(actual\ value, predicted\ value)$.

If a normal curve is an adequate model for the data set, the actual values and predicted values should be close to each other. In that case the plot will show a closely linear pattern of points. Data that are not nearly normal in shape will have many discrepancies between actual and predicated values. In those cases the graph will show a non-linear pattern of points.

Example: Here are data on eruptions of the Old Faithful geyser in Yellowstone National Park.[1] After each eruption, the park rangers estimate when the next one will happen and these estimates are pretty good, typically within a few minutes of the next eruption time. In this example, we'll consider the estimation errors (time – estimated time). Negative errors indicate that the eruption was earlier than predicted; eruptions occurring later than predicted have positive errors. The distribution of the errors is nicely symmetric and seems closely normal. The mean and standard deviation are 0.590 minutes and 4.645 minutes, respectively.

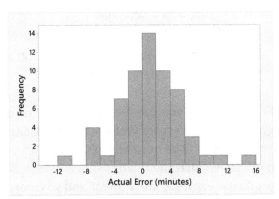

To construct a normal probability plot, we first sort the eruptions by the estimation error from low to high. The address of each eruption is just the rank of that eruption as ordered by error from 1 = smallest negative error to 61 = largest positive error. Let's use the eruption on 7/29/2007 (7 minutes earlier than estimated) as an example:

Date	Actual Error (minutes)	Address (k)	Percentile Rank (p)	Normal z-score	Predicted Difference	Point on Normal Probability Plot
7/28/2007	-11	1	1.61	-2.14	-9.4	(-11, -9.4)
8/4/2007	-8	2	3.23	-1.85	-8.0	(-8, -8.0)
7/30/2007	-8	3	4.84	-1.66	-7.1	(-8, -7.1)
8/4/2007	-8	4	6.45	-1.52	-6.5	(-8, -6.5)
7/29/2007	-7	5	8.06	-1.40	-5.9	(-7, -5.9)
⋮	⋮	⋮	⋮	⋮	⋮	⋮
8/3/2007	14	61	98.39	2.14	10.5	(14, 10.5)

- It has the fifth smallest error so its address is $k = 5$ and its percentile rank is

$$p = 100 \left(\frac{5}{62} \right) = 8.06$$

which means that the error of -7 minutes is the 8.06th percentile in the data set.

- Now use the normal curve model with $E(X) = 0.590$ and $SD(X) = 4.645$ to find the 8.06th percentile in that model:

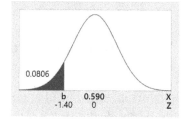

Use Table I to find the z-score that has an area of 0.0806 to its left (-1.40) and then convert the z-score to an error:

$$-1.40 = \frac{b - 0.590}{4.645} \quad \Rightarrow \quad b = 0.590 - 1.40(4.645) = -5.9$$

- Use the actual error and predicted error for this eruption as the ordered pair (-7, -5.9) on the normal probability plot.

- Repeat the process for the other 60 eruptions to get 61 ordered pairs.

From the table you can see that the normal curve gives predicted errors that are fairly close to the actual errors. The normal probability plot of all 62 pairs shows that the relationship between actual and predicted errors is fairly linear and so a normal curve model is an appropriate model for the estimation errors data.

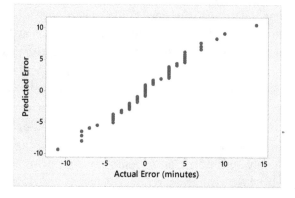

Example: Consider the top 20 U.S. institutions of higher education ranked by tuition and fees.[2] The distribution of the percent full-time faculty is clearly left skewed. The mean and standard deviation are 80.38% and 19.59%, respectively.

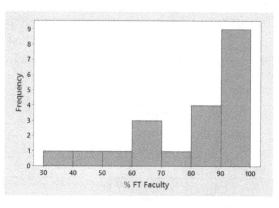

To construct a normal probability plot, we first sort the schools by the percent of faculty that are full-time from low to high. The address of each school is just the rank of that school from 1 = lowest percent full-time faculty to 20 = highest percent full-time faculty.

Name	Actual % FT Faculty	Address (k)	Percentile Rank (p)	Normal z-score	Predicted % FT Faculty	Point on Normal Probability Plot
Sarah Lawrence College	35.2	1	4.76	-1.67	47.7	(35.2, 47.7)
Columbia University in the City of New York	46.9	2	9.52	-1.31	54.7	(46.9, 54.7)
Jewish Theological Seminary of America	52.9	3	14.29	-1.07	59.4	(52.9, 59.4)

University of Southern California	60.5	4	19.05	-0.88	63.1	(60.5, 63.1)
Tufts University	63.5	5	23.81	-0.71	66.5	(63.5, 66.5)
⋮	⋮	⋮	⋮	⋮	⋮	⋮
Colgate University	100	20	95.24	1.67	113.1	(100, 113.1)

Let's use Tufts University as an example:

- It has the fifth lowest percent full-time faculty so its address is $k = 5$ and its percentile rank is

$$p = 100\left(\frac{5}{21}\right) = 23.81$$

  which means that 63.5% full-time faculty is the 23.81st percentile in the data set.

- Now use the normal curve model with $E(X) = 80.38$ and $SD(X) = 19.59$ to find the 23.81st percentile in that model:

  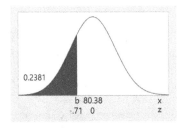

  Use Table I to find the z-score that has an area of 0.2381 to its left (-0.71) and then convert the z-score to a cost:

$$-0.71 = \frac{b-80.38}{19.59} \quad \Rightarrow \quad b = 80.38 - 0.71(19.59) = 66.5$$

- Use the actual and predicted percent full-time faculty for Tufts University as the ordered pair (63.5, 66.5) on the normal probability plot.

- Repeat the process for the other 19 schools to get 20 ordered pairs.

From the table you can see that the normal curve gives predicted percents that generally are not close to the actual percents. The normal probability plot of all 20 pairs shows that the relationship between actual and predicted percent full-time faculty is not linear and so a normal curve model is not appropriate for these data.

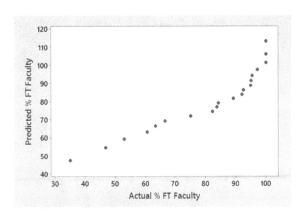

For data sets having more than just a few data values, constructing a normal probability plot by hand is rather tedious. Fortunately, most statistical software packages will easily construct one for you. Those plots may differ slightly from the procedure described here but the graphs will look very similar to ours. Here are normal probability plots made by Minitab® Statistical Software for the two examples above. Notice the difference in the y-axis scale but the similarity of the pattern of points. Minitab adds a straight reference line and two curves that indicate how close to a straight line pattern is close enough.

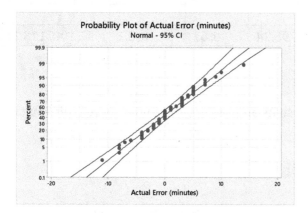

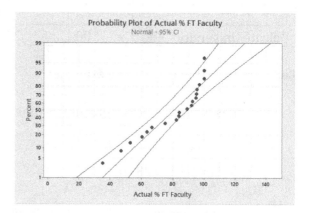

Section 4.4 Exercises

1. In a data set with 50 values, find the percentile ranks for the 3rd, 17th, and 42nd data values in order.

2. In a data set with 17 values, find the percentile ranks for the 1st, 8th, and 17th data values in order.

3. In a data set with 275 values, find the percentile ranks for the 69th, 138th, and the 207th data values in order.

4. In a normal distribution with $E(X) = 100$ and $SD(X) = 16$, find the predicted

    a.   8th percentile.
    b.   22nd percentile.
    c.   median.
    d. $Q3$.
    e.   95th percentile.

5. In a normal distribution with a mean of 550 and a standard deviation of 84, find the predicted

    a.   10th percentile.
    b. $Q1$.
    c.   55th percentile.
    d.   86th percentile.
    e.   99th percentile.

6. In a data set, $n = 38$, $\bar{x} = 110.4$, and $s = 20.9$. In a normal distribution having the same features, find

    a.   the predicted 30th percentile.
    b.   the predicted 70th percentile.
    c.   the predicted percentile corresponding to the 20th ordered data value.
    d.   the predicted percentile corresponding to the 6th ordered data value.

7. In a data set, $n = 22$, $\bar{x} = 53.6$, and $s = 5.1$. In a normal distribution having the same features, find

    a.   the predicted 20th percentile.
    b.   the predicted 80th percentile.
    c.   the predicted percentile corresponding to the 5th ordered data value.

    d.   the predicted percentile corresponding to the 17th ordered data value.

8. Here are the number of fouls committed and the points scored for the players on a basketball team during one year.

    a.   Find the mean and standard deviation of the number of fouls committed.
    b.   Find the percentile rank for player 2's fouls.
    c.   Find the predicted number of fouls in a normal distribution corresponding to player 2's fouls.
    d.   Give the coordinates of the point in the normal probability plot for player 2's fouls.

Player	Fouls	Points
1	1	0
2	22	75
3	3	3
4	14	62
5	4	9
6	10	20
7	5	29
8	22	56
9	3	5
10	3	2
11	24	56

9. Refer to exercise 8.

    a.   Find the mean and standard deviation of the points scored.
    b.   Find the percentile rank for player 9's points.
    c.   Find the predicted number of points in a normal distribution corresponding to player 9's points.
    d.   Give the coordinates of the point in the normal probability plot for player 9's points.

10. Here are the responses (in millions) to a survey question that asked, "How many people do you think live in Canada?"

    220   15   26   35   50

    a.   Find the mean and standard deviation.
    b.   Make a normal probability plot.
    c.   What does your plot indicate about the adequacy of the normal model?

11. Here are the responses to a survey question that asked, "How fast have you ever driven a car (in mph)?"

    60   90   120   100   80   100   115   155

    a.   Find the mean and standard deviation.
    b.   Make a normal probability plot.
    c.   What does your plot indicate about the adequacy of the normal model?

12. Stock screeners are strategies used to select a portfolio of stocks in which to invest. Different screeners use different criteria (value vs. growth, large vs. small companies, etc.). The data shown here are the returns for a particular month for all stock screeners.[3]

-7.47	-6.52	-3.43	-3.26	-2.28	-1.84	-1.74	-1.59	-1.53	-1.46	-0.40	-0.24
-0.20	-0.04	0.65	1.40	1.56	1.66	1.70	1.75	1.78	2.02	2.17	2.21
2.21	2.40	2.44	2.58	2.99	3.22	3.28	3.45	3.45	3.47	3.47	3.47
3.57	3.64	3.74	3.94	4.08	4.10	4.19	4.48	4.55	4.65	4.84	4.86
4.86	4.99	4.99	5.74	5.75	6.24	6.37	6.53	7.04	7.48	7.70	8.51
8.53	8.68	10.65									

    a.   Shown here is Minitab's normal probability plot of these returns. What does the plot suggest?
    b.   Find the predicted return corresponding to the return of 6.24%.
    c.   Give the coordinates in a normal probability plot corresponding to the stock screener with a return of 1.40%. Why doesn't the y-coordinate match Minitab's plot?

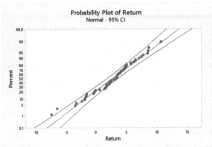

13. Here is a random sample of credit card transaction amounts from the set of all such amounts for a school district in a recent fiscal year.[4] The amounts are shown in order from low to high.

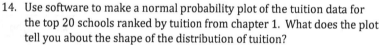

-135.00	-61.32	4.11	14.95	15.10	22.85	25.00	25.00	25.00	33.33	35.70
35.74	37.94	37.98	40.00	43.70	44.63	51.40	55.00	56.72	60.00	60.00
60.00	67.34	79.80	79.95	90.00	95.00	126.97	140.50	148.75	150.08	150.08
178.88	345.00	349.95	373.20	468.95	494.00	500.00	504.00	589.00	625.00	812.46
819.00	845.00	845.00	861.43	3500.00	4865.00					

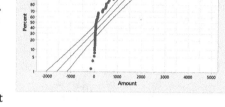

a. Shown here is Minitab's normal probability plot of these amounts. What does the plot suggest?

b. Find the predicted amount corresponding to the amount of $40.00.

c. Give the coordinates in a normal probability plot corresponding to the transaction amount of $3500. Why doesn't the y-coordinate match Minitab's plot?

14. Use software to make a normal probability plot of the tuition data for the top 20 schools ranked by tuition from chapter 1. What does the plot tell you about the shape of the distribution of tuition?

15. Use software to make a normal probability plot of the data on percent of faculty that are full-time for the top 20 schools ranked by tuition from chapter 1. What does the plot tell you about the shape of the distribution of full-time faculty rates?

4.5 Linear Combinations of Independent Random Variables

In section 4.2 you saw how to change the units of $E(X)$ and $SD(X)$ of a random variable. In this section we'll expand on those tools by considering how to combine expected values and standard deviations of several random variables. That means we'll be looking at several random variables at a time which we have not done up to this point. The tools that follow in this section are appropriate for random variables that are *independent*. Two random variables (let's call them X and Y) are **independent** if knowing the value of one of them does not change the probability distribution of the other. The chance that Y is a certain value is not related to the value of X. For example, we'd be just as likely to find Y above 10 whether X was 5 or 50. In that (somewhat technical) sense, X and Y are unrelated.

Example: You randomly select a day during the year and let X = total snowfall that day, Y = high temperature that day, and Z = closing price of Microsoft stock that day. Which pairs of variables are likely independent? Which pairs are not independent?

Solution: Let's consider each pair of random variables:

X and Y: These two variables are very likely *not independent*. If you knew that $Y < 32°F$, there's a higher chance that $X > 0$ inches than if $Y > 32°F$. The chance of snow is very much related to temperature.

X and Z: These two variables are independent. The probability that $Z > \$200$ would likely be the same whether $X = 0$ or $X > 0$. The chance that Microsoft stock closes above $\$200$/share isn't related to snowfall amount.

Y and Z: Like X and Z, these two variables are also independent. The probability that $Z > \$200$ would likely be the same whether $Y > 90°F$ or $Y < -10°F$. Weather conditions and the price of a stock like Microsoft are likely completely unrelated. (However, weather conditions and, say, the price of wheat may be related.)

A linear combination of random variables is a linear function of the form

$$Y = a_1 X_1 + a_2 X_2 + \cdots + a_n X_n$$

where the X's are random variables and the a's are constants. What follows are the features of the probability distribution of Y:

Center: The expected value of Y is: $E(Y) = a_1 E(X_1) + a_2 E(X_2) + \cdots + a_n E(X_n)$

Variability: If the X's are all independent, the variance and standard deviation of Y are

$$Var(Y) = a_1^2 Var(X_1) + a_2^2 Var(X_2) + \cdots + a_n^2 Var(X_n)$$

$$SD(Y) = \sqrt{Var(Y)}$$

Shape: If all the X's have a normal distribution and are all independent, then Y also has a normal distribution.

Note that the formula for $E(Y)$ doesn't require independence but the shape features as well as the formulas for $Var(Y)$ and $SD(Y)$ do.

Example: A small service elevator has a weight capacity of 600 pounds. Suppose the weight of an adult person has a normal probability distribution with expected value of 140 pounds and standard deviation of 15 pounds. How likely is it that the total weight of 4 adults will exceed the elevator's capacity?

Solution: If we let X_1, X_2, X_3, X_4 be the independent weights of the 4 adults, then the total weight is

$$Y = X_1 + X_2 + X_3 + X_4 = 1X_1 + 1X_2 + 1X_3 + 1X_4$$

a linear combination where each $a_i = 1$. So we know Y has a normal distribution with

$$E(Y) = 1E(X_1) + 1E(X_2) + 1E(X_3) + 1E(X_4) = 140 + 140 + 140 + 140 = 560$$

$$Var(Y) = 1^2 Var(X_1) + 1^2 Var(X_2) + 1^2 Var(X_3) + 1^2 Var(X_4) = 15^2 + 15^2 + 15^2 + 15^2 = 900$$

$$SD(Y) = \sqrt{Var(Y)} = \sqrt{900} = 30$$

Sketch the normal curve centered at 560, mark 600 and shade the area to its right; that's the probability you want. The z-score for 600 is

$$z = \frac{600 - 560}{30} = 1.33$$

and use Table I to find the area:

$$P(Y > 600) = P(Z > 1.33) = 0.0918$$

There's about a 9.18% chance that 4 adults will exceed the elevator's capacity. I wouldn't try it!

Example: Suppose that reasonable models for the times a college student spends on educational activities and leisure/sports on a weekday are random variables having expected values of 3.3 hours and 4.1 hours, respectively.[5] Also suppose that both random variables have a standard deviation of 1.6 hours. Nita and Jorge are two completely unrelated college students. Find the expected value and standard deviation of the difference between Nita's time spent on leisure/sports and Jorge's time spent on educational activities on one day.

Solution: If we let X_1 be Nita's time spent on leisure/sports and X_2 be Jorge's time spent on educational activities, then the difference in times is

$$Y = X_1 - X_2 = 1X_1 + (-1)X_2$$

a linear combination where each $a_1 = 1$ and $a_2 = -1$. We know the distribution of Y is centered at

$$E(Y) = 1E(X_1) + (-1)E(X_2) = 4.1 - 3.3 = 0.8$$

Since Nita and Jorge are unrelated, it's reasonable to assume X_1 and X_2 are independent, so

$$Var(Y) = 1^2 Var(X_1) + (-1)^2 Var(X_2) = 1.6^2 + 1.6^2 = 5.12$$

$$SD(Y) = \sqrt{Var(Y)} = \sqrt{5.12} \approx 2.26$$

This last example illustrates a not-so-intuitive results about differences between independent random variables: the variability of the difference is larger than it is for the two variables separately. Perhaps it will help to think of variation as measured by the range instead of variance or standard deviation. Suppose you take the difference between two variables, X_1 which varies between 40 and 55 (range = 15) and X_2 which varies between 30 and 50 (range = 20). What is the range of possibilities for $Y = X_1 - X_2$, the difference between the two? If you take the largest possible value for X_1 (55) and the smallest possible value for X_2 (30), you get the largest possible difference (55 – 30 = 25). On the other hand, if you take the smallest possible value for X_1 and the largest possible value for X_2, you get the smallest possible difference (30 – 55 = -25). So the range of possible values for the difference is 25 – (-25) = 50, larger than either range separately, and certainly *not* the difference in the ranges!

Example: A life insurance company provides group life insurance for the employees of a small private school. Employees fall into one of three classes based on their risk. The table below summarizes the differences between the classes in terms of the expected value and variance of the number of deaths in one year and the benefit paid to a beneficiary should an employee die.

Class	Expected Number of Deaths	Variance of Number of Deaths	Benefit Amount
A	0.1	0.099	$100,000
B	0.75	0.7275	$50,000
C	0.4	0.38	$30,000

The insurance company wants to collect a premium from the school that is equal to the expected value of the total benefits paid plus two standard deviations. (The premium is the amount paid by the school to the insurance company for the life insurance coverage.) What should the premium be?

Solution: Let X_1, X_2, and X_3 be the numbers of deaths in each of the classes A, B, and C, respectively. Since there are different people in the different classes, we can treat X_1, X_2, and X_3 as independent random variables. For example, the chance that $X_1 > 0$ should be the same when $X_2 = 0$ as when $X_2 = 2$. The total amount of benefits paid is the number of deaths in each class multiplied by the benefit amount:

$$Y = 100{,}000X_1 + 50{,}000X_2 + 30{,}000X_3$$

$$\begin{aligned} E(Y) &= 100{,}000E(X_1) + 50{,}000E(X_2) + 30{,}000E(X_3) \\ &= 100{,}000(0.1) + 50{,}000(0.75) + 30{,}000(0.4) \\ &= \$59{,}500 \end{aligned}$$

$$\begin{aligned} Var(Y) &= 100{,}000^2 Var(X_1) + 50{,}000^2 Var(X_2) + 30{,}000^2 Var(X_3) \\ &= 100{,}000^2(0.099) + 50{,}000^2(0.7275) + 30{,}000^2(0.38) \\ &= 3{,}150{,}750{,}000 \end{aligned}$$

$$SD(Y) = \sqrt{Var(Y)} = \sqrt{3{,}150{,}750{,}000} \approx \$56{,}131.54$$

So the premium should be

$$E(Y) + 2SD(Y) = \$59{,}500 + 2(\$56{,}131.54) = \$171{,}763.08$$

Example: A class of 12 students takes an exam. Students' scores on the exam are normally-distributed with expected value 75 and standard deviation 14.4. Judah is a student in the class. Find

    a. the probability that Judah will score above 90.
    b. the probability that the class average will be above 90.

Solution: Let $X_1, \ldots, X_{12}$ be the scores for the 12 students. Let's say X_4 is Judah's score. It's reasonable that they're all independent since the chance that Judah does well shouldn't be related to whether another student, say Simeon, does well or not.

a. You want $P(X_4 > 90)$. Sketch the normal curve centered at 75, mark 90 and shade the area to its right; that's the probability you want. The z-score for 90 is

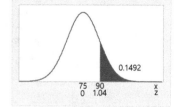

$$z = \frac{X_4 - E(X_4)}{SD(X_4)} = \frac{90 - 75}{14.4} = 1.04$$

and use Table I to find the area:

$$P(X_4 > 0) = P(Z > 1.04) = 1 - P(Z \le 1.04) = 1 - 0.8508 = 0.1492$$

There's a 14.92% chance that Judah scores above 90.

b. The class average score can be written as a linear combination:

$$Y = \bar{X} = \frac{X_1 + X_2 + \cdots + X_{12}}{12} = \frac{1}{12}X_1 + \frac{1}{12}X_2 + \cdots + \frac{1}{12}X_{12}.$$

$$\begin{aligned} E(Y) &= \frac{1}{12}E(X_1) + \frac{1}{12}E(X_2) + \cdots + \frac{1}{12}E(X_{12}) \\ &= \frac{1}{12}(75) + \frac{1}{12}(75) + \cdots + \frac{1}{12}(75) \\ &= 75 \end{aligned}$$

$$\begin{aligned} Var(Y) &= \left(\frac{1}{12}\right)^2 Var(X_1) + \left(\frac{1}{12}\right)^2 Var(X_2) + \cdots + \left(\frac{1}{12}\right)^2 Var(X_{12}) \\ &= \left(\frac{1}{12}\right)^2 (14.4)^2 + \left(\frac{1}{12}\right)^2 (14.4)^2 + \cdots + \left(\frac{1}{12}\right)^2 (14.4)^2 \\ &= 17.28 \end{aligned}$$

$$SD(Y) = \sqrt{Var(Y)} = \sqrt{17.28} \approx 4.157$$

You want $P(Y > 90)$. Sketch the normal curve centered at 75, mark 90 and shade the area to its right; that's the probability you want. The z-score for 90 is

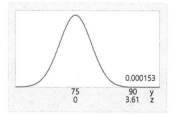

$$z = \frac{Y - E(Y)}{SD(Y)} = \frac{90 - 75}{4.157} = 3.61$$

and use Table I to find the area:

$$P(Y > 90) = P(Z > 3.61) = 1 - P(Z \le 3.61) = 1 - 0.999847 = 0.000153$$

There's very little chance that the class average will be above 90.

This example illustrates a very important concept: *averages vary less than individual values do!* It's somewhat likely that an individual student (like Judah) scores above 90 on the exam, but it's very unlikely that enough students in the class score high enough to push the class average above 90. When was the last time you saw a class average that high? We'll explore averages much more in later sections so stay tuned.

Section 4.5 Exercises

1. You take the subway home from work each day. You get on at the stop near the baseball stadium. Let Y be the number of people waiting with you to get on the subway and let X be the high temperature that day. Do you think X and Y are independent? Explain.

2. Let Y be 1 if a corn plant produces more than one ear and 0 otherwise. Let X be defined the same way for the corn plant next to it in the row. Do you think X and Y are independent? Explain.

3. Let Y be 1 if a corn plant in a field in Kansas produces more than one ear and 0 otherwise. Let X be defined the same way for a corn plant in China. Do you think X and Y are independent? Explain.

4. Let X be the closing value of the Dow Jones Industrial Average and let Y be the amount of rainfall in New York City on the same day. Do you think X and Y are independent? Explain.

5. Let X be your course average in college algebra and let Y be your course average in calculus. Do you think X and Y are independent? Explain.

6. Let X be course average in college algebra for a student selected at random and let Y be the course average in college algebra for another student selected at random. Do you think X and Y are independent? Explain.

7. Let $X_1, X_2, X_3, X_4, X_5, X_6$ be independent random variables each having $E(X) = 50$ and $SD(X) = 10$. Let $Y = \frac{1}{6}X_1 + \frac{1}{6}X_2 + \frac{1}{6}X_3 + \frac{1}{6}X_4 + \frac{1}{6}X_5 + \frac{1}{6}X_6$. Note that $Y = \bar{X}$.

   a. Find $E(Y)$.
   b. Find $Var(Y)$ and $SD(Y)$.

8. Let X_1, X_2, X_3, X_4 be independent random variables having expected values $100, 150, 125,$ and 175, respectively and having standard deviations $11, 9, 8,$ and 12, respectively. Let $Y = 5.5X_1 + 6.5X_2 + 7.5X_3 + 9.5X_4$.

   a. Find $E(Y)$.
   b. Find $Var(Y)$ and $SD(Y)$.

9. Let $X_1, X_2, X_3, X_4, X_5, X_6$ be independent random variables. Let X_1, X_2, X_3 have standard deviation 6 and let X_4, X_5, X_6 have standard deviation 5.

   a. Let $A = \frac{1}{3}X_1 + \frac{1}{3}X_2 + \frac{1}{3}X_3$. Find the variance and standard deviation of A.
   b. Let $B = \frac{1}{3}X_4 + \frac{1}{3}X_5 + \frac{1}{3}X_6$. Find the variance and standard deviation of B.
   c. Let $T = \frac{1}{2}A + \frac{1}{2}B$. Find the variance and standard deviation of T.

10. Let $X_1, X_2, \ldots, X_{400}$ be independent random variables each having standard deviation 10. Let $R = \frac{1}{400}X_1 + \frac{1}{400}X_2 + \cdots + \frac{1}{400}X_{400}$. Find the variance and standard deviation of R.

11. Let X_1 and X_2 be independent, normally-distributed random variables with $(X_1) = 45, E(X_2) = 40, SD(X_1) = 5$, and $SD(X_2) = 6$. Let $Y = X_1 - X_2$.

    a. Find $E(Y)$ and $SD(Y)$.
    b. Find $P(Y > 0)$.
    c. Find $P(|Y - 5| > 10)$.

12. Let X_1 and X_2 be independent, normally-distributed random variables with $E(X_1) = E(X_2) = 30$ and $SD(X_1) = SD(X_2) = 5$. Let $Y = X_1 - X_2$.

    a. Find $E(Y)$ and $SD(Y)$.
    b. Find $P(Y > 5)$.
    c. Find $P(|Y| > 5)$.

13. Let $X_1, X_2, \ldots, X_{100}$ be independent random variables each having standard deviation 6. Let $X_{101}, X_{102}, \ldots, X_{200}$ be independent random variables each having standard deviation 4. Let $X_{201}, X_{202}, \ldots, X_{300}$ be independent random variables each having standard deviation 9. Let $X_{301}, X_{302}, \ldots, X_{400}$ be independent random variables each having standard deviation 7.

    a. Let $A = \frac{1}{100}X_1 + \frac{1}{100}X_2 + \cdots + \frac{1}{100}X_{100}$. Find the variance and standard deviation of A.
    b. Let $B = \frac{1}{100}X_{101} + \frac{1}{100}X_{102} + \cdots + \frac{1}{100}X_{200}$. Find the variance and standard deviation of B.
    c. Let $C = \frac{1}{100}X_{201} + \frac{1}{100}X_{202} + \cdots + \frac{1}{100}X_{300}$. Find the variance and standard deviation of C.
    d. Let $D = \frac{1}{100}X_{301} + \frac{1}{100}X_{302} + \cdots + \frac{1}{100}X_{400}$. Find the variance and standard deviation of D.
    e. Let $T = \frac{1}{4}A + \frac{1}{4}B + \frac{1}{4}C + \frac{1}{4}D$. Find the variance and standard deviation of T.

14. A restaurant sells four different sizes of soft drinks: kids (12 ounces for $0.85), small (14 ounces for $1.35), medium (20 ounces for $1.59), and large (32 ounces for $1.85). In one day, the restaurant expects to sell 20, 35, 300, and 150 of each size, respectively. Suppose the numbers of each size sold are independent with standard deviations 10, 15, 50, and 60, respectively.

    a. Find the expected amount of soft drink (in ounces) sold in one day.
    b. Find the standard deviation of the amount of soft drink (in ounces) sold in one day.
    c. Find the expected revenue from soft drink sales in one day.
    d. Find the standard deviation of soft drink sales in one day.

15. The free-response portions of standardized tests are scored by trained personnel called "readers." A "reading" is a period of time (usually a week) where many readers gather to score free-response questions. Suppose the number of questions scored during the reading by a single reader is well-modeled by a normal distribution with expected value of 1100 and standard deviation 250, independently of other readers. For a particular reading there are 1 million questions to be scored and the scoring company is considering hiring 900 readers.

    a. Find the expected value, variance, and standard deviation for the total number of questions scored during the reading by the 900 readers.
    b. Find the probability that all 1 million questions will be scored during the reading.
    c. The scoring company just did the calculation you did in part b. and realizes it should hire more readers to get the job done on time. What is the minimum number of additional readers the company should hire to have a 95% chance of getting the job done on time? Hint: Try a guess-and-check approach.

16. A student takes a 10 multiple-choice question quiz. Each question has 5 choices. She knows the answers to 7 of them but has to guess, independently, at the rest. Let Y be the number of questions she gets correct out of 10.

    a. Find and interpret $E(Y)$.
    b. Now suppose she can eliminate all but two choices for each of the questions she doesn't know for sure. What is her expected score now?

17. A company provides group life insurance for its employees. Employees are classified by age. Characteristics of each age group and the benefit amounts are shown below.

Age Group	Expected Number of Deaths	Variance of Number of Deaths	Benefit Amount
18 – 40	2	1.98	$150,000
41 – 60	4	3.84	$100,000
60 – 75	5	4.5	$50,000
over 75	3	2.4	$10,000

Let X_1, X_2, X_3, X_4 be, independently, the numbers of deaths in each age group.

a. Find the expected value of the total benefit amount.
b. Find the standard deviation of the total benefit amount.
c. If the company pays a premium of $900,000, is it likely that the insurance company will incur a loss because they need to pay out more in benefits than they get from the premium? Explain.
d. The insurance company wants to change the premium to be the expected total benefit amount plus two standard deviations. What should the premium be?

18. Ten mid-sized sedans and five SUVs are driven on a 100-mile test course. Let $X_1, X_2, \ldots, X_{10}$ be the gasoline consumptions in gallons for the sedans and let $X_{11}, X_{12}, \ldots, X_{15}$ be the gasoline consumptions in gallons for the SUVs. For the sedans, $E(X) = 3.4$ and $SD(X) = 0.2$. For the SUVs, $E(X) = 4.0$ and $SD(X) = 0.3$. All consumptions have a normal distribution and are independent. To compare the data from the two groups, we can consider the difference in average gasoline consumption

$$Y = \frac{X_1 + X_2 + \cdots + X_{10}}{10} - \frac{X_{11} + X_{12} + \cdots + X_{15}}{5}.$$

a. Find $E(Y)$ and $SD(Y)$. Hint: Write Y in standard linear combination form.
b. Find $P(Y < 0)$, i.e. that the SUVs use more gas, on average, than the sedans.
c. Find $P(Y < -1)$. Explain in context what probability you just calculated.

4.6 The Binomial Distribution

Just like the normal distribution is probably the most famous continuous probability distribution, the binomial random variable is a special case of a discrete random variable that is important enough to merit its own section. It's appropriate in many experiments that are comprised of a sequence of smaller experiments called **trials**. Here's an example:

Example: Roulette is a casino game where a wheel with several slots is spun and a marble is dropped onto the spinning wheel, landing in one of the slots. Gamblers bet on the number or color of the slot. In American roulette, the wheel has 18 red slots, 18 black slots, and 2 green slots. Typically gamblers play more than one spin of the wheel so we're interested in what happens in a sequence of spins.

Let X be the number of times a player wins a bet on color in a sequence of 10 spins. The range of X is $\{0, 1, 2, \ldots, 10\}$; a player can lose on all spins, win only once, win twice, and so on up to the best outcome for the gambler: winning their bets on all 10 spins.

The roulette example has several characteristics common to all binomial experiments:

1. The experiment consists of n trials.
2. The outcome of each trial can be classified as either "success" or "failure." These terms don't necessarily imply one outcome is good and the other bad; it's just the names we give to the two outcomes in general.
3. The probability of success on each trial, denoted p, is constant from trial to trial.
4. The random variable X is defined as the number of successes in the n trials.

We then say that X is a binomial random variable or that X has a binomial probability distribution. Here's how the roulette example maps to these characteristics.

Binomial Characteristics	Roulette
n trials	trial = single spin/bet; $n = 10$
success/failure outcomes on each trial	success = winning a bet on a single spin
constant p	$p = 18/38$ and remains constant
X = number of successes in n trials	X = number of wins in 10 spins

Be careful; not every sequence of success/failure trials has all the binomial characteristics.

Example: You shuffle a standard deck of 52 playing cards and deal out 5 cards. Let X = number of red cards dealt. Is X a binomial random variable?

Solution: Let's check to see if all five binomial characteristics hold.

Binomial Characteristics	Cards
n trials	trial = single card; $n = 5$
success/failure outcomes on each trial	success = card is red
constant p	$p = \frac{1}{2}$, but only for the first card
X = number of successes in n trials	X = number of red cards out of 5

Since cards are dealt without replacement from the deck, the probability of any one card being red depends on the colors of the cards already dealt; it does not remain constant. For example, suppose you know that the first card is red. What's the probability that the second card is red? Let's draw the tree diagram for just a two-stage experiment.

If the first card is red, there are only 25 red cards remaining in the deck out of 51. So the probability that the second card is red is $\frac{25}{51} \approx 0.49$. If the first card is black, there are 26 red cards remaining in the deck out of 51. So the probability that the second card is red is $\frac{26}{51} \approx 0.51$. The probability that a card is red changes from the first card to the second card and depends on the color card dealt first. Therefore X is not a binomial random variable.

For experiments that are binomial in nature, there are several shortcuts we can use to describe the behavior of the experiment. There is a formula for the $p(x)$'s and much simpler formulas for $E(X)$ and $SD(X)$.

Binomial Formula

Let's go back to Aaron shooting free-throws. Recall that he is a 60% free-throw shooter from the foul line and he attempts 3 free-throws, observing whether or not he makes each shot. The tree diagram looks like this:

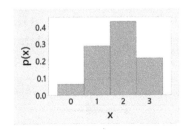

Now let's get a random variable involved: let $X = $ number of shots he makes out of 3. Notice that this is a binomial experiment:

Binomial Characteristics	Free Throws
n trials	trial = single free-throw; $n = 3$
success/failure outcomes on each trial	success = making a single shot
constant p	$p = 0.60$ and remains constant
$X = $ number of successes in n trials	$X = $ number of shots made out of 3

Here's the probability distribution:

x	Outcomes in S	$p(x)$
0	OOO	0.064
1	XOO, OXO, OOX	$0.096 + 0.096 + 0.096 = 0.288$
2	XXO, XOX, OXX	$0.144 + 0.144 + 0.144 = 0.432$
3	XXX	0.216

Wouldn't it be nice to jump right to the $p(x)$'s without having to construct the tree diagram? Would you want to draw the tree diagram for 20 attempts? I wouldn't! Fortunately there is a simpler way and it involves taking advantage of the structure of the experiment.

Notice how the probability of each outcome in the sample space is just the product of some 0.6's and some 0.4's where the order doesn't matter. The number of 0.6's in the product is just the number of made shots and the number of 0.4's in the product is the number of missed shots. So let's generalize this pattern:

p = probability of a success $1 - p$ = probability of a failure
x = number of successes $n - x$ = number of failures

$p^x(1 - p)^{n-x}$ = probability of an outcome with x successes

That last expression simply describes how we multiply the probabilities along the branches of the tree diagram to get the probability of any particular sample space outcome. Let's rewrite the $p(x)$ column in the probability distribution to illustrate this structure:

x	Outcomes in S	$p(x)$
0	OOO	$1 \cdot 0.6^0 \cdot 0.4^3 = 0.064$
1	XOO, OXO, OOX	$3 \cdot 0.6^1 \cdot 0.4^2 = 0.288$
2	XXO, XOX, OXX	$3 \cdot 0.6^2 \cdot 0.4^1 = 0.432$
3	XXX	$1 \cdot 0.6^3 \cdot 0.4^0 = 0.216$

If only we could know how many outcomes in S there are that result in 1 success, 2 successes, etc. without having to draw the tree diagram. Good news! There is a way to do that. In n trials, the number of possible outcomes having exactly x successes is

$$\binom{n}{x} = \frac{n!}{x!(n-x)!}$$

and is read "n choose x." Remember that $0! \equiv 1$. Let's consider the $n = 3$ case:

$x = 0$: $\binom{3}{0} = \frac{3!}{0!3!} = \frac{3 \cdot 2 \cdot 1}{1 \cdot 3 \cdot 2 \cdot 1} = 1$; Only 1 outcome has no successes.

$x = 1$: $\binom{3}{1} = \frac{3!}{1!2!} = \frac{3 \cdot 2 \cdot 1}{1 \cdot 2 \cdot 1} = 3$; There are 3 outcomes having exactly 1 success.

$x = 2$: $\binom{3}{2} = \frac{3!}{2!1!} = \frac{3 \cdot 2 \cdot 1}{2 \cdot 1 \cdot 1} = 3$; There are 3 outcomes having exactly 2 successes.

$x = 3$: $\binom{3}{3} = \frac{3!}{3!0!} = \frac{3 \cdot 2 \cdot 1}{3 \cdot 2 \cdot 1 \cdot 1} = 1$; Only 1 outcome has 3 successes.

Those numbers $(1, 3, 3, 1)$ are just the ones we need in front of the $p^x(1 - p)^{n-x}$ terms that tells us how many times those terms will appear in the tree diagram. Putting all the pieces together gives

$$p(x) = \binom{n}{x}p^x(1 - p)^{n-x}.$$

That's the formula for the probabilities of getting exactly x successes in a binomial experiment. No tree diagram required.

Example: Find the probability that a gambler will win exactly half of her bets on color in 10 spins in roulette. Also find the probability distribution of X = number of winning bets in 10 spins.

Solution: Good thing we don't need a tree diagram: there are $2^{10} = 1{,}024$ possible outcomes! Since we know this is a binomial experiment and that X has a binomial distribution, we can use the binomial formula with $n = 10$ and $p = \dfrac{18}{38}$.

Winning half of her 10 bets means $x = 5$ so

$$p(5) = \binom{10}{5}\left(\frac{18}{38}\right)^5\left(1-\frac{18}{38}\right)^{10-5} = \frac{10!}{5!5!}\left(\frac{18}{38}\right)^5\left(\frac{20}{38}\right)^5 \approx 0.2427 \ .$$

We can fill in the other $p(x)$'s in a similar way and show the distribution of X in a table or graph:

x	$p(x)$	
0	$\binom{10}{0}\left(\frac{18}{38}\right)^0\left(1-\frac{18}{38}\right)^{10-0} = \frac{10!}{0!10!}\left(\frac{18}{38}\right)^0\left(\frac{20}{38}\right)^{10}$	≈ 0.0016
1	$\binom{10}{1}\left(\frac{18}{38}\right)^1\left(1-\frac{18}{38}\right)^{10-1} = \frac{10!}{1!9!}\left(\frac{18}{38}\right)^1\left(\frac{20}{38}\right)^9$	≈ 0.0147
2	$\binom{10}{2}\left(\frac{18}{38}\right)^2\left(1-\frac{18}{38}\right)^{10-2} = \frac{10!}{2!8!}\left(\frac{18}{38}\right)^2\left(\frac{20}{38}\right)^8$	≈ 0.0595
3	$\binom{10}{3}\left(\frac{18}{38}\right)^3\left(1-\frac{18}{38}\right)^{10-3} = \frac{10!}{3!7!}\left(\frac{18}{38}\right)^3\left(\frac{20}{38}\right)^7$	≈ 0.1427
4	$\binom{10}{4}\left(\frac{18}{38}\right)^4\left(1-\frac{18}{38}\right)^{10-4} = \frac{10!}{4!6!}\left(\frac{18}{38}\right)^4\left(\frac{20}{38}\right)^6$	≈ 0.2247
5	$\binom{10}{5}\left(\frac{18}{38}\right)^5\left(1-\frac{18}{38}\right)^{10-5} = \frac{10!}{5!5!}\left(\frac{18}{38}\right)^5\left(\frac{20}{38}\right)^5$	≈ 0.2427
6	$\binom{10}{6}\left(\frac{18}{38}\right)^6\left(1-\frac{18}{38}\right)^{10-6} = \frac{10!}{6!4!}\left(\frac{18}{38}\right)^6\left(\frac{20}{38}\right)^4$	≈ 0.1820
7	$\binom{10}{7}\left(\frac{18}{38}\right)^7\left(1-\frac{18}{38}\right)^{10-7} = \frac{10!}{7!3!}\left(\frac{18}{38}\right)^7\left(\frac{20}{38}\right)^3$	≈ 0.0936
8	$\binom{10}{8}\left(\frac{18}{38}\right)^8\left(1-\frac{18}{38}\right)^{10-8} = \frac{10!}{8!2!}\left(\frac{18}{38}\right)^8\left(\frac{20}{38}\right)^2$	≈ 0.0316
9	$\binom{10}{9}\left(\frac{18}{38}\right)^9\left(1-\frac{18}{38}\right)^{10-9} = \frac{10!}{9!1!}\left(\frac{18}{38}\right)^9\left(\frac{20}{38}\right)^1$	≈ 0.0063
10	$\binom{10}{10}\left(\frac{18}{38}\right)^{10}\left(1-\frac{18}{38}\right)^{10-10} = \frac{10!}{10!0!}\left(\frac{18}{38}\right)^{10}\left(\frac{20}{38}\right)^0$	≈ 0.0006

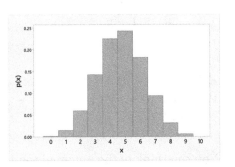

Notice how the probability distribution is fairly symmetric but slightly right-skewed, showing the slight house edge when betting on color.

Example: Find the probability that a gambler will leave the table with more money than they started with when betting on color in 10 spins in roulette.

Solution: A gambler will break if they win exactly 5 times; they'll lose as much money as they win. To leave with more money than they started with means winning more than 5 bets. Since we've already found the $p(x)$'s for winning exactly 6, 7, 8, 9, and 10 times, we just add them together:

$$P(X > 5) = p(6) + p(7) + p(8) + p(9) + p(10) \approx 0.3141$$

Expected Value and Standard Deviation

In general, we calculate the expected value and standard deviation for a discrete random variable like this:

x	$p(x)$
x_1	p_1
x_2	p_2
$\vdots$	$\vdots$
x_k	p_k

$$E(X) = x_1 p_1 + x_2 p_2 + \cdots + x_k p_k$$

$$SD(X) = \sqrt{[x_1 - E(X)]^2 p_1 + [x_2 - E(X)]^2 p_2 + \cdots + [x_k - E(X)]^2 p_k}$$

For a binomial random variable, there's a much easier way since we can exploit the structure of the experiment. Suppose T is a random variable representing the outcome on a single trial: 1 = success and 0 = failure. Then the probability distribution of T is

t	$p(t)$
0	$1 - p$
1	p

$$E(T) = 0(1 - p) + 1p = p$$
$$Var(T) = (0 - p)^2 (1 - p) + (1 - p)^2 p = p(1 - p)$$
$$SD(T) = \sqrt{p(1 - p)}$$

You can think of the binomial random variable X as the sum of n independent T's:

$$X = T_1 + T_2 + \cdots + T_n$$

The T's for the successes will be 1 and the T's for the failures will be 0. The sum will be how many T's are 1's which is exactly how many successes there are in the n trials. Pretty cool, huh? X is a linear combination of n independent random variables where all the a_i's are 1. Since the success probability is the same for each trial, the expected values of the T's are all the same and so are the variances and standard deviations. Using our tools to get the expected value and variance of a linear combination of independent random variables, we have

$$E(X) = 1E(T_1) + 1E(T_2) + \cdots + 1E(T_n) = p + p + \cdots + p = np$$

$$Var(X) = 1^2 Var(T_1) + 1^2 Var(T_2) + \cdots + 1^2 Var(T_n) = np(1 - p)$$

$$SD(X) = \sqrt{np(1 - p)}$$

For a binomial random variable, np and $\sqrt{np(1 - p)}$ are the expected value and standard deviation, respectively. You don't even need to calculate all the $p(x)$'s first! All you need to know are n and p.

Example: Find the z-score corresponding to breaking even in 10 bets on color in roulette.

Solution: To get the z-score for $X = 5$, we need both $E(X)$ and $SD(X)$.

$$E(X) = np = 10(\tfrac{18}{38}) \approx 4.737$$

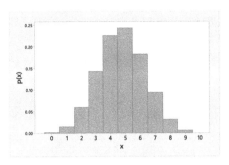

$$SD(X) = \sqrt{np(1-p)} = \sqrt{10(\tfrac{18}{38})(1 - \tfrac{18}{38})} \approx 1.579$$

$$z = \frac{X - E(X)}{SD(X)} = \frac{5 - 4.737}{1.579} = 0.1\overline{6}$$

Breaking even is not at all unusual for 10 bets on color.

Example: Now let's consider things from the casino's perspective. Consider 200 gamblers who each place 10 bets on color in roulette. Find the z-score corresponding to the 200 gamblers collectively breaking even on the 2,000 bets.

Solution: We can consider all $n = 2,000$ bets having constant $p = \frac{18}{38}$. Then $X = $ number of winning bets out of 2,000 is still binomial. To break even, the gamblers would need to collectively win 1,000 times.

$$E(X) = np = 2000(\tfrac{18}{38}) \approx 947.368$$

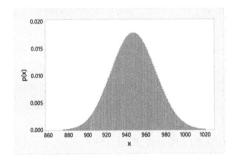

$$SD(X) = \sqrt{np(1-p)} = \sqrt{2000(\tfrac{18}{38})(1 - \tfrac{18}{38})} \approx 22.330$$

$$z = \frac{X - E(X)}{SD(X)} = \frac{1000 - 947.368}{22.330} \approx 2.36$$

I love this example because it illustrates several things: First, notice how much more unusual the break-even condition is for 2,000 bets than for 10 bets. Second, it is much more likely for the gamblers to collectively lose money ($X < 1000$) and for the casino to come out ahead. That small house edge that seems pretty insignificant for an individual gambler on a few bets is the casino's insurance against losses across many gamblers. Third, are you surprised at the shape of the binomial distribution? Looks pretty normal, doesn't it? We'll explore that in the next section.

Example: Do you think you could tell the difference between a cup of tea where the tea was poured into the milk compared to a cup where the milk was poured into the tea? There's the story of an English woman who claimed she could![6] Suppose you secretly prepare a cup of tea by tossing a fair coin. If it comes up heads, you pour the tea into the milk. If tails, you pour the milk into the tea. You then stir it well and present the cup to the woman to see if she can correctly identify how it was prepared. If you did this for 15 cups of tea and she correctly identified your preparation method 9 times, what would you think?

Solution: Okay, I know what you're thinking: "She was wrong on 6 cups so she can't really tell the difference." Would you be that strict? What if she had only missed 1? Would you give her the benefit of the doubt then? My guess is that you would because you know *it would be very unlikely to get 14 out of 15 right just by guessing*. This is exactly the kind of logic we'll use to assess her performance of 9 out of 15 correct: assume instead that she's just guessing (i.e. using a coin toss to determine what the result of your coin toss was!) and determine how likely her actual performance was under that assumption.

If we assume she was just guessing, we have a binomial experiment where she'd have a 50% chance of guessing correctly on any one cup:

Binomial Characteristics	Tea Cups
n trials	trial = single cup; $n = 15$
success/failure outcomes on each trial	success = correct guess on one cup
constant p	$p = 0.50$ and remains constant
X = number of successes in n trials	X = number of correct cups out of 15

Now we can calculate the probability that she gets 9 right by just guessing:

$$P(X = 9) = \binom{15}{9}(0.50)^9(1 - 0.50)^{15-9} = \frac{15!}{9!6!}(0.50)^9(0.50)^6 \approx 0.1527$$

Would you consider that a low probability? If we're talking about the chances of rain today, you might think a 15.27% chance of rain is low enough to not carry an umbrella. However, in the binomial distribution, considering probabilities like this for single values of X can be misleading. Here's a picture of the distribution. Notice that the highest single probabilities ever get is about 0.20. Also notice that 9 is fairly close to the center of the distribution; in the picture, 9 doesn't seem all that unusual even though it's individual probability $p(9)$ seems small.

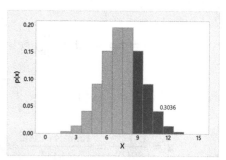

A better way to evaluate whether binomial outcomes are unusual or not is to calculate a "tail probability." That is, a probability that X is either *greater than or equal to the outcome* or *less than or equal to the outcome*. In this example, let's calculate the probability that she'd get greater than or equal to 9 cups correct.

$$\begin{aligned} P(X \geq 9) &= p(9) + p(10) + p(11) + p(12) + p(13) + p(14) + p(15) \\ &= \binom{15}{9}(0.50)^9(1 - 0.50)^{15-9} + \binom{15}{10}(0.50)^{10}(1 - 0.50)^{15-10} + \binom{15}{11}(0.50)^{11}(1 - 0.50)^{15-11} \\ &\quad + \binom{15}{12}(0.50)^{12}(1 - 0.50)^{15-12} + \binom{15}{13}(0.50)^{13}(1 - 0.50)^{15-13} \\ &\quad + \binom{15}{14}(0.50)^{14}(1 - 0.50)^{15-14} + \binom{15}{15}(0.50)^{15}(1 - 0.50)^{15-15} \approx 0.3036 \end{aligned}$$

Now that's a larger probability which better reflects the fact that 9 is not all that unusual in this distribution. *Since getting 9 cups (or more) right just by guessing is not all that unusual, we would not get too excited about this woman's claim that she's able to tell the difference.*

Let's again review the logic:
- Assume the woman is just guessing (i.e. tossing a coin) for each cup.
- We can use the binomial distribution to describe the possible results of the experiment.
- Comparing her actual performance to this "guessing distribution" shows that it's not really surprising.
- Therefore, we wouldn't rule out that she's just guessing. Guessing is still a somewhat reasonable explanation of her performance.

We'll use that general logical sequence a lot in statistics:
- Assume that a random mechanism used to obtain the data is solely responsible for any patterns or differences we see.
- Obtain a distribution of possible results which could be obtained by the random mechanism alone.
- Compare the actual data to the "random mechanism alone" distribution.
- Rule out the "random mechanism alone" explanation if the data look surprising or unusual.

How unusual is "unusual?" Statisticians typically consider probabilities less than 0.05 as unusual. Why 0.05 and not 0.06 or 0.04? Granted, it's an arbitrary number but many trace it's origin to a statement made by English statistician Sir Ronald Fisher where he said, to paraphrase, that results are surprising if we'd expect them to happen only 1 time out of 20.[7] So $\frac{1}{20} = 0.05$ stuck.

Interestingly enough, Fisher was present at the tea party where the woman claimed to be able to tell the difference and he designed a similar experiment to assess her claim.

Example: Your friend comes to you distraught after taking a multiple choice exam. "I didn't know anything," she claimed. "I had to guess at every question." It was a 25-question exam and each question had 5 choices. A week later she comes to you again but this time is elated. She exclaims, "Guess what? You know that exam I took last week that I thought I failed? Well, I got a 72%! My guesses must have been really lucky!" What would you think?

Solution: Let's apply our logic.
- Assume your friend guessed at every question.
- Then we can use the binomial distribution to describe the possible results from guessing alone:

Binomial Characteristics	Multiple Choice Exam
n trials	trial = single question; $n = 25$
success/failure outcomes on each trial	success = correct guess on one question
constant p	$p = \frac{1}{5} = 0.2$ and remains constant
X = number of successes in n trials	X = number of correct questions out of 25

Since all the conditions are satisfied, X has a binomial distribution. Here's a picture. We can use our shortcuts to get the expected number of correct questions and the standard deviation:

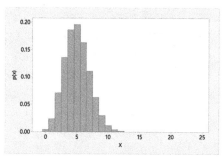

$E(X) = np = 25(0.2) = 5$

$SD(X) = \sqrt{np(1-p)} = \sqrt{25(0.2)(1-0.2)} = 2$

Over many guessers taking this same kind of exam (25 questions, 5 choices each), the average score would be close to 5 correct and scores would differ from 5 by around 2, on average.

- To compare your friend's score of 72% to the binomial distribution with $n = 25$ and $p = 0.2$, we'll calculate the probability of getting a 72% *or higher* just by guessing. That means getting at least 18 correct out of 25:

$P(72\% \ or \ higher) = P(X \geq 18) = p(18) + p(19) + \cdots + p(25)$

$= \binom{25}{18}(0.2)^{18}(1-0.2)^{25-18} + \binom{25}{19}(0.2)^{19}(1-0.2)^{25-19} + \cdots + \binom{25}{25}(0.2)^{25}(1-0.2)^{25-25}$

$= 0.0000000291$

That's very small. Another way to compare your friend's score to what we'd expect for a guesser is to calculate a z-score:

$z = \frac{X - E(X)}{SD(X)} = \frac{18 - 5}{2} = 6.5$

Getting a 72% by guessing is not only higher than we'd expect, it is 6.5 standard deviations higher!

- Getting a 72% (or anything higher) is highly unlikely if she was just guessing. The probabilities of all those scores are so small they don't even show up in the graph! We can easily rule out the guessing mechanism as an explanation for her performance. Perhaps your friend tends to have a penchant for the dramatic and underestimated her abilities.

In the last two examples, we found probabilities of outcomes that involved several values of X. To do that required adding up the binomial $p(x)$'s which is tedious, especially if there are many of them. In some cases, we can use the normal distribution as a good approximation and save ourselves some work. That's the topic of the next section.

Section 4.6 Exercises

For exercises 1 – 11,
    i.   give the range of the random variable X.
    ii.   determine whether the random variable X has a binomial distribution or not by checking the four characteristics.
    iii. give the values of n and p if X has a binomial distribution.

1. Consider the experiment of tossing a fair coin 3 times and recording the sequence of heads and tails you get. Let X be the number of tails you get.

2. In Monopoly, you move around the board by rolling two fair dice. Let's say that one is green and one is blue. Let X be the number of sixes shown on the dice.

3. One drawer in our kitchen is the "junk drawer." Among other things in there are batteries. Let's say there are 10 batteries in the drawer but 3 of them are totally dead. I select 2 batteries at random for my son's train. Let X be the number of dead batteries I get.

4. Darryl is a baseball player who has a 0.250 batting average. That means that over some period of time, Darryl had a hit in 1 out of 4 at-bats. Let X be the number of hits Darryl gets in a 4 at-bat game.

5. A department store runs a promotion where customers spin a wheel at the checkout to determine their discount. The wheel has 10 equal-sized slices: six are labeled "2%," three are labeled "10%," and one is labeled "20%." Let X be the number of times a customer has to spin the wheel before getting 20%.

6. There are two traffic lights on Maria's way to work. If both lights are green, she can get to work in 20 minutes. Stopping at the first light adds 1 minute to her commute time and stopping at the second light adds 2 minutes. The probabilities that she stops at the first and second lights are 0.3 and 0.2, respectively. Let X be the number of lights she stops at on her way to work.

7. Brian, Art, Sam, and Phil audition for America's Got Luck. The audition process consists of tossing a fair coin for each person. Heads they get on; tails they don't. Let X be how many of these guys get on the show.

8. Brian, Art, Sam, and Phil audition for America's Got More Luck. The audition process consists counting the number of folks in line to audition and putting half that number of red marbles in a bucket and half that number of green marbles in the bucket. As people come in for their "audition," a marble is selected from the bucket without replacement. Red they get on; green they don't. Suppose that these four guys are the only ones in line. Let X be how many of these guys get on the show.

9. Micah purchases a car insurance policy with a $3000 annual premium. The table below gives the possible annual costs that the insurance company may have to pay on Micah's behalf along with the probabilities.

Cost	$0	$1500	$20,000	$300,000
Probability	0.80	0.15	0.045	0.005

    Let X be the net gain for the insurance company per year on Micah's policy.

10. Consider the game of spinning the spinner shown here twice and observing the random variable X = sum of the numbers on both spins.

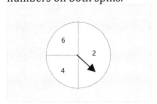

11. You want to dunk your friend in the dunk tank at the carnival. Your friend sits on a ledge above a tank of water and you throw baseballs at a target. If you hit the target, the ledge collapses and your friend gets dunked. You figure you have a 20% chance of hitting the target with each throw. It costs $2 per throw and you only have $8. Let X be the number of baseballs you throw.

12. Evaluate the following.

    a. $\binom{5}{2}$ b. $\binom{5}{3}$ c. $\binom{9}{7}$ d. $\binom{6}{4}$

13. Evaluate the following.

    a. $\binom{20}{10}$ b. $\binom{18}{7}$ c. $\binom{13}{12}$ d. $\binom{25}{15}$

14. Evaluate $\binom{6}{x}$ for $x = 0, 1, 2, 3, 4, 5, 6$. What patterns do you notice?

15. Evaluate $\binom{5}{x}$ for $x = 0, 1, 2, 3, 4, 5$. What patterns do you notice?

16. Find a value of n so that there is only one largest value of all the values of $\binom{n}{x}$ for $x = 0, 1, 2, ..., n$.

17. Find a value of n so that there are two largest values of all the values of $\binom{n}{x}$ for $x = 0, 1, 2, ..., n$.

18. What is the smallest possible of value $\binom{n}{x}$? When does it occur?

19. Let X have a binomial distribution with $n = 5$ and $p = 0.3$. Find

    a. $P(X = 2)$.
    b. $p(4)$.
    c. $P(X \le 1)$.
    d. the probability that X is more than 2.
    e. $P(1 \le X < 3)$.
    f. the probability that X is more than 1 but at most 3.
    g. $E(X)$.
    h. $Var(X)$ and $SD(X)$.

20. Let X have a binomial distribution with $n = 10$ and $p = 0.7$. Find

    a. $p(3)$.
    b. the probability that X is exactly 8.
    c. $P(X < 2)$.
    d. the probability that X is at least 9.
    e. $P(5 < X < 8)$.
    f. the probability that X is at least 5 but at most 8.
    g. $E(X)$.
    h. $Var(X)$ and $SD(X)$.

21. Let X have a binomial distribution with $n = 16$ and $p = 0.25$. Find

    a. $P(X = 6)$.
    b. the probability of no successes.
    c. $P(X \ge 12)$.
    d. the probability of at most 4 successes.
    e. $P(0 < X \le 3)$.
    f. the probability of more than two but no more than 5 successes.
    g. $E(X)$.
    h. $Var(X)$ and $SD(X)$.

22. Let X have a binomial distribution with $n = 9$ and $p = 0.6$. Find

    a. $p(9)$.
    b. the probability of exactly 5 successes.
    c. $P(X > 6)$.
    d. the probability of fewer than 7 successes.
    e. $P(6 \le X \le 8)$.
    f. the probability of at least 5 but at most 7 successes.

    g. $E(X)$.

    h. $Var(X)$ and $SD(X)$.

23. Let X be the number of tails in three tosses of a fair coin. Use a tree diagram to get the probability distribution of X and then verify your $p(x)$'s, with the binomial formula.

24. Let X be the number of dead batteries I get when I randomly select two for my son's train from my junk drawer having 7 good batteries and 3 dead ones. Use a tree diagram to get the probability distribution of X and compare it to the $p(x)$'s you get from the binomial formula with $n = 2$ and $p = 0.3$. Why don't they match?

25. Let X be the number of fours you roll on two fair dice (a green one and a blue one). Use a tree diagram to get the probability distribution of X and then verify your $p(x)$'s using the binomial formula.

26. Brian, Art, Sam, and Phil audition for America's Got More Luck. The audition process consists counting the number of folks in line to audition and putting half that number of red marbles in a bucket and half that number of green marbles in the bucket. As people come in for their "audition," a marble is selected from the bucket without replacement. Red they get on; green they don't. Let X be how many of these guys get on the show. Use a tree diagram to get the probability distribution of X and compare it to the $p(x)$'s you get from the binomial formula with $n = 4$ and $p = 0.5$. Why don't they match?

27. Consider the experiment of tossing a fair coin 3 times and recording the sequence of heads and tails you get.

    a.   Find the probability of getting exactly 2 tails.

    b.   Find the probability of getting at least 1 tail.

    c.   Calculate and interpret the expected number of tails.

    d.   Calculate the variance and standard deviation of the number of tails.

28. In Monopoly, you move around the board by rolling two fair dice. Let's say that one is green and one is blue.

    a.   Find the probability of rolling exactly 1 six.

    b.   Find the probability of rolling at least 1 six.

    c.   Calculate and interpret the expected number of sixes.

    d.   Calculate the variance and standard deviation of the number of sixes.

    e.   Would your answers above change if you considered fours instead of sixes? Explain.

29. Consider the group life insurance problem from section 4.5. A more realistic problem would be to give you the table below instead of the table of expected values and variances of the numbers of deaths in each class.

Class	Number in Class	Probability of Death	Benefit Amount
A	10	0.01	$100,000
B	25	0.03	$50,000
C	8	0.05	$30,000

    a.   Verify the expected number of deaths in each class given in the example in section 4.5.

    b.   Interpret the expected number of deaths in class B.

    c.   Verify the variance of the number of deaths in each class given in the example in section 4.5.

    d.   Calculate the standard deviation of the number of deaths in each class.

30. Darryl is a baseball player who has a 0.250 batting average. That means that over some period of time, Darryl had a hit in 1 out of 4 at-bats. In a game where Darryl has 4 at-bats

    a.   find the probability that he gets only 1 hit.

    b.   find the probability that he gets fewer than 2 hits.

    c.   calculate and interpret the expected number of hits.

    d.   calculate the variance and standard deviation of the number of hits.

31. In Europe, roulette is played with a wheel having 18 red slots, 18 black slots, and 1 green slot. Consider betting on red for each of three spins of a European roulette wheel.

    a. Find the probability of winning exactly 2 bets.
    b. Find the probability of winning more bets than you lose.
    c. Calculate and interpret the expected number of wins.
    d. Calculate the variance and standard deviation of the number of wins.

32. Consider an American roulette wheel with 18 red slots, 18 black slots, and 2 green slots. Consider betting on red on each spin.

    a. What are your chances of winning on one spin?
    b. What are your chances of winning more bets than you lose in a set of 3 spins?
    c. What are your chances of winning more bets than you lose in a set of 7 spins.
    d. As the gambler, how many times should you spin to maximize your chances of winning more than you lose?
    e. What do you think the casino hopes you do?

33. You are a scout for a baseball team. Someone told you that Javier is a 0.275 hitter (that is, he's gotten a hit in 27.5% of his at-bats over some time period). You observe Javier hit across several games. During that time, he got 3 hits in 30 at-bats.

    a. Calculate the z-score for Javier's performance.
    b. What is the probability that Javier gets 3 or fewer hits in 30 at-bats?
    c. What would you conclude based on your answer to part a.?

34. In exercise 33, what would you think if Javier got 6 hits in 30 at-bats? Explain.

35. A test for extrasensory perception (ESP) involves presenting a subject 25 cards, one at a time, face down on a table. The subject concentrates intently on each card and then responds with what shape is on the card. The shapes on the cards are chosen from a set of 5 different shapes and each shape is equally-like to show up on each card. Your friend Tom takes the test and gets 12 correct.

    a. If Tom doesn't have ESP and is just guessing at each card, calculate the z-score for Tom's performance.
    b. If Tom doesn't have ESP and is just guessing at each card, what are the chances that he gets at least 12 correct?
    c. What would you conclude based on your answers to parts a. and b.?

36. In exercise 35, what would you think if Tom got 7 correct? Explain.

37. Tanya always uses her favorite coin to determine who goes first when you both play checkers. She always calls heads and tells you that it doesn't matter that you always get tails since the coin is fair. Recently, you've noticed that Tanya is getting to go first a lot so you decide to collect some data. In the last 28 games, she went first in 16 of them.

    a. If Tanya's coin is fair, calculate the z-score for this outcome.
    b. If Tanya's coin is fair, what are her chances of going first in at least 16 games?
    c. What would you conclude based on your answers to parts a. and b.?

38. In exercise 37, what would you think if Tanya went first in 20 games? Explain.

39. The Standard and Poor's 500 (S&P500) index of the performance of 500 large U.S. companies dates back to 1957. (The index of 90 such stocks was started by the Standard Statistics Company in 1926.) Since 1957, the S&P500 has increased in weekly value for 56.5% of weeks. Evidence suggests that treating the status of one week (increased or decreased) as independent of status of the next week is very reasonable. Let X be how many weeks in the past 5 weeks the S&P500 index has increased in weekly value.

    a. Use the binomial formula to find the probability distribution of X.
    b. Draw a graph showing the distribution.

40. Consider a common voting practice where a simple majority of "yes" votes (more than half) is needed for some action to be adopted. You and eight other people sit on a board of directors for an organization. Your vote matters if exactly 4 of the others vote "yes" since then your vote would decide the matter. Assume each member votes independently of the others.

    a. What is the probability that your vote matters if each of the other members uses a fair coin to cast their vote?
    b. What is the probability that your vote matters if each of the other members uses a fair six-sided die to cast their vote?

41. Rework exercise 40 for a board of 21 people (you and 20 others).

4.7 The Normal Approximation to the Binomial

In the last section, you saw that for 2,000 roulette bets on color, X = the number of winning bets has a binomial distribution that looks much like a normal distribution. Below is a graph of the binomial distribution ($n = 2000, p = \frac{18}{38}$) with a normal curve overlaid. The normal curve fits very well! However, in Aaron's free-throw shooting experiment, the binomial distribution ($n = 3, p = 0.60$) of X = number of shots made is not very normal-looking.

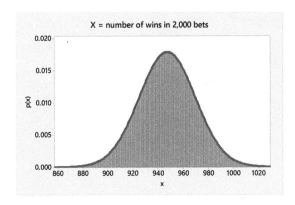
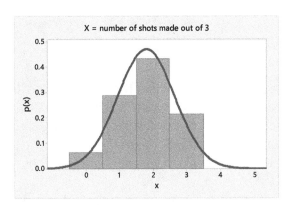

In a probability histogram, not only the heights, but also the *areas* of the bars are the $p(x)$'s. That's because the width of each bar is 1 so the area of each bar is just the height:

$$area = height \cdot width = p(x) \cdot 1 = p(x)$$

For example, in the free-throw shooting problem, the area of the bar at $X = 2$ is 0.432. So $P(X = 2) = p(2) = 0.432$. The total area of the bars for $X = 0$ and $X = 1$ is $0.064 + 0.288 = 0.352$. So $P(X \leq 1) = p(0) + p(1) = 0.352$.

To get the areas under the normal curve to match up well to the areas of the binomial probability histogram bars, the binomial experiment must have enough trials to allow the probability to get spread out like a bell curve. Here are several binomial distributions with normal curves overlaid.

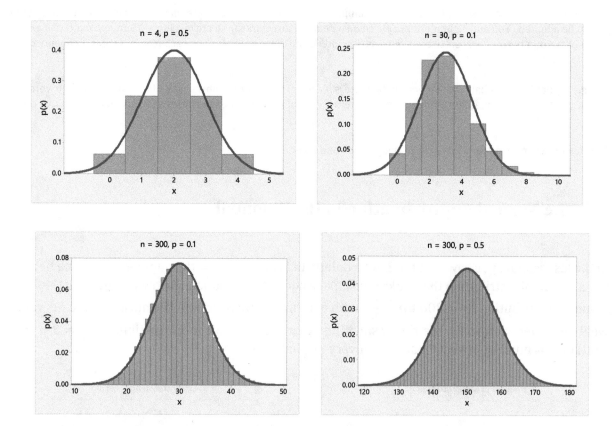

Even when $p = 0.5$ so that the binomial distribution is symmetric, if there aren't enough trials, the areas won't match up well. On the other hand, the success probability might be too large or too small so that the binomial distribution is too skewed to match a normal curve well. It's like Goldilocks; everything has to be "just right." Here, the combination of n and p have to be "just right." The rule of thumb we'll use is the following:

> If $np \geq 15$ and $n(1 - p) \geq 15$, the normal distribution areas/probabilities will be good approximations to the binomial areas/probabilities.

Note that this rule is satisfied for the bottom two pictures but not for the top two pictures.

If the rule of thumb is satisfied, use the binomial expected value and standard deviation as the $E(X)$ and $SD(X)$ of the approximating normal distribution. So you would use a normal distribution with

$$E(X) = np$$
$$SD(X) = \sqrt{np(1 - p)}.$$

Example: A tennis player has a 60% chance of a successful first serve that remains constant from serve to serve. How likely is he to make at most 110 first serves in 200 attempts?

Solution: Let X = number of successful first serves out of 200. X has a binomial distribution with $n = 200$ and constant $p = 0.6$. Finding $P(X \leq 110)$ would involve adding up $p(0), p(1), \ldots, p(110)$

and calculating 111 binomial probabilities! Let's check the rule of thumb conditions to see if we can use a normal distribution as an approximation:

$$np = 200(0.6) = 120 \geq 15$$
$$n(1-p) = 200(1-0.6) = 80 \geq 15$$

Both conditions are satisfied so we'll use a normal distribution with

$$E(X) = np = 200(0.6) = 120$$
$$SD(X) = \sqrt{np(1-p)} = \sqrt{200(0.6)(1-0.6)} \approx 6.9282$$

and use our normal distribution approach: find the z-score and use the table.

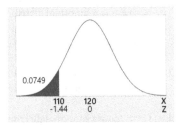

$$z = \frac{110-120}{6.9282} = -1.44$$

$$P(X \leq 110) = P(Z \leq -1.44) = 0.0749$$

There's about a 7.49% chance that he'll make 110 first serves or fewer in 200 attempts.

Example: Suppose at a certain casino, there are 500 bets on color placed at the roulette wheels every day. In a week's time, how likely is it that the casino makes money on those bets?

Solution: In 7 days' time, there would be 3,500 color bets placed. We already know that $X = $ number of winning (from the gamblers' perspective) bets out of 3,500 has a binomial distribution with $n = 3,500$ and $p = \frac{18}{38}$. Let's check the rule of thumb conditions:

$$np = 3500\left(\frac{18}{38}\right) \approx 1657.9 \geq 15$$
$$n(1-p) = 3500\left(1 - \frac{18}{38}\right) \approx 1842.1 \geq 15$$

Both conditions are easily satisfied so we can use the normal distribution with

$$E(X) = np = 3500\left(\frac{18}{38}\right) \approx 1657.9$$
$$SD(X) = \sqrt{np(1-p)} = \sqrt{3500(\frac{18}{38})(1 - \frac{18}{38})} \approx 29.5394.$$

The casino makes money if, collectively, the gamblers win less than half of their bets, so we want $P(X < 1750)$:

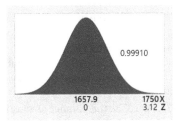

$$z = \frac{1750-1657.9}{29.5394} = 3.12$$

$$P(X < 1750) = P(Z < 3.12) = 0.99910$$

It's almost a given that the casino makes money on those bets. The "action" in the distribution is almost entirely below the break-even spot for the gamblers. While it's somewhat likely for individual gamblers to make money on a handful of bets, the casino is almost guaranteed to be ahead because it can take advantage of how the game works for a large number of bets.

Section 4.7 Exercises

1. For each of the following binomial distributions, say whether a normal distribution would be a good approximation.

    a. X has a binomial distribution with $n = 5$ and $p = 0.4$.
    b. X has a binomial distribution with $n = 100$ and $p = 0.01$.
    c. X has a binomial distribution with $n = 100$ and $p = 0.98$.
    d. X has a binomial distribution with $n = 40$ and $p = 0.5$.
    e. X has a binomial distribution with $n = 40$ and $p = 0.2$.

2. For each part of exercise 1 where the normal distribution would be a good approximation, find the center and standard deviation of the normal curve.

3. Let X have a binomial distribution with $n = 6$ and $p = 0.25$.

    a. Use the binomial formula to find $P(X \leq 1)$.
    b. Is the normal distribution a good approximation? Explain.
    c. Find $E(X)$ and $SD(X)$.
    d. Use the normal distribution (anyway) to approximate $P(X \leq 1)$.
    e. Is your answer to d. close to your answer to b.?

4. Let X have a binomial distribution with $n = 9$ and $p = 0.8$.

    a. Use the binomial formula to find $P(X \geq 8)$.
    b. Is the normal distribution a good approximation? Explain.
    c. Find $E(X)$ and $SD(X)$.
    d. Use the normal distribution (anyway) to approximate $P(X \geq 8)$.
    e. Is your answer to d. close to your answer to b.?

5. Let X have a binomial distribution with $n = 250$ and $p = 0.7$.

    a. Use the binomial formula to find $P(160 \leq X \leq 170)$.
    b. Is the normal distribution a good approximation? Explain.
    c. Find $E(X)$ and $SD(X)$.
    d. Use the normal distribution to approximate $P(160 \leq X \leq 170)$.
    e. Is your answer to d. close to your answer to b.?

6. Let X have a binomial distribution with $n = 500$ and $p = 0.1$.

    a. Use the binomial formula to find $P(60 \leq X \leq 70)$.
    b. Is the normal distribution a good approximation? Explain.
    c. Find $E(X)$ and $SD(X)$.
    d. Use the normal distribution to approximate $P(60 \leq X \leq 70)$.
    e. Is your answer to d. close to your answer to b.?

7. Let X be the number of wins in 10,000 \$1 bets on color placed on roulette wheels in a casino. Approximate the probability that the casino makes money on those bets.

8. Let X be the number of wins in n bets on color placed on roulette wheels in a casino. What is the smallest number of bets so that there would be at least a 90% chance that the casino would make money on those bets?

9. Approximate the probability of getting 460 or fewer tails in 1,000 tosses of a fair coin.

10. A company's flat screen TV has a 1.5% chance that it will fail and need to be replaced within a year. The company anticipates selling 10,000 of these TV's. Each comes with a 1-year full-replacement warrantee.

    a. Find the approximate probability that the company will have to replace more than 180 TV's.
    b. Fill in the blank: There's a 65% chance that the company will have to replace at most _____ TV's.
    c. Find the expected number of replacements.
    d. Find the standard deviation of the number of replacements.

11. Refer to exercise 10. The company figures that replacing a customer's TV would cost the company $250.

    a. Find the expected total replacement cost.
    b. Find the standard deviation of the total replacement cost.
    c. What is the approximate probability that the total replacement cost will be more than $35,000?
    d. The company has reserves of $50,000 designated for TV replacement costs. Will this be enough? Explain.

12. The germination rate for a type of grass seed is 88%. If a bag of grass seed contains 400,000 seeds,

    a. what is the probability that more than 351,500 of them germinate?
    b. what is the probability that more than 88.1% of them germinate?
    c. there is a 20% chance that more than _____% of them germinate.

13. The normal distribution can be used to approximate single binomial $p(x)$'s by adding and subtracting 0.5 to the x value and then finding the area under the curve between the two numbers. For example, to approximate $p(32) = P(X = 32)$, find $P(31.5 \leq X \leq 32.5)$ in the normal distribution. Doing this approximates the area of one binomial rectangle centered at x. Use the normal distribution this way to approximate each of the following binomial probabilities. Compare your approximation to the exact probability you get by using the binomial formula.

    a. $p(55) = P(X = 55), n = 300, p = 0.2$
    b. $p(40) = P(X = 40), n = 75, p = 0.45$

14. Refer to exercise 13. This method can be extended to any range of x's. First decide which x's are included in your event. Then add or subtract 0.5 to the extremes so that you capture their entire rectangle widths (and not just half of them) in the normal approximation. This adjustment is called the *continuity correction*. Here are some examples for a binomial distribution with $n = 100$ and $p = 0.4$.

	x's included	approximate normal
$P(30 \leq X \leq 35)$	$30, 31, 32, 33, 34, 35$	$P(29.5 \leq X \leq 35.5)$
$P(30 < X < 35)$	$31, 32, 33, 34$	$P(30.5 \leq X \leq 34.5)$

    a. Find the exact $P(32 \leq X \leq 35)$ using the binomial formula.
    b. Approximate $P(32 \leq X \leq 35)$ using the normal distribution both with and without using the continuity correction.
    c. Find the exact $P(32 < X < 35)$ using the binomial formula.
    d. Approximate $P(32 < X < 35)$ using the normal distribution both with and without using the continuity correction.

Notes

[1] Old Faithful Visitor Center log books (www.geyserstudy.org/ofvclogs.aspx), transcribed by Lynn Stephens. Many thanks to the park rangers, volunteers, and Lynn for collecting and compiling these data.

[2] U.S. Department of Education College Scorecard (as of 5/21/2019).

[3] American Association of Individual Investors, www.aaii.com.

[4] data.delaware.gov.

[5] American Time Use Survey (2010 – 2014), Bureau of Labor Statistics.

[6] Salsburg, D. (2001), *The Lady Tasting Tea: How Statistics Revolutionized Science in the Twentieth Century*, W. H. Freeman and Company, New York.

[7] Fisher, R. A. (1935), *The Design of Experiments*, Oliver and Boyd, Edinburgh.

Chapter 5: Probability Distributions for Random Sampling

5.1 Introduction to Sampling Distributions

Remember the components of the scientific method that we looked at in chapter 3?

1. Pose a Question
2. Make a Hypothesis
3. Collect Data
4. Analyze the Data
5. Make a Conclusion

This method can help us answer questions about a large population based on a relatively small sample from that population. We called that process *generalization inference* and saw that *random sampling* is the best way to collect the data in order to answer those kinds of questions. Samples that are not random samples can easily suffer from *selection bias* and prohibit us from making valid generalization inferences.

In this chapter, we're going to use the probability tools in chapter 4 to give us insight into how we might analyze the data in order to answer the question. Random sampling involves structured uncertainty; we don't know exactly what the results will be ahead of time but we can describe the distribution of the possibilities. We'll learn what we can expect to see in a random sample taken from a large population.

First a few new terms: A summary of data observed for every object in a *population* is called a **parameter**. A summary of data observed for only the objects in a *sample* is called a **statistic**. Parameters come from populations; statistics come from samples. The alliteration will help you remember which is which. The distinction between parameters and statistics is very important for us right now because we'll be using all kinds of data summaries (means, standard deviations, proportions, etc.) but sometimes we'll need to refer to the population and sometimes to the sample. Here's a list of common summaries and the different notation used for them.

Summary	Variable Type	Population Parameter Notation	Sample Statistic Notation
mean	numeric	μ	$\bar{X}$
median	numeric	η	M
standard deviation	numeric	σ	S
variance	numeric	σ^2	S^2
proportion	categorical	p	$\hat{P}$

Note that for categorical variables, appropriate summaries are often proportions.

Example: A survey was sent to all 1,187 faculty members at a large university. One of the questions asked faculty about their satisfaction with their experience as a faculty member. Of the 433 faculty who responded to the question, 220 (50.8%) were either "satisfied" or "very satisfied."[1] What is the population, sample, parameter, and statistic in this scenario? Comment on the validity of the results.

Solution: The <u>population</u> is all 1,187 faculty members at the university. The <u>sample</u> is the 433 faculty members who responded to the question. The variable (satisfaction level) is categorical so the appropriate parameter would be a proportion. Since the "satisfied" and "very satisfied" categories are combined and reported together, p = proportion of all faculty members at the university who *would have* reported that they are "satisfied" or "very satisfied" with their experience as a faculty member is the <u>parameter</u> of interest. Its value, however, is unknown since not all 1,187 faculty members responded. The <u>statistic</u> is $\hat{P}$, the proportion of the 433 faculty members who reported that they are "satisfied" or "very satisfied" with their experience as a faculty member. Its value is $220/433 \approx 0.508$.

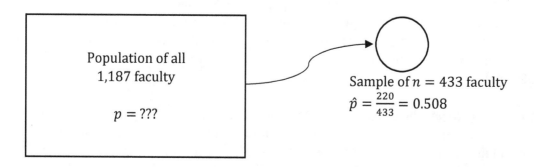

The statistic $\hat{p} \approx 0.508$ is an estimate of the unknown parameter p. Do you think that's a valid estimate? You might say "no, because 433 is much smaller than 1,187." Be careful. We know that a randomly-selected sample, even one much smaller than the population size, can be used to make valid generalizations. The problem here is that the 433 faculty members *are not a random sample*. Since less than half of the faculty surveyed actually responded (433 out of 1,187 is only 36.5%), there could be substantial *non-response bias*. Perhaps only those faculty members who were particularly disgruntled tended to respond and those who were fairly satisfied chose not to respond. If so, the population proportion p would be larger than 0.508 since the less satisfied faculty would be over-represented. On the other hand, if those faculty who were satisfied with their experience were more likely to respond, the population proportion p would be smaller than 0.508 since the satisfied faculty would be over-represented. Because of the potential for substantial non-response bias, we should be very careful not to use the sample to *generalize* to the population of all 1,187 faculty.

Even if samples are obtained randomly, the potential for non-response bias exists when dealing with human beings. That's why reputable polling organizations work hard to reduce the non-response rate.

So, we know *random sampling* is good for doing *generalization inference*, but you might be wondering how we can trust the outcome of a random mechanism. Just like in any experiment that involves a random mechanism (like tossing a coin or rolling dice), we can describe the distribution of the possible outcomes. The **sampling distribution** of a statistic based on a sample from a population is the probability distribution of all possible values of the statistic. It shows how frequently the possible values would occur if the sampling process were repeated many times.

So far, we've been primarily focused on describing variability in a single data set. For random sampling, there is another kind of variability: the variability that comes from repeating the random sampling process many times:

1. Take a random sample of size n from a population.
2. Calculate a statistic that summarizes the sample data.
3. Repeat steps 1-2 many times.

At the end of this process, you'd have a bunch of values of the statistic. The distribution of those values is the *sampling distribution* of the statistic.

Does any of this sound familiar? Back in chapter 3 we explored several random sampling methods in detail and assessed each one in exactly this way. Let's go back to one of those examples involving a comparison of the simple random sample (SRS) and cluster random sample (CRS).

Example: Ethnicity is a variable that tends not to vary much within a household: individuals within a household tend to be all of the same ethnicity. Suppose we wanted to estimate the percentage of Hispanic individuals in a certain community. There are 5,000 households in the community with 4 individuals per household for a total of 20,000 individuals. Let's say that there are 1,000 purely Hispanic households giving 4,000 (20%) Hispanic individuals in the population. Let's take both a SRS of 400 individuals and a CRS of 100 households (which will also give us 400 individuals). In each sample we'll compute the percentage of Hispanics.

Here are the results from repeating each sampling method 1,000 times. The statistic for each sample is the sample percentage of Hispanics (out of 400). The histograms show the two sampling distributions of the sample percentages: one for a SRS and one for a CRS.

	SRS Sample %	CRS Sample %
1	19.00	14
2	19.25	21
3	19.00	14
...	...	...
1,000	22.50	24

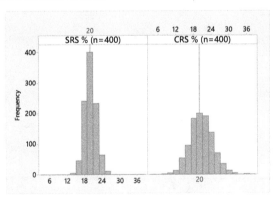

It was only through sampling distributions that we were able to explore the characteristics of the SRS and CRS. Both methods produce sample percentages that are centered at the population percent of Hispanics (20%) but the SRS sample percentages are less variable; they tend to be closer to 20% on average than the ones produced by the CRS.

Repeating the sampling process like we did for the example above is tedious and computationally intensive. Practically, you need to know how to write computer code. There are other ways to get the sampling distribution of a statistic. If the population and sample sizes are quite small, a tree diagram approach might work.

Example: Here are the annual salaries (in thousands of dollars) for all employees at a small company.

$$100 \quad 80 \quad 20 \quad 40$$

Two common parameters are the population mean and standard deviation. Here, those are

$$\mu = \frac{100+80+20+40}{4} = 60$$

and

$$\sigma = \sqrt{\frac{(100-60)^2+(80-60)^2+(20-60)^2+(40-60)^2}{4}} = \sqrt{1000} \approx 31.623.$$

The population mean and standard deviation are $60,000 and $31,623, respectively.

Let's find the sampling distribution of the sample mean $\bar{X}$ for a SRS of 2 employees. A tree diagram will give us all possible sample results, their probabilities, and help to get the sampling distribution of $\bar{X}$:

1st	2nd	Sample	Probability	$\bar{X}$
100	80	100, 80	$\frac{1}{4} \cdot \frac{1}{3} = \frac{1}{12}$	90
	20	100, 20	"	60
	40	100, 40	"	70
80	100	80, 100	"	90
	20	80, 20	"	50
	40	80, 40	"	60
20	100	20, 100	"	60
	80	20, 80	"	50
	40	20, 40	"	30
40	100	40, 100	"	70
	80	40, 80	"	60
	20	40, 20	"	30

Notice that each possible sample result has the same probability. That's what a SRS means. Now just treat $\bar{X}$ as the random variable and record the possible values and their probabilities:

This table is the sampling distribution of the sample mean $\bar{X}$ for a SRS of 2 employees. It shows how likely the possible means are. Once you have this, you can find out more about how a SRS behaves.

$\bar{x}$	$p(\bar{x})$
30	$1/12 + 1/12 = 1/6$
50	$1/12 + 1/12 = 1/6$
60	$1/12 + 1/12 + 1/12 + 1/12 = 1/3$
70	$1/12 + 1/12 = 1/6$
90	$1/12 + 1/12 = 1/6$

Let's get the expected value of $\bar{X}$. Multiply the $\bar{x}$'s by their probabilities and add the products:

$$E(\bar{X}) = 30\left(\tfrac{1}{6}\right) + 50\left(\tfrac{1}{6}\right) + 60\left(\tfrac{1}{3}\right) + 70\left(\tfrac{1}{6}\right) + 90\left(\tfrac{1}{6}\right) = 60$$

Over many SRS's of 2 employees, the average salary would be close to \$60,000. Notice anything? That's the population mean. One really nice property of a SRS is that it produces sample means that hang out around the population mean.

What about variability? Let's get the standard deviation of $\bar{X}$:

$$SD(\bar{X}) = \sqrt{(30-60)^2\left(\tfrac{1}{6}\right) + (50-60)^2\left(\tfrac{1}{6}\right) + (60-60)^2\left(\tfrac{1}{3}\right) + (70-60)^2\left(\tfrac{1}{6}\right) + (90-60)^2\left(\tfrac{1}{6}\right)}$$
$$= \sqrt{333.\overline{3}} \approx 18.257$$

This is much less than 31.623, the population standard deviation. The difference is that 31.623 measures variation in the *salaries* but 18.257 measures variation in the *sample means*. A very important concept that is illustrated here is that *sample means vary less than the data values.* Even the ranges show this! The salaries go from 20 to 100 (range = 80) but the possible sample means go from 30 to 90 (range = 60). It's impossible to get a sample mean of 20 because, even if you got 20 in your sample, the other number would be higher, pulling the mean up. Likewise it's impossible to get a sample mean of 100 because, even if you got 100 in your sample, the other number would be lower, pulling the mean down. So the sample average can't vary as much as the individual salaries do. Stay tuned for more about sample means in section 5.3.

Example: Leon and Sharon invite Bob and Trisha over for dinner. Unfortunately, the dinner conversation turns toward politics and the couples begin arguing over whether there should be term limits for U.S. Supreme Court justices. Leon and Sharon favor term limits but Bob and Trisha do not.

Don't worry; I'm not going to get political on you. Let's stick to statistics. The population proportion who favor term limits is $p = 2/4 = 1/2$. Let's find the sampling distribution of $\hat{P}$, the sample proportion who favor term limits in a SRS of 2 people from this dinner party. A tree diagram will give us all possible sample results, their probabilities, and help to get the sampling distribution of $\hat{P}$:

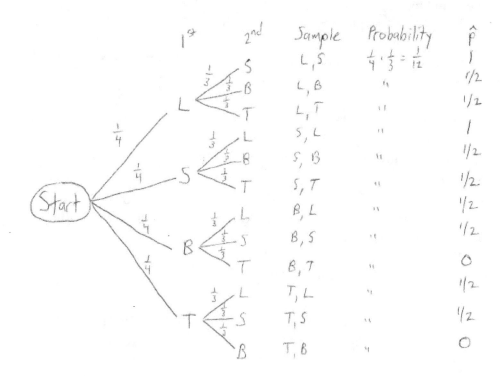

Notice that each possible sample result has the same probability. That's what a SRS means. Now just treat $\hat{P}$ as the random variable and record the possible values and their probabilities:

This table is the sampling distribution of $\hat{P}$, the sample proportion who favor term limits, for a SRS of 2 people. It shows how likely the possible proportions are.

$\hat{p}$	$p(\hat{p})$
0	$2(1/12) = 1/6$
1/2	$8(1/12) = 2/3$
1	$2(1/12) = 1/6$

In the next sections, we'll see how to describe the sampling distributions of several common statistics when a SRS is done. Except for the examples we've considered, describing sampling distributions in general for other sampling methods like the CRS and the stratified random sample (STRS) is beyond the scope of this book.

Section 5.1 Exercises

For exercises 1 – 9, define the population, sample, parameter, and value of the statistic in context. Use correct notation for the parameter and statistic.

1. You want to determine the fraction of people in the U.S. who use alternative medical treatments that improve in their condition. You ask 10 of your friends who use alternative medical treatments about whether their condition has improved. Seven of them said it has.

2. In order to determine the proportion of the daily transactions at a bank that are processed correctly, the bank selects 100 transactions at the end of the day on Friday. It finds that 92 of those transactions were processed correctly.

3. A school board wants to find out how many parents in the district are in favor of a new policy being considered. A poll is posted on the school district's website and only 68 of the 678 parents who responded were in favor of the new policy.

4. A potato chip manufacturer wants to monitor the average weight of product in bags of chips. To do so it selects a SRS of 30 bags from the production line each hour. The average weight of the most recent sample was 12.05 ounces.

5. Mastitis in cows is an inflammation of the udder caused by bacteria. Veterinarians will often use antibiotics to treat cows with mastitis. Milk from cows with mastitis is not to be mixed with milk from healthy cows until the cow has recovered. To determine whether the average antibiotic level in batches of raw milk collected from local farmers is at an acceptable level, a SRS of 10 local farmers is selected and the most recent batch of milk from each is tested. The mean antibiotic level was 22 μg/kg.

6. What fraction of the trees in a forest are ready for harvesting? The landowner selects a SRS of 100 latitude/longitude coordinates in her forest and determines whether the nearest tree to each location is ready for harvest. 45% of the trees selected are ready.

7. A car dealership wants to estimate the proportion of students at a local university who own their own car. There are 10,000 freshman, 9,000 sophomores, 8,000 juniors, and 7,000 seniors. The dealership takes a STRS of 100 freshman, 90 sophomores, 80 juniors, and 70 seniors. Of those in the sample, 15 freshman, 18 sophomores, 68 juniors, and 67 seniors owned their own car.

8. A business analyst wants to average debt-to-income ratio for all businesses in the state. There are 200 product-oriented businesses and 800 service-oriented businesses in the state. The analyst takes a STRS of 10 product-oriented businesses and 40 service-oriented businesses. The average debt-to-income ratio for the product-oriented businesses was 0.44 and was 0.32 for the service-oriented businesses.

9. A wireless service provider wants to estimate the proportion of people (including children) in the country who have their own smartphone. The project manager for the study takes a CRS of 1000 U.S. households. Out of the 1693 people, 1405 own their own smartphone.

10. Which of these have sampling distributions? $\mu, S^2, \hat{P}, \sigma, M$

11. Which of these have sampling distributions? $\eta, p, \bar{X}, S, \sigma^2$

12. Identify the correct statements. Fix the incorrect ones.
    a. A sampling distribution describes the distribution of data values.
    b. A sampling distribution shows the behavior of a sampling process over many samples.
    c. The probability distribution of a parameter is called a sampling distribution.
    d. Sampling distributions describe the values of a data summary for many samples.

13. Identify the correct statements. Fix the incorrect ones.
    a. A sampling distribution describes the distribution of values of a statistic.
    b. A sampling distribution shows the distribution of a variable for many objects in a population.
    c. The probability distribution of a statistic is called a sampling distribution.
    d. Sampling distributions describe the data in a single sample.

14. Why is random sampling important?

15. What is a potential problem with non-random sampling?

16. Here are the annual salaries for everyone at a small company from exercise 13 in section 3.2.

ID Number	Position	Salary
15938	Management	$100,000
94587	Management	$80,000

ID Number	Position	Salary
82400	Laborer	$30,000
28401	Laborer	$40,000
38427	Laborer	$35,000
58339	Laborer	$15,000
48927	Laborer	$25,000
60113	Laborer	$40,000

a.   A SRS of 4 salaries is drawn and the sample mean salary is calculated. This process is repeated 20 times and the 20 sample means are shown here:

43750   50000   63750   35000   40000   51250   46250   42500   30000   57500
43750   55000   26250   46250   61250   60000   45000   30000   30000   47500

Describe the sampling distribution of the sample mean salary for a SRS of size 4.

b.   A STRS of 4 salaries is drawn (1 salary randomly selected from the managers and 3 salaries randomly selected from the laborers) and the sample mean salary is calculated. This process is repeated 20 times and the 20 sample means are shown here:

40000   46250   40000   45000   43750   47500   45000   46250   42500   53750
41250   46250   46250   45000   48750   43750   47500   47500   47500   45000

Describe the sampling distribution of the sample mean salary for a STRS of size 4.

17.   Consider these two sampling scenarios from exercise 16 in section 3.2.

<u>Scenario I</u>: A beverage company makes ginger beer and bottles it in 12-oz. bottles. The bottles are shipped in cases of 24 bottles. Recently customers have been reporting that some bottles have not been sealed properly and the company wants to investigate by randomly selecting bottles to check from the most recent 1000 cases.

<u>Scenario II</u>: In order to estimate the proportion of the 24,000 people in a certain town who attend weekly religious services, an organization randomly selects people and surveys them.

a.   For one of the scenarios (I'm not telling you which one) the results of 1000 SRSs and 1000 CRSs are shown here. The relevant percentage was calculated for each sample and the distributions of the 1000 percentages are displayed for each sampling method. Compare the two sampling distributions.

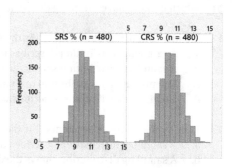

b.   For the other scenario, the results of 1000 SRSs and 1000 CRSs are shown here. The relevant percentage was calculated for each sample and the distributions of the 1000 percentages are displayed for each sampling method. Compare the two sampling distributions.

c.   Which scenario goes with part a.? Which scenario goes with part b.? Why?

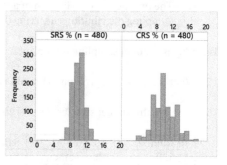

18. A large university wants to send an exit survey to a SRS of 500 of its 3,827 graduating seniors. One question is, "Would you recommend to others that they should attend the university?" The response choices are: not sure, not likely, likely, definitely, most definitely. Describe how you would (in theory) use repeated sampling to obtain the sampling distribution of the sample proportion who would "definitely" or "most definitely" recommend the university to others.

19. "Do you favor or oppose abortion in the third trimester of pregnancy?" A survey organization plans to ask a SRS of 1000 U.S. adults this question. Describe how you would (in theory) use repeated sampling to obtain the sampling distribution of the sample proportion who oppose third trimester abortion.

20. Here are the durations (in seconds) for a class of students to complete an online survey. Consider this the population.

129	178	156	288	116	150	479	112	152	187
146	191	175	200	166	269	145	271	180	383
326	198	312	137	157					

a. Find μ and σ and construct a histogram of the population durations.
b. Consider a SRS of 5 durations. How many possible SRS's are there?
c. A SRS of 5 durations is drawn and the sample mean duration is calculated. This process is repeated 30 times and the 30 sample means are shown here:

219.8	144.2	215.6	144.4	189.2	218.8	188.2	213.4	183.6	210.2
214.4	228.6	270.8	205.4	197.2	213.6	239.2	161.6	271.4	214.4
208.2	187.0	180.2	185.0	152.4	205.6	238.8	175.0	158.6	274.4

Construct a histogram of the simulated sampling distribution of the sample mean duration using these 30 sample means. Compare this distribution to the population distribution.

d. Now consider a SRS of 10 durations. How many possible SRS's are there?
e. A SRS of 10 durations is drawn and the sample mean duration is calculated. This process is repeated 30 times and the 30 sample means are shown here:

210.1	205.0	183.1	206.0	256.8	241.4	165.9	189.8	221.0	250.2
158.6	199.6	215.8	215.3	229.7	202.1	210.6	176.6	174.0	208.0
224.0	229.5	181.4	211.5	218.6	235.8	208.2	189.7	188.4	218.1

Construct a histogram of the simulated sampling distribution of the sample mean duration using these 30 sample means. Compare this distribution to the population distribution and to the sampling distribution in part c.

21. Here are the reported number of people texted since midnight for a class of students who responded to survey. Consider this the population.

2	6	3	1	3	1	5	10	1	5
2	1	2	8	2	2	2	4	4	5
6	1	10							

a. Find η, the population median, and construct a histogram of the population data.
b. Consider a SRS of 3 values. How many possible SRS's are there?
c. A SRS of 3 is drawn and the sample median is calculated. This process is repeated 30 times and the 30 sample medians are shown here:

8	3	8	1	6	3	2	2	1	2
4	2	2	2	2	3	3	2	2	4
1	3	3	5	2	5	2	1	2	2

Construct a histogram of the simulated sampling distribution of the sample median number of people texted using these 30 sample medians. Compare this distribution to the population distribution.

d.   Now consider a SRS of 14 values. How many possible SRS's are there?

e.   A SRS of 14 is drawn and the sample median is calculated. This process is repeated 30 times and the 30 sample medians are shown here:

2.5	2.5	2.5	2.0	3.5	2.5	3.0	2.5	2.5	4.0
3.0	4.5	3.0	3.0	3.0	3.0	3.0	2.0	3.0	2.0
2.0	2.5	2.5	3.0	2.5	2.5	2.5	3.5	2.5	2.0

Construct a histogram of the simulated sampling distribution of the sample median number of people texted using these 30 sample medians. Compare this distribution to the population distribution and to the sampling distribution in part c.

22. A box contains 4 bills in different denominations: $1, $5, $10, $20. Consider a SRS of 2 bills from the box.

a.   Verify that $\mu = 31$ and $\sigma \approx 34.785$.

b.   Use a tree diagram to find the sampling distribution of $\bar{X}$, the sample mean bill amount.

c.   Find and interpret $E(\bar{X})$ and compare to μ.

d.   Find $SD(\bar{X})$ and compare to σ.

23. Consider exercise 22 and a SRS of 3 bills from the box.

a.   Use a tree diagram to find the sampling distribution of M, the sample median bill amount.

b.   Find and interpret $E(M)$ and compare to η (the population median) and to μ,

24. Consider exercise 22.

a.   Verify that $\sigma^2 = 1210$.

b.   Use a tree diagram to find the sampling distribution of S^2, the sample variance of the bill amounts.

c.   Find $E(S^2)$ and compare to σ^2.

d.   Use a tree diagram to find the sampling distribution of S^{*2}, the sample variance of the bill amounts calculated by using n in the denominator instead of $n - 1$.

e.   Find $E(S^{*2})$ and compare to σ^2.

f.   Use your analysis to explain why we use $n - 1$ instead of n when calculating the sample variance.

25. Alana, Bill, Chuck, Denise, and Emily are in line at the post office. Alana, Chuck, and Emily need to purchase stamps; Bill and Denise are mailing packages. Consider a SRS of 3 people from the line and observe whether each one needs to purchase stamps or not.

a.   Use a tree diagram to find the sampling distribution of $\hat{P}$, the sample proportion of people who need stamps.

b.   Find and interpret $E(\hat{P})$.

c.   Find $SD(\hat{P})$.

26. Consider exercise 25 and a SRS of only 2 people.

a.   Use a tree diagram to find the sampling distribution of $\hat{P}$, the sample proportion of people who need stamps.

b.   Find $E(\hat{P})$ and compare to exercise 25.

c.   Find $SD(\hat{P})$ and compare to exercise 25.

27. Of the 14,000 U.S. stocks available at the time of this writing, about 9,000 are considered "small cap" companies whose market capitalization (total dollar value of outstanding shares) is under $2 billion. Consider a SRS of 3 U.S. stocks and observe whether each one is small cap or not.

a.   Use a tree diagram to find the sampling distribution of $\hat{P}$, the sample proportion of small cap stocks. Hint: The probabilities on the branches change from one selection to the next.
b.   Find $E(\hat{P})$ and $SD(\hat{P})$.
c.   Repeat parts a. and b. keeping the probabilities on the branches the same from one selection to the next.
d.   Did it matter much whether the branch probabilities stayed constant or not? Why do you think this is so?

28.   Consider exercise 27. Let's say there are only 14 U.S. stocks in the population, 9 of which are "small cap" stocks. Consider a SRS of 3 U.S. stocks from this much smaller population.

a.   Use a tree diagram to find the sampling distribution of $\hat{P}$, the sample proportion of small cap stocks. Hint: The probabilities on the branches change from one selection to the next.
b.   Find $E(\hat{P})$ and $SD(\hat{P})$.
c.   Repeat parts a. and b. keeping the probabilities on the branches the same from one selection to the next.
d.   Did it matter much whether the branch probabilities stayed constant or not? Why do you think this is so?

29.   Consider the example of the four employee salaries from the text. Suppose the first two ($100k and $80k) are for two managers and the other two ($20k and $40k) are for laborers. Instead of a SRS, consider a STRS of 2 employees, one manager and one laborer.

a.   Find the sampling distribution of $\bar{X}$, the sample mean salary. Hint: Use a tree diagram.
b.   Find $E(\bar{X})$ and $SD(\bar{X})$.
c.   Compare the expected value and standard deviation to those in the example for a SRS. What does this tell you about how a STRS and a SRS compare in this situation?

30.   Consider the example of the dinner party from the text. Instead of a SRS, consider a CRS of one couple to get a random sample of 2 people.

a.   Find the sampling distribution of $\hat{P}$, the sample proportion who favor term limits. Hint: Use a tree diagram.
b.   Find $E(\hat{P})$ and $SD(\hat{P})$.
c.   Compare the expected value and standard deviation to those in the example for a SRS. What does this tell you about how a CRS and a SRS compare in this situation?

31.   Consider the example of the dinner party from the text. Suppose, instead that Leon favors term limits but Sharon does not and that Bob favors term limits but Trisha does not.

a.   Find the sampling distribution of $\hat{P}$, the sample proportion who favor term limits, for a SRS of 2 people.
b.   Find $E(\hat{P})$ and $SD(\hat{P})$.
c.   Repeat parts a. and b. for a CRS of one couple to get a random sample of 2 people.
d.   What do your results tell you about how a CRS and a SRS compare in this situation?

5.2 Sampling Distributions of Counts and Proportions

When categorical variables are observed on objects, the natural summaries are counts (frequencies) and proportions (similar to percentages). In this section, we'll explore the sampling distributions of counts and proportions for random samples using the SRS method.

Sample Counts

Take a SRS of n objects from a population and observe whether or not each object has some feature of interest. Count how many do and let $X =$ number of objects in the sample with the feature.

Sound familiar? This scenario possesses some of the characteristics of the binomial experiment! Here are the binomial characteristics again:

Binomial Characteristics
n trials
success/failure outcomes on each trial
constant p
X = number of successes in n trials

The n trials are the n objects in the SRS. A single object is considered a "success" if it has the feature of interest and a "failure" otherwise. The proportion of objects in the population that have the feature is denoted p. It is the probability of selecting a single object at random from the population having the feature of interest. The random variable X is defined in just the right way: the number of objects in the sample with the feature.

The only slight problem in applying the binomial distribution to a SRS is that p technically doesn't stay the same from one object to the next. Since a SRS is done without replacement, the composition of the population (and hence p) changes from one object to the next. Fortunately, the binomial distribution is a very good approximation as long as the sample size (n) is very small compared to the population size (N). Our rule of thumb is that X will have a binomial distribution as long as $\frac{n}{N} < 0.05$. When this is the case, the changes in p from one object to the next will be negligible.

As we saw in chapter 4, when np and $n(1 - p)$ are both at least 15, the binomial distribution is "normal enough" to use the normal distribution as an approximation. (Since the binomial distribution is itself an approximation to the sampling distribution of X for a SRS, using the normal distribution would be an approximation of an approximation!) Don't worry, in those cases it's a very good approximation.

Example: Trisomy 18 (T18) is a genetic abnormality where a baby has an extra copy of chromosome 18 in some or all cells in the body. According to the Trisomy 18 Foundation (www.trisomy18.org), T18 occurs in about 1 in every 2500 pregnancies.[2] Suppose there are 6 million pregnant women in the U.S. That gives 2,400 pregnant women who are carrying a baby with T18. Take a SRS of 50,000 U.S. pregnant women. Find the sampling distribution of X = number who have a T18 baby out of the 50,000 and use it to find the probability that 30 or more women are carrying a T18 baby.

Solution: The binomial conditions are well-satisfied:

Binomial Characteristics	T18
n trials	trial = single pregnant woman; $n = 100$
success/failure outcomes on each trial	success = carrying a baby with T18
constant p	$p = 0.0004$ and remains *almost* constant
X = number of successes in n trials	X = number who have a T18 baby out of 100

Since $\frac{n}{N} = \frac{50,000}{6,000,000} = 0.008\bar{3} < 0.05$, X has (almost) a binomial sampling distribution with $n = 50,000$ and $p = \frac{2,400}{6,000,000} = 0.0004$. To find $P(X \geq 30)$, we could do this:

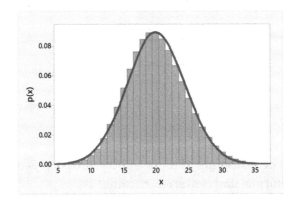

$$P(X \geq 30) = p(30) + p(31) + \cdots + p(50,000)$$

calculating each $p(x)$ using the binomial formula. That's too many calculations! To save some time we could do this instead:

$$P(X \geq 30) = 1 - P(X < 30) = 1 - [p(0) + p(1) + \cdots + p(29)]$$

but that still means doing 30 binomial formula calculations. To get around doing that, notice that the binomial distribution in this case looks fairly bell-shaped. Since $np = 50,000(0.0004) = 20 \geq 15$ and $n(1 - 0.0004) = 49,980 \geq 15$, X has an approximately normal sampling distribution with

$$E(X) = np = 50,000(0.0004) = 20$$

and

$$SD(X) = \sqrt{np(1 - p)} = \sqrt{50,000(0.0004)(1 - 0.0004)} \approx 4.47124,$$

so we can use the normal distribution and find $P(X \geq 30)$ as an area under the curve:

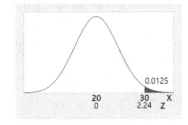

$$z = \frac{30 - 20}{4.47124} = 2.24$$

$$P(X \geq 30) = P(Z \geq 2.24) = 1 - 0.9875 = 0.0125$$

There is only a 1.25% chance of finding more than 30 T18 babies in a SRS of 50,000 pregnant women in the U.S.

Just to convince you that the normal model is a good one to describe the sampling distribution of X, here's a simulation of 1,000 SRS's of 50,000 each from a population of 6,000,000 pregnant women where 2,400 have babies with T18. For each sample the number with T18 babies is observed.

Sample #	X
1	14
2	15
3	13
...	...
1,000	19

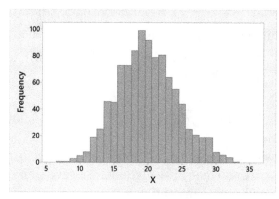

The histogram of these 1,000 values of X is nearly normal and centered at around 20.

The normal distribution is not always a good approximation to the sampling distribution of X as the next example illustrates.

Example: Consider the Trisomy 18 example with a SRS of only 100 pregnant women in the U.S. Find the sampling distribution of X = number who have a T18 baby out of the 100 and use it to find the probability that at least one woman is carrying a T18 baby.

Solution: Notice that since $\frac{n}{N} = \frac{100}{6,000,000} < 0.05$, X has (almost) a binomial sampling distribution with $n = 100$ and $p = \frac{2,400}{6,000,000} = 0.0004$. But since $np = 100(0.0004) = 0.04 < 15$, we know a normal distribution will not be a good approximation. Here's a picture of the sampling distribution of X. Since T18 is very rare, there are relatively few pregnant women with T18 babies in the population, so p is very small. For a sample size of 100, the distribution is severely right skewed; in fact most of the "action" is on values of 0 and 1.

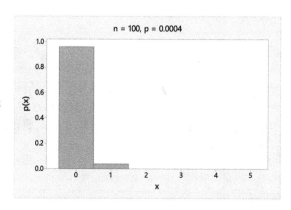

To find the probability that at least one woman is carrying a T18 baby in the SRS, we shouldn't use the normal distribution, but we can use the binomial formula:

$$P(X \geq 1) = 1 - P(X = 0) = 1 - \binom{100}{0}(0.0004)^0(1 - 0.0004)^{100} = 1 - 0.9608 = 0.0392.$$

There is only a 3.92% chance of finding at least one T18 baby in a SRS of 100 pregnant women in the U.S.

Sample Proportions

Here's a quick science quiz for you:

What does a light-year measure?

    A. Brightness   B. Time   C. Distance   D. Weight

If you said "Distance," you are correct, along with an estimated 72% of all U.S. adults. (A light-year is the distance light travels in a year. At a speed of about 186,000 miles per second, light will travel about 6 trillion miles in a year!) The survey results are based on a Pew Research Center study of 3,278 randomly-selected U.S. adults that was conducted between August 11 and September 3, 2014.[3] The 72% figure is the value of a statistic ($\hat{p} = 0.72$) that was based on the random sample of 3,278 adults. The parameter of interest was p = proportion of all U.S. adults who would have gotten the light-year question correct. At the time of the study, the U.S. population was around 319,000,000[4] and while not all of those were adults, the adult population was still in the hundreds of millions.

Do you think that exactly 72% of all U.S. adults would have gotten the light-year question right? If you answered "no, not exactly," then how far off is the 72% as an estimate? The answer lies in the sampling distribution of $\hat{P}$. What if the Pew Research Center took another survey of 3,278 randomly-selected U.S. adults at the same time? Would you expect them to find exactly 72% answering the light-year question correctly in that sample? The answer is likely "no" because there would be different people in that new sample. How different would the second sample's $\hat{p}$ be from 0.72? Again, the answer lies in the sampling distribution of $\hat{P}$. Exactly what does this sampling distribution look like? Let's find out.

Suppose there were 255,000,000 U.S. adults in the population as of August 2014. Also suppose that exactly 70% of them (178,500,000) would get the light-year question correct. Now take a simple random sample (SRS) of 3,278 of them and ask them the light-year question. Write down the sample proportion $\hat{p}$ who get it right out of 3,278. Repeat this process 1,000 times so that you end up with 1,000 $\hat{p}$'s. Here's a partial list of the results along with a histogram showing the distribution of the 1,000 $\hat{p}$'s.

Sample #	# who got the light-year question correct	$\hat{p}$
1	2276	0.694
2	2275	0.694
3	2282	0.696
⋮	⋮	⋮
1,000	2309	0.704

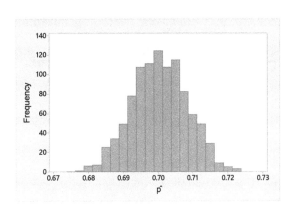

Notice that these sample proportions are centered at 0.70 and appear nearly normal. Did you expect that? Consider how you calculate a sample proportion: it's the number of objects with some feature (call that number X) divided by the number of objects in the sample:

$$\hat{P} = \frac{X}{n} = \frac{1}{n}X = aX + b$$

where $a = \frac{1}{n}$ and $b = 0$.

This is just a change of units of the binomial random variable X from a number between 0 and n to a proportion between 0 and 1. Using the rules for units changes (chapter 4) we can get the features of the sampling distribution of $\hat{P}$:

$$E(\hat{P}) = E\left(\frac{1}{n}X\right) = aE(X) = \frac{1}{n}E(X) = \frac{1}{n}(np) = p$$

$$Var(\hat{P}) = Var\left(\frac{1}{n}X\right) = a^2 Var(X) = \left(\frac{1}{n}\right)^2 Var(X) = \frac{1}{n^2}np(1-p) = \frac{p(1-p)}{n}$$

$$SD(\hat{P}) = \sqrt{Var(\hat{P})} = \sqrt{\frac{p(1-p)}{n}}$$

In words, these facts say that for SRS's, $\hat{p}$'s want to "hang out" around the population proportion of successes. For example, in the light-year example, sample proportions for a SRS of $n = 3{,}278$ will stack up around

$$E(\hat{P}) = p = 0.70$$

and vary with a standard deviation of

$$SD(P) = \sqrt{\frac{p(1-p)}{n}} = \sqrt{\frac{0.70(1-0.70)}{3278}} = 0.0080.$$

While the sampling distribution of $\hat{P}$ is not binomial, questions about $\hat{P}$ can be framed as questions about X which has a binomial distribution as long as $\frac{n}{N} < 0.05$. In addition, if np and $n(1-p)$ are both at least 15, $\hat{P}$ will have a nearly normal sampling distribution.

Here, $\frac{n}{N} = \frac{3{,}278}{255{,}000{,}000} \approx 0.00001 < 0.05$ and both

$np = 3278(0.70) = 2294.6 \geq 15$
$n(1-p) = 3278(1-0.70) = 983.4 \geq 15$

so the normal distribution with $E(\hat{P}) = 0.70$ and $SD(\hat{P}) = 0.0080$ will describe the sampling distribution of $\hat{P}$ very well.

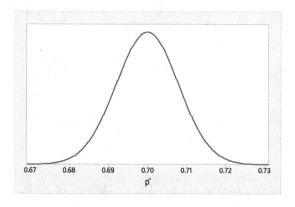

I want to point out something here before we move to the next example. A SRS of 3,278 U.S. adults that results in 72% getting the light-year question right is not very likely if we assume only 70% of all U.S. adults would get it right. (Note where 0.72 is in the sampling distribution of $\hat{P}$.) Perhaps our 70% assumption about the population is too low. What if we change the assumption from 70% to 71%? Now $p = 0.71$ but both

$$np = 3278(0.71) = 2327.38 \geq 15$$
$$n(1 - p) = 3278(1 - 0.71) = 950.62 \geq 15$$

so $\hat{P}$ will still have a nearly normal sampling distribution with

$$E(\hat{P}) = p = 0.71$$

and

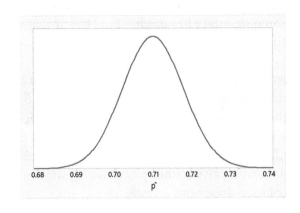

$$SD(\hat{P}) = \sqrt{\frac{p(1-p)}{n}} = \sqrt{\frac{0.71(1-0.71)}{3278}} = 0.0079.$$

Now a 72% sample result is not so unusual. In light of the sample data, it seems that $p = 0.71$ is a better choice for the population proportion than is $p = 0.70$. That last sentence is a generalization inference about the population parameter p and the sampling distribution of the statistic $\hat{P}$ was at the heart of it.

Summary

Shown below are the features of the sampling distributions of sample counts and sample proportions. These features are based on SRS's of size n from a population where p is the proportion of successes.

	X = number of successes	$\hat{P} = \frac{X}{n}$ = proportion of successes
Center†	$E(X) = np$	$E(\hat{P}) = p$
Variance*	$Var(X) = np(1 - p)$	$Var(\hat{P}) = \dfrac{p(1 - p)}{n}$
Standard Deviation*	$SD(X) = \sqrt{np(1 - p)}$	$SD(\hat{P}) = \sqrt{\dfrac{p(1 - p)}{n}}$
Notes*	Binomial Nearly normal if $np \geq 15$ and $n(1 - p) \geq 15$	Nearly normal if $np \geq 15$ and $n(1 - p) \geq 15$

* These assume that $\frac{n}{N} < 0.05$, that the sample size is less than 5% of the population size.

† Also holds for CRS's and certain STRS's.

To compute probabilities, here's my recommendation: first check the np and $n(1 - p)$ conditions. If they are both satisfied, you can use the normal distribution to compute probabilities using either the sampling distribution of X or $\hat{P}$, whichever one makes more sense to you. If the np and $n(1 - p)$ conditions are not both satisfied, use X and the binomial distribution to compute probabilities.

Example: Suppose that 56% of all hotel guests from luxury hotels to economy hotels receive complimentary breakfast.[5] Find the probability that in a SRS of 150 hotel guests, a majority got complimentary breakfast.

Solution: We have a SRS of $n = 150$ from a large population of hotel guests where the proportion who get complimentary breakfast is $p = 0.56$. Let X = number who got complimentary breakfast in the SRS of 150 hotel guests and let $\hat{P}$ = proportion who got complimentary breakfast in the SRS of 150 hotel guests. Both of these statistics are nearly normal because $np = 150(0.56) = 84$ and $n(1 - p) = 150(1 - 0.56) = 66$ are both 15 or more. A majority means $X > 75$ or $\hat{P} > 0.5$, depending on whether you think in terms of counts or proportions, respectively. Here is the solution done both ways:

X has a nearly normal distribution with

$$E(X) = np = 150(0.56) = 84$$

and

$$SD(X) = \sqrt{np(1 - p)} = \sqrt{150(0.56)(1 - 0.56)} = 6.07947$$

so we can use the normal distribution and find $P(X > 75)$ as an area under the curve:

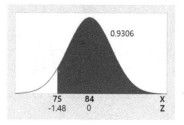

$$z = \frac{75 - 84}{6.07947} = -1.48$$

$P(X > 75) = P(Z > -1.48) = 1 - P(Z \leq -1.48) = 1 - 0.0694 = 0.9306.$

Here's an equivalent solution in terms of proportions. $\hat{P}$ has a nearly normal distribution with

$$E(\hat{P}) = p = 0.56$$

and

$$SD(\hat{P}) = \sqrt{\frac{p(1 - p)}{n}} = \sqrt{\frac{0.56(1 - 0.56)}{150}} = 0.04053$$

so we can use the normal distribution and find $P(\hat{P} > 0.5)$ as an area under the curve:

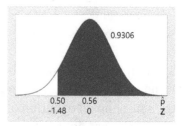

$$z = \frac{0.50 - 0.56}{0.04053} = -1.48$$

$P(\hat{P} > 0.50) = P(Z > -1.48) = 1 - P(Z \leq -1.48) = 1 - 0.0694 = 0.9306.$

Example: Consider the complimentary breakfast example. The 56% figure came from looking at all kinds of hotels, but suppose we assume that it also applies to *economy hotel* guests. Find the probability that, in a SRS of 10 economy hotel guests, 20% or fewer got complimentary breakfast. What would you conclude if you observed this kind of sample result?

Solution: We have a SRS of $n = 10$ from a large population of economy hotel guests where we assume the proportion who get complimentary breakfast is $p = 0.56$. Let X = number who got complimentary breakfast in the SRS of 10 economy hotel guests and let $\hat{P}$ = proportion who got complimentary breakfast in the SRS of 10 economy hotel guests. Neither of these statistics is nearly normal because $np = 10(0.56) = 5.6$ and $n(1 - p) = 10(1 - 0.56) = 4.4$ are not both 15 or more (although the distribution looks "bell-shaped," it's a bit too "blocky" for a smooth normal curve to fit well), so we'll use X and the binomial distribution. "20% or fewer" means $X \leq 2$.

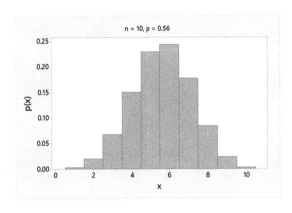

$$P(X \leq 2) = p(0) + p(1) + p(2)$$
$$= \binom{10}{0}(0.56)^0(1 - 0.56)^{10} + \binom{10}{1}(0.56)^1(1 - 0.56)^9 + \binom{10}{2}(0.56)^2(1 - 0.56)^8$$
$$= 0.0236$$

There's only a 2.36% chance of getting 2 or fewer guest who got complimentary breakfast in a SRS of 10 economy hotel guests. That's assuming the SRS comes from the population of all economy hotel guests where 56% get complimentary breakfast. If we were to see a sample like this, we would have reason to believe the overall 56% figure doesn't apply to economy hotels and that for economy hotels, fewer than 56% of all guests get complimentary breakfast.

Section 5.2 Exercises

1. For which of these scenarios would p be almost constant for a SRS?

    a. $N = 52, n = 5$
    b. $N = 1500, n = 15$
    c. $N = 200, n = 100$
    d. $N = 300,000,000, n = 2,000$

2. For which of these scenarios would p be almost constant for a SRS?

    a. $N = 100, n = 50$
    b. $N = 1,000,000, n = 600,000$
    c. $N = 100, n = 3$
    d. $N = 50,000, n = 1,000$

3. Let X = number of successes in a SRS of n objects from a large population. For each of the following, find $E(X)$, $Var(X)$, $SD(X)$ and indicate whether a normal distribution would be a good approximation to the shape.

    a. $p = 0.5, n = 10$
    b. $p = 0.5, n = 30$
    c. $p = 0.1, n = 50$
    d. $p = 0.9, n = 150$
    e. $p = 0.99, n = 2{,}000$

4.   Let X = number of successes in a SRS of n objects from a large population. For each of the following, find $E(X)$, $Var(X), SD(X)$ and indicate whether a normal distribution would be a good approximation to the shape.

    a. $p = 0.25, n = 5$
    b. $p = 0.75, n = 100$
    c. $p = 0.05, n = 200$
    d. $p = 0.85, n = 90$
    e. $p = 0.95, n = 300$

5.   Let $\hat{P}$ = proportion of successes in a SRS of n objects from a large population. For each of the following, find $E(\hat{P})$, $Var(\hat{P}), SD(\hat{P})$ and indicate whether a normal distribution would be a good approximation to the shape.

    a. $p = 0.3, n = 40$
    b. $p = 0.5, n = 100$
    c. $p = 0.7, n = 1{,}500$
    d. $p = 0.05, n = 50$
    e. $p = 0.8, n = 50{,}000$

6.   Let $\hat{P}$ = proportion of successes in a SRS of n objects from a large population. For each of the following, find $E(\hat{P})$, $Var(\hat{P}), SD(\hat{P})$ and indicate whether a normal distribution would be a good approximation to the shape.

    a. $p = 0.01, n = 5{,}000$
    b. $p = 0.2, n = 80$
    c. $p = 0.6, n = 55$
    d. $p = 0.75, n = 81$
    e. $p = 0.97, n = 1{,}000$

7.   Let X = number of successes in a SRS of 100 objects from a large population.

    a.   For each of these population proportions (0.1, 0.2, 0.3, 0.4, 0.5, 0.6, 0.7, 0.8, 0.9) calculate $Var(X)$.
    b.   Plot the $Var(X)$ versus the value of p.
    c.   For what population proportion is the sample number of successes most variable?

8.   Let $\hat{P}$ = proportion of successes in a SRS of 64 objects from a large population.

    a.   For each of these population proportions (0.05, 0.15, 0.25, 0.35, 0.45, 0.55, 0.65, 0.75, 0.85, 0.95) calculate $Var(\hat{P})$.
    b.   Plot the $Var(\hat{P})$ versus the value of p.
    c.   For what population proportion is the sample proportion of successes most variable?

9.   In a SRS of $n = 150$ objects from a large population having 80% successes, give the center, standard deviation, and shape (binomial or approximately normal) of the sampling distribution of

    a. X.
    b. $\hat{P}$.

10.   In a SRS of $n = 120$ objects from a large population having 40% successes, give the center, standard deviation, and shape (binomial or approximately normal) of the sampling distribution of

    a. X.

b. $\hat{P}$.

11. In a SRS of $n = 20$ objects from a large population having 10% successes, give the center, standard deviation, and shape (binomial or approximately normal) of the sampling distribution of

   a. X.
   b. $\hat{P}$.

12. In a SRS of $n = 150$ objects from a large population having 2% successes, give the center, standard deviation, and shape (binomial or approximately normal) of the sampling distribution of

   a. X.
   b. $\hat{P}$.

13. Suppose 60% of all people in the U.S. who use alternative medical treatments improve in their condition. You ask 10 of your friends who use alternative medical treatments about whether their condition has improved. Describe the sampling distribution of the sample number who say their condition improved out of 10.

14. Suppose 98% of all daily transactions at a bank are processed correctly. The bank selects the last 100 transactions at the end of the day on Friday. Describe the sampling distribution of the sample number of correctly-processed transactions out of the 100.

15. Let's say that 70% of all parents in a school district are in favor of a new policy being considered by the school board. A poll is posted on the school district's website and 678 parents respond. Describe the sampling distribution of the sample proportion in favor out of 678.

16. In a SRS of $n = 5$ objects from a large population having 60% successes, find

   a. $P(X \le 115)$.
   b. $P(\hat{P} > 0.87)$.
   c. $P(0.55 \le \hat{P} \le 0.65)$.
   d. the probability that the sample contains a majority of successes.

17. In a SRS of $n = 120$ objects from a large population having 40% successes, find

   a. $P(X \ge 50)$.
   b. $P(X < 35)$.
   c. $P(\hat{P} \le 0.50)$
   d. $P(0.39 \le \hat{P} \le 0.41)$

18. In a SRS of $n = 20$ objects from a large population having 10% successes, find

   a. $P(X < 3)$.
   b. $P(X > 3)$.
   c. $P(0.05 \le \hat{P} \le 0.15)$.
   d. $P(\hat{P} = 0.10)$.

19. In a SRS of $n = 50$ objects from a large population having 60% successes, find

   a. $P(X \ge 35)$.
   b. $P(0.44 \le \hat{P} \le 0.56)$.
   c. $P(\hat{P} \ge 0.52)$
   d. the probability that the sample does not contain a majority of successes.

20. Suppose that 40% of all the trees in a forest are ready for harvesting. The landowner selects a SRS of 100 latitude/longitude coordinates in her forest and determines whether the nearest tree to each location is ready for harvest.

    a. Describe the sampling distribution of the number of trees ready for harvesting in the SRS.
    b. Describe the sampling distribution of the proportion of trees ready for harvesting in the SRS.
    c. Find the probability that there are fewer than 28 trees in the sample are ready for harvesting.
    d. Find the probability that 49% or more of the trees in the sample are ready for harvesting.

21. Let's assume that 74% of all students at a university own their own car. Consider a SRS of 340 students.

    a. Describe the sampling distribution of the number of students who own their own car in the SRS.
    b. Describe the sampling distribution of the proportion of students who own their own car in the SRS.
    c. Find the probability that at least 240 students in the sample own their own car.
    d. Find the probability that at most 65% of the students in the sample own their own car.

22. Suppose 1 out of every 700 babies born in the U.S. has Down syndrome. Consider a SRS of 1,500 U.S. newborns.

    a. Describe the sampling distribution of the number of newborns with Down syndrome in the SRS.
    b. Describe the sampling distribution of the proportion of newborns with Down syndrome in the SRS.
    c. Find the probability that at most 2 newborns in the sample have Down syndrome.
    d. Find the probability that the proportion of newborns in the sample that have Down syndrome is between 0.002 and 0.004.

23. A variety of perennial ryegrass is advertised to have a 90% germination rate. Consider a one-pound bag of perennial ryegrass to be a SRS of 250,000 seeds. What is the probability that

    a. over 90.1% of the seeds in the bag will germinate?
    b. between 89.88% and 90.12% of the seeds in the bag will germinate?

24. Suppose that 39% of individual taxpayers in the U.S. prepare their own tax return.

    a. In a SRS of 250 U.S. taxpayers, what is the probability that fewer than 75 prepare their own tax return?
    b. If you found that only 40 out of a SRS of 250 students at a certain university prepare their own tax return, what would you think of the assumption that the 39% figure applies to that university? Explain.

25. According to the U.S. Department of Education, 10.8% of all borrowers in the federal student loan programs' fiscal year 2015 cohort defaulted on their loan. Consider a SRS of 300 federal student loan borrowers in the most recent cohort.

    a. Assuming that the 10.8% rate has not changed, what is the probability that more than 15% of the borrowers in the SRS default?
    b. What would you conclude if you actually found that only 15 borrowers in the SRS defaulted? Explain.

26. Consider exercise 25 and the SRS of 300 federal student loan borrowers in the most recent cohort. Let's say your SRS results in 15 borrowers who defaulted. Calculate a z-score for this sample result when the population default rate is

    a. 0.01.
    b. 0.02.
    c. 0.04.
    d. 0.06.
    e. 0.08.
    f. 0.10.
    g. A population default rate is considered reasonable if it results in $|z| < 2$ for the sample statistic. Using this criterion, which population default rates (a – f) are reasonable based on the data?

27. Suppose 85% of all eligible U.S. employees are currently contributing to a 401(k) retirement plan. You record the 401(k) contribution status (contributing or not) for a SRS of 275 workers at your company.

    a. Assuming that the national rate applies to your company, what is the probability that fewer than 220 of the employees in the SRS contribute?
    b. What would you conclude if you found that 215 employees in the SRS contribute? Explain.

28. Consider exercise 27 and the SRS of 275 employees from your company. Let's say your SRS resulted in 215 employees who contribute. Calculate the z-score for this sample result when the contribution rate for all employees at your company is

    a. 0.60.
    b. 0.70.
    c. 0.75.
    d. 0.80.
    e. 0.85.
    f. 0.90.
    g. A population default rate is considered reasonable if it results in $|z| < 2$ for the sample statistic. Using this criterion, which population default rates (a – f) for your company are reasonable based on the data?

29. Consider exercise 25 but with a SRS of only 30 federal student loan borrowers in the most recent cohort.

    a. Assuming that the 10.8% rate has not changed, what is the probability that 2 or fewer of the borrowers in the SRS default?
    b. What would you conclude if you actually found that 10 borrowers in the SRS defaulted? Explain.

5.3 Sampling Distribution of the Mean

In the last section, we considered sampling distributions of sample counts and sample proportions, statistics that are used to summarize categorical data. In this section, we'll shift the focus from categorical data to numeric data and do a deep dive into the properties of $\bar{X}$, the sample mean, for SRS's.

What values of $\bar{X}$ are typical for a SRS? What values are considered unusually high or low? These are the questions that we'll answer in this section.

How fast have you ever driven a car? I asked this question in a survey distributed to students on the first day of a statistics class one year and got 96 responses in miles per hour (mph). A histogram of the responses is shown here. Considering this set of speeds as a population, the mean and standard deviation are

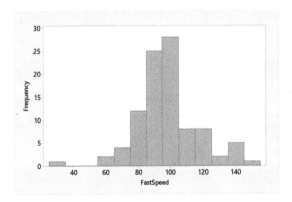

$\mu = 96.4375$

$\sigma \approx 19.0444.$

Now suppose you take a SRS of $n = 2$ of these speeds, calculate $\bar{x}$ for those two speeds, and repeat that process 1,000 times. Here's a diagram showing the process, a partial listing of the results and a histogram of the 1,000 sample means.

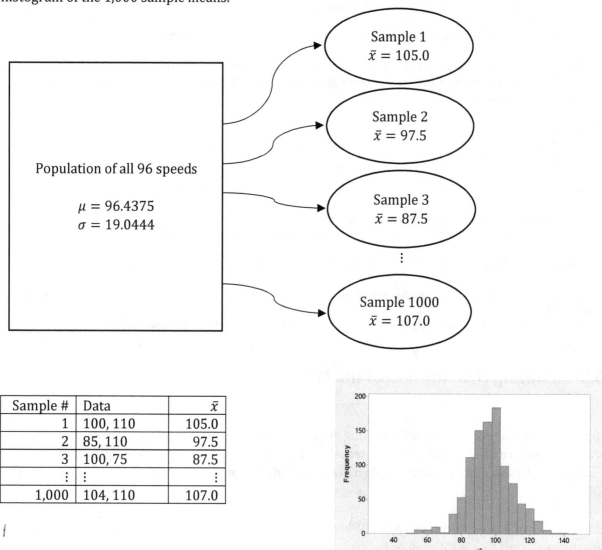

Sample #	Data	$\bar{x}$
1	100, 110	105.0
2	85, 110	97.5
3	100, 75	87.5
⋮	⋮	⋮
1,000	104, 110	107.0

The mean and standard deviation of these 1,000 sample means are 95.9 and 13.1, respectively. Notice that the sample means tend "hang out" around 96mph or so, where the population data are centered. However, the sample means vary a bit less than the population speeds. The smallest speed in the population is 30mph. (I wonder if that's really true or if someone just wanted to get their data value noticed!) Since there is only one of those, it is impossible to get a sample mean as small as 30; if 30 is chosen, the other value in the sample would have to be higher than 30 and therefore the sample mean would be higher than 30. The largest value in the population is 150. Since there is only one of those, it is impossible to get a sample mean as high as 150; if 150 is chosen, the other value in the sample would have to be lower than 150 and therefore the sample mean would be lower than 150. This is an important fact: *sample means vary less than the data values.*

Suppose we take samples of size $n = 10$ and repeat the process. Here is a partial listing of the results and a histogram of the 1,000 sample means.

Sample #	Data	$\bar{x}$
1	110, 105, 98, 140, 130, 95, 90, 80, 90, 125	106.3
2	85, 65, 100, 120, 140, 100, 98, 101, 100, 120	102.9
3	85, 120, 90, 150, 100, 100, 115, 100, 90, 102	105.2
⋮	⋮	⋮
1,000	90, 90, 75, 90, 80, 85, 95, 85, 120, 102	91.2

The mean and standard deviation of these 1,000 sample means are 96.4 and 5.6, respectively. Notice that the sample means still tend "hang out" around 96mph or so, where the population data are centered. But now the sample means vary even less! This illustrates another important concept: *the variability in sample means decreases as the sample size increases*. This concept may make more sense if you think about the extreme case: what if we took SRS's of $n = 96$ speeds? Each sample would be the same as the population data set so all the sample means would be 96.4375 and the standard deviation of the 1,000 sample means would be 0; there would be no variation from one $\bar{x}$ to the next.

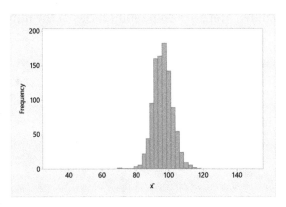

It turns out there is a nice model that describes the sampling distribution of $\bar{X}$ for SRS's. In the context of a SRS, denote the population mean as μ, the population variance as σ^2, and the population standard deviation as σ. If the sample size is small compared to the population size, each X_i randomly selected has expected value of $E(X_i) = \mu$, variance $Var(X_i) = \sigma^2$, and will be independent of other X_i's. That is, we let the population features of center and variability be the features of the distribution of single values randomly selected from the population. The sample mean is a linear combination of n of these random variables:

$$\bar{X} = \frac{X_1 + X_2 + \cdots + X_n}{n} = \frac{1}{n}X_1 + \frac{1}{n}X_2 + \cdots + \frac{1}{n}X_n$$

Using the tools from chapter 4 for linear combinations we have

$$E(\bar{X}) = \frac{1}{n}E(X_1) + \frac{1}{n}E(X_2) + \cdots + \frac{1}{n}E(X_n) = \frac{1}{n}\mu + \frac{1}{n}\mu + \cdots + \frac{1}{n}\mu = n\frac{1}{n}\mu = \mu$$

$$Var(\bar{X}) = \left(\frac{1}{n}\right)^2 Var(X_1) + \left(\frac{1}{n}\right)^2 Var(X_2) + \cdots + \left(\frac{1}{n}\right)^2 Var(X_n) = \frac{1}{n^2}\sigma^2 + \frac{1}{n^2}\sigma^2 + \cdots + \frac{1}{n^2}\sigma^2$$
$$= n\frac{1}{n^2}\sigma^2 = \frac{\sigma^2}{n}$$

and

$$SD(\bar{X}) = \sqrt{Var(\bar{X})} = \sqrt{\frac{\sigma^2}{n}} = \frac{\sigma}{\sqrt{n}}.$$

Those results confirm what we saw in the simulations: sample means are centered at the population mean, vary less than the population data, and vary less for larger sample sizes than for smaller ones.

In terms of shape, notice that the fastest speeds in the population are nearly normal and so are the sampling distributions of $\bar{X}$. You might wonder whether a normal model works in all cases...

Zillow.com is a website which displays homes for sale and lots of other goodies related to residential real estate. At Zillow's website, you can download county-level data on every county in the United States. One of the variables observed on each county is the number of homes listed on Zillow. Here are some summaries of the number of homes variable as of October 24, 2016.

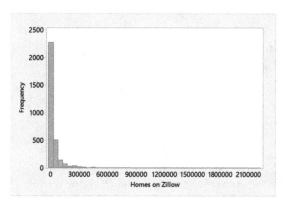

The first features that jumps out at me is the shape of this distribution: extremely right-skewed. There are a few counties with over a million listings on Zillow, corresponding to highly-populated metropolitan areas. The largest is 2,208,716 for Los Angeles county, California. The mean and standard deviation for this population of counties is

$$\mu = 36{,}890$$
$$\sigma = 97{,}235.$$

For a SRS of $n = 9$ counties, we can use our model to describe some features of the sampling distribution of $\bar{X}$, the mean number of homes on Zillow for those 9 counties:

$$E(\bar{X}) = \mu = 36{,}890$$

$$SD(\bar{X}) = \frac{\sigma}{\sqrt{n}} = \frac{97235}{\sqrt{9}} = 32411.667$$

For 1,000 SRS's of $n = 9$ counties, the shape of the sampling distribution of $\bar{X}$ is still quite right-skewed. Clearly a normal model would not work well. So does the shape of the sampling distribution of $\bar{X}$ always reflect the shape of the population data? The answer may surprise you.

Shown below are $\bar{X}$'s sampling distributions for SRS's of $n = 100$ and $n = 1{,}500$.

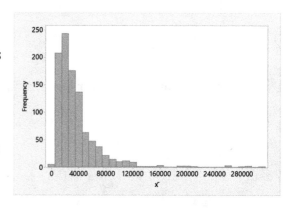

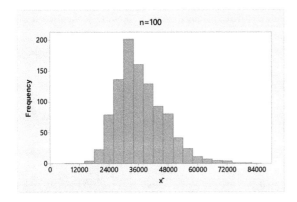
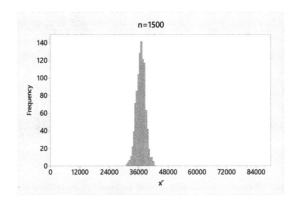

For sample sizes of 100, the sample means still show some right-skewness, but not nearly as much as the population data. For samples of size 1,500, not only is the distribution very symmetric, it is also nearly normal! The principle at work here is the same one that allowed us to use a normal distribution as an approximation to the binomial distribution in section 4.7. It is called the **Central Limit Theorem**: "central" because it is one of the more important results in probability and "limit" because we're seeing what happens to the shape as n gets larger and larger, the limit of the shape as n goes to infinity. The theorem states that the larger the sample size of a SRS, the more normal the shape of the distribution of $\bar{X}$ will get.

Of course, if the data values themselves are nearly-normal, then sample means have a normal distribution regardless of sample size. But *even if the data values themselves are not nearly-normal*, the Central Limit Theorem says that *sample means* stack up in an approximately normal shape over many SRS's as long as the sample size is large enough.

How large does the sample size have to be before the normal distribution works well as a model? That's a hard question to answer because it depends on how non-normal the data values are. The more skewed the data are, the larger the sample size has to be before the normal model works well. For fairly symmetric, unimodal, numeric data distributions, the convergence to normality happens pretty quickly so a sample size of 5 to 10 may be sufficiently large. For mildly-skewed populations, a sample size of 10 or 20 may be sufficiently large. For severely-skewed populations, the sample size may need to be several hundred or even over 1,000. The short answer is that there is no short answer.

Summary

Shown below are the features of the sampling distribution of sample means. These features are based on SRS's of size n from a population having mean μ and standard deviation σ.

	$\bar{X}$ = sample mean
Center†	$E(\bar{X}) = \mu$
Variance*	$Var(\bar{X}) = \dfrac{\sigma^2}{n}$
Standard Deviation*	$SD(\bar{X}) = \dfrac{\sigma}{\sqrt{n}}$
Shape*	Normal if: 1) population data are normal 2) n is sufficiently large

† Also holds for CRS's and certain STRS's.

* These assume that $\frac{n}{N} < 0.05$, that the sample size is less than 5% of the population size.

Example: Two-liter soda bottles are filled with 2 liters of soda. Actually, there are small differences in the fill amounts from bottle to bottle; you just don't notice them. Suppose the fill amounts for all soda bottles filled at a certain bottling plant are normally-distributed with a mean of 2000ml and a standard deviation of 2ml. In a SRS of 16 soda bottles from this plant, what is the probability that the mean fill amount is 1998.5ml or less? What would you conclude if, in a SRS of 16 bottles, you found the sample mean fill amount to be 1998.25ml?

Solution: Supposing the population of fill amounts is normal, the sample mean, $\bar{X}$, is also normal. If we suppose, as given, that $\mu = 2000$ and $\sigma = 2$ in the population, then

$E(\bar{X}) = \mu = 2000$

$SD(\bar{X}) = \dfrac{\sigma}{\sqrt{n}} = \dfrac{2}{\sqrt{16}} = 0.5$

so we can use the normal distribution and find $P(\bar{X} \leq 1998.5)$ as an area under the curve:

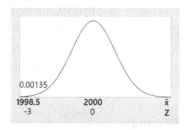

$z = \dfrac{1998.5 - 2000}{0.5} = -3.00$

$P(\bar{X} \leq 1998.5) = P(Z \leq -3.00) = 0.00135.$

There is only a 0.135% chance of getting a sample mean of 1998.5ml or less in a SRS from this population of bottles. If we were to actually observe something this unusual, like $\bar{x} = 1998.25$, we would question the assumptions about the population. Perhaps μ is less than 2000ml. Perhaps σ is greater than 2ml. Perhaps the fill amounts are not normal. The point is that *unusual outcomes from a SRS lead us to look for explanations other than simple random sampling.*

Example: Suppose that, at the end of last year, the average time employees in the U.S. had been with their current employer was 4.5 years and that the standard deviation of these times was 3.3 years. In a SRS of 1,100 U.S. employees last year, what is the probability that the average time spent with their current employer will be between 4.3 and 4.7 years?

Solution: Given that $\mu = 4.5$ years, $\sigma = 3.3$ years, and that times cannot be negative, the distribution of times is likely right-skewed. However, because of the large sample size, the distribution of the sample mean should be nearly normal. Also

$$E(\bar{X}) = \mu = 4.5$$

$$SD(\bar{X}) = \frac{\sigma}{\sqrt{n}} = \frac{3.3}{\sqrt{1100}} \approx 0.0995$$

so we can use the normal distribution and find $P(4.3 \leq \bar{X} \leq 4.7)$ as an area under the curve:

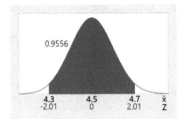

$$z = \frac{4.3-4.5}{0.0995} = -2.01 \quad \text{and} \quad z = \frac{4.7-4.5}{0.0995} = 2.01$$

$$P(4.3 \leq \bar{X} \leq 4.7) = P(-2.01 \leq Z \leq 2.01) = 0.9778 - 0.0222 = 0.9556.$$

It's a pretty good bet (95.56% chance) that the sample mean time spent with their current employer for a SRS of 1,100 U.S. employees last month will be between 4.3 and 4.7 years.

Section 5.3 Exercises

1. Consider this small population of responses to the question, "How much cash (in dollars) do you have on you right now?"

    100   10   50   0   4

    a. Find the population mean (μ) and standard deviation (σ) for these amounts.
    b. Make a histogram of the population of amounts.
    c. Complete this table showing all possible samples of size $n = 2$ from this population of amounts.

Sample	100,10	100,50	100,0	100,4	10,100					
$\bar{x}$	55	75								
Sample										
$\bar{x}$										

    d. Find the mean and standard deviation for the population of the 20 possible sample means.
    e. Why isn't the standard deviation of the sample means equal to $\sigma/\sqrt{2}$?
    f. Make a histogram of the 20 possible sample means.
    g. Compare the features of the population to the features of the sampling distribution of the mean.

2. Consider this small population of responses to the question, "How many siblings do you have?"

    2   1   3   0

    a. Find the population mean (μ) and standard deviation (σ) for these data.
    b. Make a histogram of the population.
    c. Complete this table showing all possible samples of size $n = 2$ from this population.

Sample	2,1	2,3	2,0	1,2		
$\bar{x}$	1.5	2.5				
Sample						
$\bar{x}$						

    d.   Find the mean and standard deviation for the population of the 12 possible sample means.

    e.   Why isn't the standard deviation of the sample means equal to $\sigma/\sqrt{2}$?

    f.   Make a histogram of the 12 possible sample means.

    g.   Compare the features of the population to the features of the sampling distribution of the mean.

3.   Consider the following sampling scenarios.

        I.   SRS of 2 from a population of 15

        II.   SRS of 2,000 from a population of 300,000,000

        III. SRS of 2,000 from a population of 10,000

        IV. STRS of 25 from a population of 100,000

        V.   CRS of 1,500 from a population of 300,000,000

    a.   For which of these scenarios would $E(\bar{X}) = \mu$?

    b.   For which of these scenarios would $SD(\bar{X}) = \frac{\sigma}{\sqrt{n}}$?

4.   Consider the following sampling scenarios.

        I.   SRS of 10 from a population of 50

        II.   SRS of 2,000 from a population of 3,000

        III. SRS of 2,000 from a population of 10,000,000

        IV. STRS of 1,600 from a population of 250,000

        V.   CRS of 16 from a population of 500

    a.   For which of these scenarios would $E(\bar{X}) = \mu$?

    b.   For which of these scenarios would $Var(\bar{X}) = \frac{\sigma^2}{n}$?

5.   Choose the symbol from this list that matches each description: μ $\bar{X}$ σ S $\frac{\sigma}{\sqrt{n}}$

    a.   the standard deviation of sample data

    b.   the mean of population data

    c.   the standard deviation of sample means

    d.   the standard deviation of population data

    e.   the mean of sample data

6.   Choose the symbol from this list that matches each description: μ $\bar{X}$ σ S $\frac{\sigma}{\sqrt{n}}$

    a.   a measure of center for sample data

    b.   a measure of variation for population data

    c.   a measure of how sample data vary

    d.   a measure of center for population data

    e.   a measure of how sample means vary

7.   Which of these best describes the sampling distribution of $\bar{X}$ for SRS's?

    a.   The sampling distribution of $\bar{X}$ describes the distribution of sample data.

    b.   The sampling distribution of $\bar{X}$ describes the variability of all values in the population.

    c.   The sampling distribution of $\bar{X}$ describes how sample means vary.

    d.   The sampling distribution of $\bar{X}$ describes the variability of all values in the sample.

Study these maybe? (handwritten margin note)

8. Which of these best explains the Central Limit Theorem for SRS's?

    a. As the sample size increases, the sample data become more closely normally distributed.
    b. As the population size increases, the sample data become more closely normally distributed.
    c. As the sample size increases, the sample mean gets closer to the population mean.
    d. As the sample size increases, the standard deviation of the sample mean gets closer to $\frac{\sigma}{\sqrt{n}}$.
    e. As the sample size increases, the distribution of sample means becomes more closely normally distributed.
    f. As the sample size increases, the distribution of the population data becomes more closely normally distributed.

9. Determine whether each statement about SRS's is true or false. Fix the false ones.

    a. Sample data vary less for larger sample sizes.
    b. The standard deviation of sample means is equal to the sample standard deviation.
    c. The center of the distribution of sample means is equal to the population mean.
    d. When $n = N, SD(\bar{X}) = 0$.
    e. There is never a situation where $Var(\bar{X}) = \sigma^2$.

10. Let $\bar{X}$ = sample mean in a SRS of n objects from a large population. For each of the following, find $E(\bar{X}), Var(\bar{X}),$ $SD(\bar{X})$ and indicate whether a normal distribution would be a good approximation to the shape.

    a. normal population, $\mu = 3.5, \sigma = 0.8, n = 4$
    b. right-skewed population, $\mu = 50, \sigma = 42, n = 5$
    c. symmetric, unimodal population, $\mu = 22, \sigma = 4, n = 20$
    d. severely skewed population, $\mu = 97, \sigma = 410, n = 50$

11. Let $\bar{X}$ = sample mean in a SRS of n objects from a large population. For each of the following, find $E(\bar{X}), Var(\bar{X}),$ $SD(\bar{X})$ and indicate whether a normal distribution would be a good approximation to the shape.

    a. severely right-skewed population, $\mu = 183, \sigma = 106, n = 35$
    b. normal population, $\mu = 515, \sigma = 116, n = 50$
    c. symmetric, unimodal population, $\mu = 10, \sigma = 3, n = 14$
    d. mildly left-skewed population, $\mu = -16, \sigma = 22, n = 100$

12. In a sample of 52 volunteer adults, each person was given a dose of melatonin just before bedtime. The time (minutes) before each person fell asleep was recorded. Suppose that the mean time to fall asleep for all adults without taking melatonin is 21 minutes and the standard deviation is 8 minutes.

    a. Describe the sampling distribution of the sample mean time to fall asleep.
    b. Would it be valid to compare the mean time to fall asleep for a SRS of adults to 21? Explain.

13. Suppose the average age in the U.S. population of adults is 38.5 years and the standard deviation is 23.0 years. Describe the sampling distribution of the sample mean age for the sample of U.S. Senators.

14. A global rental car company assesses its fleet by periodically inspecting a sample of cars. Inspections are quantified on a scale where 100 is a perfect score. Suppose the average score for all cars in the fleet is 82 and the standard deviation is 27. The company randomly selects 9 of its locations and inspects all 210 cars currently located at those locations. Describe the sampling distribution of the sample mean score.

15. There are 200 product-oriented businesses and 800 service-oriented businesses in a state. The average debt-to-income ratio for all 1000 businesses is 0.344 and the standard deviation is 0.083. A business analyst takes a STRS of 10 product-oriented businesses and 40 service-oriented businesses. Describe the sampling distribution of the sample average debt-to-income ratio.

16. The mean GPA at a university is 3.14 and the standard deviation is 0.88. A calculus instructor takes a survey of the 86 students in her class and each student reports their current GPA. Describe the sampling distribution of the sample average GPA.

17. In a SRS of $n = 49$ from a large population having a mean of 75.2 and a standard deviation of 15.9, find

    a. $P(\bar{X} > 80)$.
    b. $P(72 \leq \bar{X} < 74)$.
    c.   the probability that the sample mean is at most 79.9.
    d. a and b so that $P(a < \bar{X} < b) = 0.68$.

18. In a SRS of $n = 9$ from a large population having a mean of -15.5 and a standard deviation of 20.3, find

    a. $P(\bar{X} < -5)$.
    b. $P(-20 \leq \bar{X} \leq 0)$.
    c.   the probability that the mean for the sample is at least 0.
    d. b so that $P(\bar{X} \leq b) = 0.975$.

19. Suppose the fastest speed someone has driven a car has a mean of 104.1 mph and a standard deviation of 26.7 mph in a large population of U.S. drivers.

    a.   Find the probability that a randomly-selected U.S. driver has driven over 90 mph.
    b.   Find the probability that the average fastest speed in a SRS of 25 U.S. drivers is over 90 mph.
    c.   Assuming that the U.K. population of drivers looks like the U.S. population, what is the probability that the average fastest speed in a SRS of 25 U.K. drivers is at most 90 mph.
    d.   What would you conclude if you observed $\bar{x} = 89.5$ in a SRS of 25 U.K. drivers? Explain.

20. The average number of children under 18 per family in the U.S. is currently 1.8. Suppose the standard deviation is 2.2.

    a.   In a SRS of 1,500 families, what is the probability that the mean number of children under 18 is more than 2?
    b.   Give an interval where we'd expect most of the possible sample means to be for SRS's of 1,500 families.
    c.   Assuming that the French population looks like the U.S. population, what is the probability that the mean number of children in a SRS of 1,500 French families is 1.6 or less?
    d.   What would you conclude if you observed a sample mean of 1.57 in a SRS of 1,500 French families?

21. A potato chip manufacturer wants to monitor the average weight of product in bags of chips. To do so it selects a SRS of 30 bags from the production line each hour. The average weight of the most recent sample was 12.05 ounces.

    a.   If the mean product weight in bags is supposed to be 12 ounces and the standard deviation is 0.3 ounces, what is the probability of getting a mean of 12.05 ounces or more in a SRS of 30 bags?
    b.   Would your answer to part a. hold if the distribution of product weights was skewed? Explain.

22. Mastitis in cows is an inflammation of the udder caused by bacteria. Veterinarians will often use antibiotics to treat cows with mastitis. Milk from cows with mastitis is not to be mixed with milk from healthy cows until the cow has recovered. Acceptable antibiotic levels in batches of milk have a mean of 20 μg/kg and a standard deviation of 3.2 μg/kg. To determine whether the average antibiotic level in batches of raw milk collected from local farmers is at an acceptable level, a SRS of 10 local farmers is selected and the most recent batch of milk from each is tested. The mean antibiotic level was 20.51 μg/kg.

    a.   What's the chance of getting a mean level at least this high in a SRS of 10 batches with acceptable levels?
    b.   What assumption did you make to obtain your answer in part a.?
    c.   Would your answer to part a. hold if the distribution of acceptable levels was skewed? Explain.

23. Suppose that passengers on flights average 185 pounds and that those weights have a standard deviation of 35 pounds.

    a.   Do you think the distribution of passenger weights is normal? Explain.

b. Find the probability that the average weight in a SRS of 162 passengers (the capacity of a Boeing 737 with a two-class configuration) is under 190 pounds?

c. Find the probability that the total weight in a SRS of 162 passengers is over 31,000 pounds?

24. Suppose the average smartphone user spends $17.73 on games per year and that the standard deviation of amounts spent on games is $20.

a. What do I really mean by "average smartphone user spends $17.73 on games per year?"

b. In a simple random sample of 100 smartphone users, what is the probability that the average amount spent on games in a year is more than $20?

c. Consider a SRS of 100 Canadian smartphone users yields $\bar{x} = \$12.62$. Suppose that the standard deviation of yearly amounts spent on games in the Canadian smartphone user population is also $20. Would $17.73 be considered a reasonable value for the Canadian population average given these data? Explain.

d. Give a reasonable range of values for the Canadian population average yearly amount spent on games given these data ($\bar{x} = \$12.62$ from a SRS of 100).

25. In chapter 3 you learned that a stratified random sample (STRS) almost always results in estimates that are less variable and more consistent than those from a SRS. Consider a population having standard deviation $\sigma = 10$. The population is divided into 4 strata. The strata have standard deviations $\sigma_1 = 6, \sigma_2 = 5, \sigma_3 = 9$, and $\sigma_4 = 7$.

a. For a SRS of size $n = 16$, let the sample mean $\bar{X} = \frac{X_1+X_2+\cdots+X_{16}}{16} = \frac{1}{16}X_1 + \frac{1}{16}X_2 + \cdots + \frac{1}{16}X_{16}$ be an estimate of the population mean based on the SRS. Find the variance and standard deviation of $\bar{X}$ for the SRS.

b. For a STRS of size $n = 16$, a SRS of size 4 is taken in each stratum. Let $\bar{X}_1, \bar{X}_2, \bar{X}_3$, and $\bar{X}_4$ be the sample means in each strata. Find the variance and standard deviations of the four sample means.

c. Let $\bar{\bar{X}} = \frac{\bar{X}_1+\bar{X}_2+\bar{X}_3+\bar{X}_4}{4} = \frac{1}{4}\bar{X}_1 + \frac{1}{4}\bar{X}_2 + \frac{1}{4}\bar{X}_3 + \frac{1}{4}\bar{X}_4$ be an estimate of the population mean based on the STRS. Find the variance and standard deviation of $\bar{\bar{X}}$ for the STRS.

26. Repeat exercise 25 using a SRS of size $n = 400$ and a STRS of size $n = 400$ (SRS of 100 in each stratum).

27. One reason I recommend that you consult a statistician when using a sampling method other than a SRS is that things can go wrong if you're not very careful. Here's a summary of the hospital patient satisfaction survey from exercise 15 in section 3.2:

A hospital conducts a patient satisfaction survey of its "critical units:" the emergency department (ED), the neo-natal intensive care unit (NICU), and the medical intensive care unit (MICU). The hospital selects a stratified random sample of 40 patients in each unit. Shown below are the actual population figures.

	ED	NICU	MICU
Number of patients	1000	300	700
Percent satisfied	20%	80%	40%

Let p be the proportion of all patients in these three critical units were satisfied overall with their level of care. Let $\hat{P} = \frac{X_{ED}+X_{NICU}+X_{MICU}}{120} = \frac{1}{120}X_{ED} + \frac{1}{120}X_{NICU} + \frac{1}{120}X_{MICU}$ be an estimate of p where the X's are the numbers of satisfied patients in each sample.

a. In this somewhat rare example, you can find the value of p. What is it?

b. It can be shown that $E(X_{ED}) = 200, E(X_{NICU}) = 240$, and $E(X_{MICU}) = 280$. Find $E(\hat{P})$.

c. Compare your answers from parts a. and b. Why are they different?

d. Let $\tilde{P} = \frac{1000}{2000}\hat{P}_{ED} + \frac{300}{2000}\hat{P}_{NICU} + \frac{700}{2000}\hat{P}_{MICU}$ be a weighted estimate of p that weights the sample proportions in each stratum according to the sizes of each stratum. Find $E(\tilde{P})$.

28. Consider exercise 15 in section 3.2 about the hospital patient satisfaction survey. Suppose that instead of selecting a STRS of 40 patients in each unit, the hospital samples proportionately to the unit sizes: a sample of 60 patients from the ED, 18 patients from the NICU, and 42 patients from the MICU.

a.   Show that the proportions of patients in each sample match the proportions of patients in each unit.
b.   Consider again the unweighted estimate $\hat{P} = \frac{X_{ED}+X_{NICU}+X_{MICU}}{120} = \frac{1}{120}X_{ED} + \frac{1}{120}X_{NICU} + \frac{1}{120}X_{MICU}$. Find $E(\hat{P})$.

29. Consider exercises 27 and 28. When using a STRS, what are two ways to ensure that the expected value of an estimator is correct?

Notes

[1] University of Delaware Faculty Climate Survey (2018).

[2] www.trisomy18.org.

[3] Funk, C. and Goo, S. K. (2015), *A Look at What the Public Knows and Does Not Know about Science*, Pew Research Center.

[4] www.census.gov/popclock.

[5] Based on the 2016 J.D. Power and Associates North American Guest Satisfaction Survey, www.jdpower.com.

Chapter 6: Hypothesis Testing

6.1 Introduction to Hypothesis Testing

Is the popularity of football ("American football" to most of the world) waning? The Gallup organization reported that, in the mid-2000's, football was the favorite sport to watch for 43% of all U.S. adults. Has that figure dropped since then? A recent Gallup poll conducted on a simple random sample of 1,049 U.S. adults showed that 388 (about 37%) said that football is their favorite sport to watch.[1]

Do you think the survey data convincingly show that football has slipped in popularity in the U.S.? The random sampling process permits a generalization inference from the sample of survey data to the population of all U.S. adults because it gives the best chance that the sample of U.S. adults represent the population of all U.S. But now there is uncertainty introduced by that random mechanism. Is the 37% just an artifact of random sampling or has the population really changed in their preferences since the mid-2000's? It turns out there is a way to evaluate these two competing descriptions about U.S. adults' preference for watching football by using the scientific method:

1. Pose a Question
2. Make a Hypothesis
3. Collect Data
4. Analyze the Data
5. Make a Conclusion

Let's consider each of these pieces. In statistics, the application of the scientific method is called a **hypothesis test**.

1. *Pose a Question*: The scientific method always begins with a question about the universe (or a part of it) and how it works. The question also indicates some hunch about how someone thinks the universe works. We'll call that someone the **researcher**. Sometimes the question is actually about the universe. Centuries ago Copernicus posed this question: "Does the Earth really revolve around the Sun and not the other way around?" Sometimes questions are about only a small part of the universe (like the group of all U.S. adults). Our hunch about football's popularity is:

   "Has the proportion of all U.S. adults whose favorite sport to watch is football decreased since the mid-2000's?"

   There is no question that is off-limits. As a researcher, you are free to ask any question you like, regardless of how it is viewed by others. The scientific method does not put restrictions on the kinds of questions that are allowed.

2. *Make a Hypothesis*: The researcher's hypothesis is a statement that reflects the researcher's question. It is a statement that the researcher hopes to support with data. We refer to this

hypothesis as the **research hypothesis** or the **alternative hypothesis**; we'll use the two terms interchangeably. For example, Copernicus' hypothesis was this: The Earth revolves around the Sun. Gallup's hypothesis about U.S. adults' sports preferences is: The proportion of all U.S. adults whose favorite sport to watch is football has decreased since the mid-2000's. The term *research hypothesis* makes sense because it reflects the *research* question. However, the term *alternative hypothesis* is a bit strange because it begs the question, "Alternative to what?" The other hypothesis is called the **null hypothesis**. It is a statement that contradicts the research hypothesis. It states what the researcher hopes to discredit. It may be a prevailing theory at the time, like in the case of Copernicus. It may, like the word "null" suggests, be a statement of no difference, no effect, or no relationship. It may be a claim someone else has made counter to what the researcher hopes to support, like in the case of your roommate's guessing claim. We often abbreviate the term null hypothesis by using the symbol H_0 and the research or alternative hypothesis with H_a. Here are both hypotheses for Copernicus:

H_0: The Sun revolves around the Earth.
H_a: The Earth revolves around the Sun.

Here they are about football:

H_0: Football is currently the favorite sport to watch for 43% of all U.S. adults.
H_a: The proportion of all U.S. adults whose favorite sport to watch is football is less than 0.43.

How do you know which is which? The alternative/research hypothesis always reflects the researcher's question. It is the statement that the researcher wants to support. Support for the alternative hypothesis comes from data and that takes us to the next step.

3. *Collect Data*: Both the kind of data and the way the data are collected should give the researcher the best opportunity of supporting the alternative hypothesis. By no means should the data collection process be rigged or fudged to favor H_a, but neither should it be haphazardly or carelessly done. If the research question is about a population, care should be taken to obtain a random sample of data so that a generalization inference is valid. If the research question is about the effectiveness of some intervention, the treatments should be randomly assigned so that a causal inference is valid. See chapter 3 for much more detail about collecting high-quality data.

In the case of football, we're trying to generalize from a sample (the 1,049 adults) to a population (all U.S. adults). The SRS of $n = 1,049$ resulted in $\hat{p} = \frac{388}{1,049} \approx 0.37$.

If it is easy to do so, we often write the hypotheses in terms of the appropriate probability model for the data. From chapter 5, we know that under certain conditions, $\hat{P}$, the sample proportion, has a nearly normal distribution centered at p, the population proportion of all U.S. adults who currently prefer football. In this case, we can write the hypotheses in terms of p:

H_0: $p = 0.43$ (no change since the mid-2000's)
H_a: $p < 0.43$ (a drop since the mid-2000's)

4. *Analyze the Data*: How much support do the data give for the research hypothesis? The answer to that question comes from a somewhat unusual procedure. The researcher assumes, for the moment, that the null hypothesis (the hypothesis that contradicts her hypothesis) is true. Under that assumption, she finds the distribution of a relevant statistic (a summary of the data). This distribution is called the **null distribution** because it reflects "null hypothesis world." It shows what values of the statistic are typical and what values are not if the null hypothesis is true. Then she compares the value of the statistic she calculated from her data to the null distribution and measures how unusual it is.

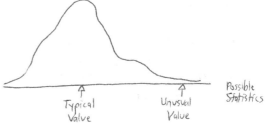

The figure here shows a generic null distribution of possible statistics. Values close to the center are considered typical. Values far from the center are considered unusual.

Let's consider the football scenario. A relevant statistic that summarizes the data is $\hat{P}$ = proportion who prefer football out of 1,049. Let's also assume, for the moment, that the null hypothesis is true (that there's been no change since the mid-2000's). We have a SRS from a large population and both $np = 1,049(0.43) = 451.07 \geq 15$ and $n(1 - p) = 1,049(1 - 0.43) = 597.93 \geq 15$. That means $\hat{P}$, the sample proportion who prefer football, has a nearly-normal distribution. Here's a picture of the null distribution of $\hat{P}$.

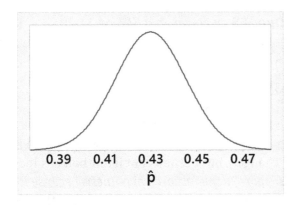

If there's been no change since the mid-2000's, you'd expect 43% in the SRS to prefer football since

$$E(\hat{P}) = p = 0.43.$$

Of course not every SRS of 1,049 would produce exactly 43% who prefer football (that would be 451.07 people...impossible!); there is variability around the center which we can quantify with the standard deviation:

$$SD(\hat{P}) = \sqrt{\frac{p(1-p)}{n}} = \sqrt{\frac{0.43(1-0.43)}{1,049}} \approx 0.0153$$

It's easy to see from the picture that getting 37% ($\hat{p} = 388/1{,}049$) who prefer football in the SRS is highly unlikely! Here are a couple of ways to measure exactly how unlikely it is. A z-score for $\hat{p} = 388/1{,}049$ is

$$z = \frac{\hat{p} - E(\hat{P})}{SD(\hat{P})} = \frac{\frac{388}{1{,}049} - 0.43}{0.0153} = -3.93.$$

If there's been no change since the mid-2000's, getting 388 who prefer football in a SRS of 1,049 is not only lower than we'd expect, it is 3.93 standard deviations lower! That is extremely unlikely if things haven't changed, but perhaps not if football has indeed lost some popularity. *The z-score gives strong support to the research hypothesis <u>because it is so unusual-looking compared to the null hypothesis</u>.* That should make you look for explanations other than "randomness under the status quo" for the survey data.

Another approach to measuring how unlikely a 37% result in a SRS of 1,049 would be under the status quo is to calculate a probability called the *p-value* ("p" for probability) that measures how likely the data are under the null hypothesis. You can think of the p-value as telling you how likely it would be to find data like the researcher's data in null hypothesis world. The **p-value** is defined as the probability of finding *at least as much evidence* for the research hypothesis as the researcher's data in a null hypothesis world. For the 37% survey result, finding the p-value would involve finding the area under the normal distribution below 0.37:

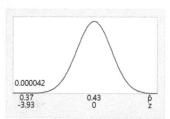

$$p - value = P(getting\ 388\ or\ fewer\ in\ a\ SRS\ of\ 1{,}049)$$
$$if\ p = 0.43$$
$$= P\left(\hat{P} \le \frac{388}{1{,}049}\right) if\ p = 0.43$$
$$= P(Z \le -3.93) = 0.000042$$

Getting 388 or fewer who prefer football in a SRS of 1,049 is a result that is not only lower than we'd expect, it would happen for only about 42 out of every million SRS's! That is extremely unlikely if things haven't changed, but perhaps not if football has indeed lost some popularity. *The p-value gives strong support to the research hypothesis <u>because it is so small under the null hypothesis</u>.* That should make you look for explanations other than "randomness under the status quo" for the survey data. *Small p-values mean the data are unlikely under the null hypothesis and therefore give support to the alternative hypothesis. The smaller the p-value, the more support there is in the data for H_a.*

5. *Make a Conclusion:* In most cases, the data show some support of the research hypothesis; the formal data analysis will tell us exactly just how much support there is. The smaller the p-value, the more support there is in the data for H_a because a small p-value means that the data are unlikely to have come about from a universe where H_0 is true. If you think the p-value is too small, then you would reject the null hypothesis and conclude that the data convincingly support the research hypothesis. How small is "too small?" A typical threshold used in many studies is 1/20 or 0.05 but you may encounter smaller or larger thresholds used depending on the study. The threshold you use is called the **significance level** of the test and is denoted by the Greek letter *alpha* (denoted α). A p-value that is less than α reflects data that are unusual under the null hypothesis. Such data are called **statistically significant**.

If you don't think the p-value is too small, then the results are inconclusive. While the data may show some support of the research hypothesis, it is also somewhat likely that they could have come about from a universe where H_0 is true.

In the football context, a p-value of 0.000042 is small enough to convince almost anyone that football has lost some popularity. You would reject the null hypothesis

H_0: $p = 0.43$ (no change since the mid-2000's)

in favor of

H_a: $p < 0.43$ (a drop since the mid-2000's).

Your conclusion might go something like this: "I have convincing evidence in the data that fewer than 43% of all U.S. adults currently choose football as their favorite sport to watch. The data show enough support (in this case very strong support!) for the research hypothesis.

Jury Trial Analogy

There is a helpful analogy between a hypothesis test and a trial by jury in the U.S. criminal justice system. A defendant is brought to trial because there is some evidence that they committed a crime. The prosecutor plays the role of the researcher and the question they are trying to answer is, "Is the defendant guilty of the alleged crime?" The hypotheses are

H_0: defendant is innocent
H_a: defendant is guilty.

Notice how the alternative hypothesis reflects the prosecutor's question. The evidence against the defendant is the data and is presented to the jury during the trial. After both the prosecution and defense finish presenting their cases, the jury deliberates – analyzes the data – to determine whether there is enough evidence against the defendant. The threshold for "enough evidence" is evidence that is convincing of guilt "beyond a reasonable doubt." If the jury determines that there is enough evidence against the defendant, H_0 is rejected and the verdict is "guilty." If not, the evidence is deemed inconclusive so the verdict is "not guilty."

Just as in a hypothesis test, the criminal justice system assumes that the null hypothesis is true and weighs the evidence to the contrary: "innocent until proven guilty." The defendant is afforded several protections to that end. A preliminary group of people called a grand jury must decide whether there is enough evidence to indict the defendant and proceed to trial. Evidence obtained by certain means (like an illegal search or a coerced confession) can be excluded by the judge. And the verdict must be unanimous; one person on the jury can stick to their minority position which results in a mistrial and the trial must start over with a new jury.

Hypothesis Test	Trial by Jury
Researcher	Prosecutor
Question	Is the defendant guilty?
Null Hypothesis	H_0: defendant is innocent
Alternative Hypothesis	H_a: defendant is guilty
Data	Evidence presented at trial
Assume H_0 is true	"Innocent until proven guilty"
Analyze data	Jury deliberations
p-value	Likelihood that the evidence presented came from an innocent defendant
Significance level (α)	"Reasonable doubt"
p-value < α	Evidence goes "beyond a reasonable doubt"
Reject H_0; statistically significant results	"Guilty" verdict
Inconclusive results	"Not guilty" verdict

Note that in a case where the jury decides that there is not enough evidence to convict the defendant, their verdict is "not guilty." You'll never hear a verdict of "innocent." That's because the defendant is presumed to be innocent at the start of the trial. If the prosecution does a lousy job of collecting and presenting evidence against the defendant, that lack of evidence does *not* support the defendant's innocence! All it means is that there is not enough evidence against the innocence hypothesis. In the same way, *we will never conclude that the null hypothesis is true in our hypothesis tests*, only that there is not enough evidence against it.

The Four Steps

In a hypothesis test, we'll use a four step approach. In practice, a researcher will pose a question and also collect the data, but for most of the problems we'll consider, the question will already be posed and the data will have already been collected. The chart below shows how the four steps correspond to the scientific method.

Scientific Method		*Our Four Steps*
1. Pose a Question		
2. Make a Hypothesis	→	Hypotheses
3. Collect Data		
4. Analyze the Data	→	Method/Conditions
	→	Mechanics
5. Make a Conclusion	→	Conclusion

Using the four step approach, the hypothesis test for the football survey would look like this:

Hypotheses: Set up H_0 and H_a based on the research question.

H_0: $p = 0.43$ (no change since the mid-2000's)
H_a: $p < 0.43$ (a drop since the mid-2000's)

p = proportion of all U.S. adults who currently prefer football

<u>Method/Conditions</u>: Select the method of analysis and identify the conditions required for your method to be valid.

> Method: one-proportion z-test for p
> Conditions:
> - SRS? Yes, the data are from a simple random sample.
> - $np = 1{,}049(0.43) = 451.07 \geq 15$ and
> $n(1 - p) = 1{,}049(1 - 0.43) = 597.93 \geq 15$

<u>Mechanics</u>: Use your method to find the p-value based on the observed data. Remember that the p-value is a measure of how likely the observed data are if the null hypothesis is true.

$$p - value = P(getting \; 388 \; or \; fewer \; in \; a \; SRS \; of \; 1{,}049)$$
$$if \; p = 0.43$$
$$= P\left(\hat{P} \leq \frac{388}{1{,}049}\right) if \; p = 0.43$$
$$= P(Z \leq -3.93) = 0.000042$$

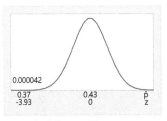

The survey data are not likely if there's been no change since the mid-2000's (the null hypothesis).

<u>Conclusion</u>: Use the p-value to make a conclusion about the alternative hypothesis in context. You have two choices:

- p-value is small
  - $\rightarrow$ data are unlikely under H_0
  - $\rightarrow$ reject H_0
  - $\rightarrow$ convincing evidence for H_a

- p-value is not small
  - $\rightarrow$ data are somewhat likely under H_0
  - $\rightarrow$ do not reject H_0
  - $\rightarrow$ inconclusive results; not convincing evidence for H_a

Your conclusion should always be in the context of the original research hypothesis. Never simply say things like, "reject H_0" or "not convincing evidence for H_a" or "the p-value is small" and stop there. Those phrases are statistical jargon that will likely not be understood by the layperson.

Here, reject H_0 since the p-value (probability of getting 37% or fewer who prefer football in a SRS of 1,049 under the "no change" hypothesis) is extremely small. Our conclusion is that fewer than 43% of all U.S. adults currently choose football as their favorite sport to watch.

Note that both possible conclusions are in terms of H_a (just like "guilty" or "not guilty" in a trial). The one piece of the trial analogy that breaks down is that, in a hypothesis test, we never use strong language like "prove." We never say that we've "proven that H_a is true," only that there is "convincing evidence for H_a." There is the chance that our conclusion is incorrect so we don't want to use the word "prove" which means there's no chance we could be wrong.

These four steps can be used in any hypothesis test. In the coming sections and chapters, we'll be exploring how to use them to answer different kinds of research questions and using different data structures.

Section 6.1 Exercises

1. What is another name for the researcher's hypothesis?

2. How do the four steps in a hypothesis test relate to the steps in the scientific method?

3. The statement that is contradictory to the researcher's hypothesis is called the _____.

For exercises 4 – 8, give the null and alternative hypotheses (in words) based on the research question.

4. "Can Elizabeth tell whether the milk was poured into the tea or the tea into the milk?"

5. "Is the average range of all electric cars currently more than 150 miles?"

6. "Do people with whorls (bulls-eye patterns) in their fingerprints have higher blood pressure on average than those who don't?"

7. "Is there a difference in music preferences between rule-focused people and people who are not rule-focused?"

8. "Would gastric surgery improve the remission rate for diabetics compared to diet and lifestyle changes?"

For exercises 9 – 16, give the null and alternative hypotheses (using symbols) based on the research question. The question involves one or more populations and so your symbols must be population parameters. Define the parameter(s) in words. Also identify the objects/subjects, the variable and type of the variable.

9. "Have more than 60% of seniors at my university attended a career fair?"

10. "Is the mean 0 – 60mph time for all electric cars less than 8 seconds?"

11. "Do more than 30% of students at my college have all required course materials within the first two weeks of the semester?"

12. "Is there a difference in average 0 – 60mph acceleration between electric and gasoline cars?"

13. "Do a majority of likely voters favor candidate X?"

14. "Is there a difference in the proportions of men and women who favor candidate X?"

15. "Is the average forced expiratory volume in one second (FEV1) for male athletes higher than 4.5?"

16. "Do a majority of people in my market segment prefer my product to my main competitor's product?"

17. Consider these four p-values: 0.999 0.50 0.01 0.00005

    a.   Which of these p-values would be convincing evidence for the research hypothesis?
    b.   For which of these p-values would the result of the test be deemed inconclusive?

c. For which of these p-values would the researcher likely reject the null hypothesis in favor of the alternative hypothesis?
d. Which of these p-values do not provide statistically convincing support for the research hypothesis?

18. Consider these four p-values: 1.00 0.004 0.00001 0.368

a. Which of these p-values would be convincing evidence for the research hypothesis?
b. For which of these p-values would the result of the test be deemed inconclusive?
c. For which of these p-values would the researcher likely reject the null hypothesis in favor of the alternative hypothesis?
d. Which of these p-values do not provide statistically convincing support for the research hypothesis?

19. Shown here is the binomial null distribution of a random variable, X = number of successes in a SRS of 6 from a large population. The hypotheses are: H_0: $p = 0.2$ vs. H_a: $p > 0.2$.

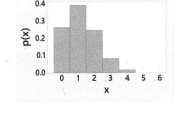

a. Which values of x would convince you that the null hypothesis was false? Explain.
b. Which values of x would somewhat (but not convincingly) support the alternative hypothesis? Explain.
c. Which values of x would not support the alternative hypothesis at all? Explain.

20. Shown here is the normal null distribution of a random variable, $\bar{X}$ = sample mean in a SRS of 50 from a large population. The hypotheses are H_0: $\mu = 54$ vs. H_a: $\mu < 54$.

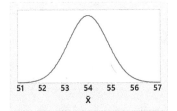

a. Which values of $\bar{x}$ would convince you that the null hypothesis was false? Explain.
b. Which values of $\bar{x}$ would somewhat (but not convincingly) support the alternative hypothesis? Explain.
c. Which values of $\bar{x}$ would not support the alternative hypothesis at all? Explain.

21. A SRS of 10 success/failure observations was selected from a large population. The data are used to test the hypotheses H_0: $p = 0.75$ vs. H_a: $p < 0.75$. The binomial null distribution of X = number of successes in the SRS out of 10 is shown here.

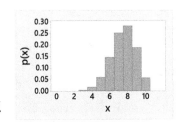

a. Assuming the null hypothesis is true, find the z-score for $X = 6$.
b. The SRS resulted in 6 successes. Find and interpret the p-value for this result.
c. Would a SRS resulting in 6 successes out of 10 be statistically convincing evidence for the research hypothesis? Explain.

22. Consider the setting of exercise 21.

a. Assuming the null hypothesis is true, find the z-score for $X = 3$.
b. The SRS resulted in 3 successes. Find and interpret the p-value for this result.
c. Would a SRS resulting in 3 successes out of 10 be statistically convincing evidence for the research hypothesis? Explain.

23. Tanya claims the coin she uses to decide who goes first in games is fair. You think it's biased toward heads and want to investigate. You toss Tanya's coin 20 times and observe X = number of heads in the 20 tosses. These data are used to test the hypotheses H_0: Tanya's coin is fair ($p = 0.5$) vs. H_a: Tanya's coin is biased toward heads ($p > 0.5$). The binomial null distribution of X is shown here.

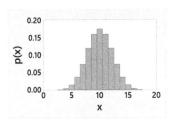

a. Assuming the null hypothesis is true, find the z-score for a result of 15 heads.

    b.   You got 15 heads out of 20 tosses. Find and interpret the p-value for this result.

    c.   Would 15 heads out of 20 tosses be statistically convincing evidence that Tanya's coin is biased toward heads? Explain.

24. Consider the setting of exercise 23.

    a.   Assuming the null hypothesis is true, find the z-score for a result of 12 heads.

    b.   You got 12 heads out of 20 tosses. Find and interpret the p-value for this result.

    c.   Would 12 heads out of 20 tosses be statistically convincing evidence that Tanya's coin is biased toward heads? Explain.

25. Consider these hypotheses: H_0: all crows are black vs. H_a: not all crows are black. You select a SRS of 15 crows.

    a.   Let's say you observe that all 15 crows are black. What is the p-value for this result?

    b.   Let's say you observe 14 black crows and 1 pink crow. What is the p-value for this result?

    c.   How many non-black crows in your SRS would be convincing evidence that the null hypothesis is false?

    d.   Does a p-value of 1 mean that the null hypothesis must be true? Explain.

26. Are sci-fi movies longer on average than romance movies? To investigate, two SRS's of 30 movies are selected in each genre and the runtimes are observed. The data are used to test the hypotheses $H_0: \mu_1 = \mu_2$ vs. $H_a: \mu_1 > \mu_2$ where μ_1 and μ_2 are the population mean runtimes for all sci-fi movies and romance movies, respectively. The null distribution of the difference in the two sample means $\bar{X}_1 - \bar{X}_2$ is shown here. It is a normal distribution with $E(\bar{X}_1 - \bar{X}_2) = 0$ and $SD(\bar{X}_1 - \bar{X}_2) = 6.45$.

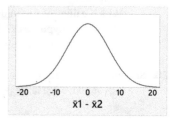

    a.   Let's say you find that the sample mean runtime for the SRS of sci-fi movies was 100.8 minutes and the sample mean runtime for the SRS of romance movies was 92.7 minutes. Find and interpret the p-value.

    b.   Would these data be statistically convincing evidence that sci-fi movies are longer on average than romance movies? Explain.

27. Consider the setting of exercise 26.

    a.   Let's say you find that the sample mean runtime for the SRS of sci-fi movies was 105.1 minutes and the sample mean runtime for the SRS of romance movies was 86.9 minutes. Find and interpret the p-value.

    b.   Would these data be statistically convincing evidence that sci-fi movies are longer on average than romance movies? Explain.

28. Consider these hypotheses: H_0: gastric surgery does not improve remission rates for diabetics compared to diet and lifestyle changes vs. H_a: gastric surgery improves remission rates for diabetics compared to diet and lifestyle changes. Here are some possible conclusions based on data. For each one, explain what is wrong and correct the error.

    a.   The data prove that gastric surgery improves remission rates for diabetics compared to diet and lifestyle changes.

    b.   The p-value is quite small.

    c.   There is no evidence that gastric surgery improves remission rates for diabetics compared to diet and lifestyle changes.

    d.   There is no difference in remission rates for diabetics who had gastric surgery compared to those who made diet and lifestyle changes.

    e.   We can reject H_0.

29. Consider these hypothesis: H_0: preference among people in my market segment for my product to my main competitor's product is 50/50 vs. H_a: a majority of people in my market segment prefer my product to my main competitor's product. Here are some possible conclusions based on data. For each one, explain what is wrong and correct the error.

a. The p-value was very large.
b. My data show that only half of people in my market segment prefer my product over my main competitor's product.
c. We have proof in the data that a majority of people in my market segment prefer my product to my main competitor's product.
d. I did not find convincing evidence for H_a.
e. The data provide absolutely no evidence of a preference among people in my market segment for either my product or my main competitor's product.

30. A recent Gallup poll conducted on a simple random sample of 1,049 U.S. adults showed that 73 said that soccer ("football" to most of the world) is their favorite sport to watch.[2] We'd like to use these data to test the hypotheses H_0: $p = 0.09$ (9% of all U.S. adults prefer soccer...the same as baseball) vs. H_a: $p < 0.09$ (soccer is less popular than baseball). The null distribution of $\hat{P}$ for a SRS of 1,049 is shown here.

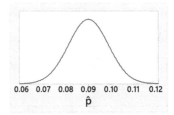

a. Assuming the null hypothesis is true, find the z-score for Gallup's result.
b. Find and interpret the p-value for Gallup's result.
c. Would 73 who prefer soccer in a SRS of 1,049 be statistically convincing evidence that soccer is less popular than baseball among all U.S. adults? Explain.

31. A company wants to find out if a majority of customers prefer its product over its closest competitor's product. The company selects a simple random sample of 2,000 customers and finds that 1,070 prefer its product over its closest competitor's product. We'd like to use these data to test the hypotheses H_0: $p = 0.5$ (half of all customers prefer the company's product over the competition's) vs. H_a: $p > 0.5$ (a majority of all customers prefer the company's product). The null distribution of $\hat{P}$ for a SRS of 2,000 is shown here.

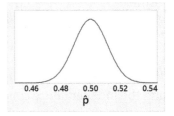

a. Assuming the null hypothesis is true, find the z-score for the sample result.
b. Find and interpret the p-value for the sample result.
c. Would 1,070 who prefer the company's product over the competition's in a SRS of 2,000 be statistically convincing evidence that a majority of all its customers do? Explain.

6.2 Test for a Population Proportion

In the last section you saw an example of a hypothesis test about a population proportion, p. In that example, p represented the proportion of all U.S. adults who currently prefer football. In this section we'll look at this particular test in more detail. The idea is to take a SRS of n objects from the population and use the sample data to make a generalization inference about p, the proportion of successes in the population.

Population
p = proportion of successes

SRS

$\hat{p} = \dfrac{x}{n}$

The test is conducted by considering the sampling distribution of either $X = $ number of successes or $\hat{P} = X/n$, the proportion of successes in the sample of n objects, under the assumption that the null hypothesis is true.

Shown below are the features of the sampling distributions of X and $\hat{P}$ based on SRS's of size n from a population where p is the proportion of successes.

	$X = $ number of successes	$\hat{P} = \frac{X}{n} = $ proportion of successes
Center	$E(X) = np$	$E(\hat{P}) = p$
Variance*	$Var(X) = np(1-p)$	$Var(\hat{P}) = \dfrac{p(1-p)}{n}$
Standard Deviation*	$SD(X) = \sqrt{np(1-p)}$	$SD(\hat{P}) = \sqrt{\dfrac{p(1-p)}{n}}$
Notes*	Binomial Nearly normal if $np \geq 15$ and $n(1-p) \geq 15$	Nearly normal if $np \geq 15$ and $n(1-p) \geq 15$

* These assume that $\frac{n}{N} < 0.05$, that the sample size is less than 5% of the population size.

The number we use for p is the one specified by the null hypothesis. As long as the SRS condition and the independence conditions are satisfied, the binomial distribution on X can be used to find p-values. We'll call this method the **one-proportion binomial test**. Note that if both np and $n(1-p)$ are at least 15, the normal distribution on either X or $\hat{P}$ can be used to find p-values. We'll call this method the **one-proportion z-test** because finding the p-value involves finding z-scores and using the standard normal distribution. That can save a lot of time compared to using the binomial formula.

One-proportion binomial test	**One-proportion z-test**
H_0: $p = p_0$ H_a: $p < p_0$ or $p > p_0$ or $p \neq p_0$ The p-value is found by comparing the value of X to a binomial distribution with $p = p_0$ under H_0. Conditions • SRS	H_0: $p = p_0$ H_a: $p < p_0$ or $p > p_0$ or $p \neq p_0$ The p-value is found by comparing the value of $z = \dfrac{x - np_0}{\sqrt{np_0(1-p_0)}}$ or $z = \dfrac{\hat{p} - p_0}{\sqrt{\dfrac{p_0(1-p_0)}{n}}}$ to a standard normal distribution under H_0. Conditions • SRS • $np_0 \geq 15$ and $n(1-p_0) \geq 15$

Note that these methods are valid only for simple random samples. As you saw in chapter 3, the sampling distribution of a statistic based on data from a fancier random sample like a cluster random sample or a stratified random sample may exhibit very different variability. The analyses of data collected by those sampling methods are beyond the scope of this book.

The research hypothesis can take several forms:

H_a: $p < p_0$
The researcher believes that the population proportion of successes is lower than a certain number. The data provide evidence for this hypothesis in the form of *low* values of x, $\hat{p}$, and z. The p-value is the probability of observing a value of X, $\hat{P}$, or Z *as low or lower* than the one observed in the data under the null hypothesis.

One-proportion binomial test

One-proportion z-test

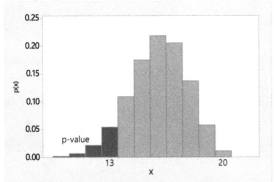

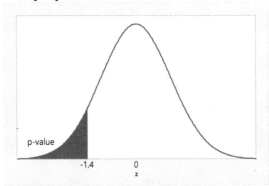

p-value based on observed value of x

p-value based on observed value of z

H_a: $p > p_0$
The researcher believes that the population proportion of successes is higher than a certain number. The data provide evidence for this hypothesis in the form of *high* values of x, $\hat{p}$, and z. The p-value is the probability of observing a value of X, $\hat{P}$, or Z *as high or higher* than the one observed in the data under the null hypothesis.

One-proportion binomial test

One-proportion z-test

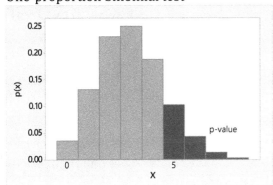

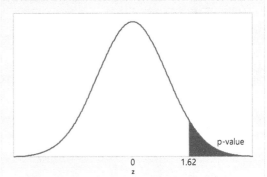

p-value based on observed value of x

p-value based on observed value of z

H_a: $p \neq p_0$
The researcher believes that the population proportion of successes differs from a certain number. The data provide evidence for this hypothesis in the form of *low or high* values of x, $\hat{p}$, and z. The p-

value is the probability of observing a value of X, $\hat{P}$, or Z *at least as far from expected* as the one observed in the data under the null hypothesis.

Getting p-values for the one-proportion binomial test in this case takes some careful thinking since the binomial distribution is not generally symmetric. However, since the one-proportion z-test uses the perfectly symmetric standard normal null distribution, you can double the "tail area" to get the p-value. If the data suggest lower values of p, the p-value is *twice* the probability of observing a value of X, $\hat{P}$, or Z *as low or lower* than the one observed in the data under the null hypothesis. If the data suggest higher values of p, the p-value is *twice* the probability of observing a value of X, $\hat{P}$, or Z *as high or higher* than the one observed in the data under the null hypothesis.

One-proportion binomial test

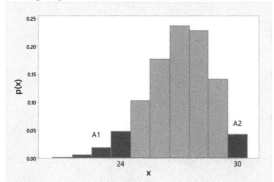

p-value = A1 + A2 based on observed value of x

One-proportion z-test

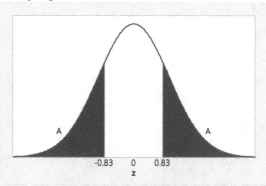

p-value = 2A based on observed value of z

Example: In an early January 2017 survey conducted by the Pew Research Center, 781 (about 52%) of those surveyed thought that China's power and influence is a major threat to the well-being of the United States.[3] The report is based on a random sample of 1,502 adults in the U.S. Is there convincing evidence in the data that a majority of all U.S. adults at the time thought that China's power and influence was a major threat to the well-being of the United States? This is not a trivial question: while the 52% figure is indeed greater than 50%, it is based on a relatively small sample of U.S. adults. What if only 50% of all U.S. adults felt that way? In that case, is it likely to get 52% in a random sample? That's the question a hypothesis test is designed to answer. Let's look at the details.

<u>Hypotheses</u>:

H_0: $p = 0.50$
H_a: $p > 0.50$

p = proportion of all U.S. adults at the time who thought that China's power and influence was a major threat to the well-being of the United States

<u>Method/Conditions</u>: We'll use the one-proportion z-test.

Conditions:

- SRS: In the Methodology section of the report, we find that the sampling procedure was a complex, multi-stage random sample. Since analyses of data from those kinds of random samples are beyond the scope of this book, we'll treat the data as if they came from a SRS in order to illustrate the method.
- $np_0 = 1{,}502(0.50) = 751 \geq 15$ and
  $n(1 - p_0) = 1{,}502(1 - 0.50) = 751 \geq 15$

Mechanics:

$$\hat{p} = \frac{781}{1{,}502} \approx 0.52 \qquad \text{from the sample data}$$

$$z = \frac{\hat{p} - p_0}{\sqrt{\frac{p_0(1-p_0)}{n}}} = \frac{\frac{781}{1{,}502} - 0.50}{\sqrt{\frac{0.50(1-0.50)}{1{,}502}}} = 1.55$$

The data provide evidence for H_a in the form of *high* values of $\hat{p}$ and z. The p-value is the probability of observing a value $\hat{P}$ *as high or higher* than 0.52 (or a value of Z *as high or higher* than 1.55) if H_0 were true.

Using the standard normal table, we get

$$p - value = P(Z \geq 1.55)$$
$$= 1 - 0.9394$$
$$= 0.0606.$$

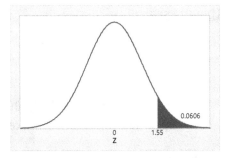

If only 50% of all U.S. adults thought that China's power and influence was a major threat to the well-being of the United States, there is a 6.06% chance that 52% or more would have thought that way in a SRS of 1,502 U.S. adults.

Conclusion:

The p-value is marginally small. If we use a 10% significance level, we can reject H_0 in favor of H_a. At the 10% significance level, the data show enough evidence that a majority of all U.S. adults at the time thought that China's power and influence was a major threat to the well-being of the United States.

It is important to note that the conclusion of this test depends upon the choice of significance level. If we use a significance level of $\alpha = 0.10$, then the p-value *is* small enough and we *can* reject H_0 in favor of H_a. At the 10% significance level, the data *do* show enough evidence that a majority of all U.S. adults at the time thought that China's power and influence was a major threat to the well-being of the United States. However, if we use the common significance level of $\alpha = 0.05$, then the p-value *is not* small enough and we *cannot* reject H_0 in favor of H_a. At the 5% significance level, the data *do not* show quite enough evidence that a majority of all U.S. adults at the time thought that China's power and influence was a major threat to the well-being of the United States.

I realize you're just starting out with hypothesis testing, so I don't want to complicate things too much. However, I also need you to not turn your brain off when making conclusions based on p-values. P-values below, say, 0.10 can start to get my attention as looking small. Choosing a cutoff below which we declare that the data convincingly support the research hypothesis is arbitrary. The choice of what is small enough can also depend on the consequences for making an incorrect decision; more on that in a later section. The bottom line is that it is sometimes not so cut-and-dry as simply comparing the p-value to the significance level.

Example: Sampling is a very useful tool in product testing because if it is done correctly, only a small fraction of products need be tested. Sampling is absolutely necessary when the testing is destructive; in those cases, if all products were tested, there would be nothing left to sell! Consider a tire manufacturer who wants to assess the failure rate of a tire when it is underinflated. A simple random sample of 150 tires are selected from the most recent production run of 10,000 tires and each is mounted on a test machine, inflated to only 15 pounds per square inch (psi), and run for 4 hours at 70 miles per hour (mph). Only 6 of the 150 tires failed. A competitor advertises that only 10% of their tires fail under these conditions. Do the manufacturer's data show convincing evidence that the proportion of all tires in the most recent production run that fail under these conditions differs from 0.10? Conduct a hypothesis test.

<u>Hypotheses</u>:

H_0: $p = 0.10$
H_a: $p \neq 0.10$

p = proportion of all 10,000 tires in the most recent production run that fail under the specified conditions (inflated to only 15psi and driven for 4 hours at 70mph)

<u>Method/Conditions</u>: We'll use the one-proportion z-test.

Conditions:
- SRS: We're told the manufacturer used a simple random sample.
- $np_0 = 150(0.10) = 15 \geq 15$ and
  $n(1 - p_0) = 150(1 - 0.10) = 135 \geq 15$

<u>Mechanics</u>:

$$\hat{p} = \frac{6}{150} = 0.04$$

$$z = \frac{\hat{p} - p_0}{\sqrt{\frac{p_0(1-p_0)}{n}}} = \frac{0.04 - 0.10}{\sqrt{\frac{0.10(1-0.10)}{150}}} = -2.45$$

The data provide evidence for H_a in the form of *low or high* values of $\hat{p}$ and z. Since the data suggest lower values of p, the p-value is *twice* the probability of observing a value of $\hat{P}$ *as low or lower* than 0.04 (or a value of Z *as low or lower* than -2.45) if H_0 were true. The idea is that we want to consider the *magnitude* of the z-score, not its *sign*. In this problem, a z-score of $z = 2.45$ (based on a $\hat{p} = 24/150 = 0.16$) would have been just as much evidence for our H_a, just in the other direction.

Using the standard normal table, we get

$$p - value = 2P(Z \leq -2.45)$$
$$= 2(0.0071)$$
$$= 0.0142.$$

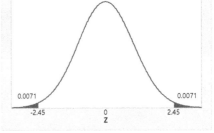

If 10% of all 10,000 tires in the most recent production run fail under the specified conditions, there is only a 1.42% chance that 4% or fewer or 16% or more in a SRS of 150 tires would fail.

Conclusion:

Since the p-value is quite small (unless your criteria for small is something below 0.0142), we can reject H_0 in favor of H_a. The data show convincing evidence that the proportion of all tires in the most recent production run that fail under these conditions differs from 0.10. In fact, the evidence points toward a lower failure rate than 10%. The manufacturer might consider advertising this as a way to compete more favorably with their competitor.

Example: According to a 2014 SAMHSA (Substance Abuse and Mental Health Services Administration) survey of full-time college students, an estimated 38% of all college students aged 18 – 22 engaged in binge drinking within the month prior to the survey.[4] Binge drinking is defined as consuming enough alcohol in 2 hours to raise one's BAC (blood alcohol level) to 0.08 g/dl. That's about 4 or 5 drinks in 2 hours. Suppose you would like to assess the extent of the binge drinking problem on your small campus of 2,500 students, but you have a hunch that the number is lower at your school. You can manage only a SRS of 30 students from your institution and find that 9 reported having engaged in binge drinking in the past month. Is this enough evidence that the binge drinking rate at your school is less than 38%? Conduct a hypothesis test.

Hypotheses:

H_0: $p = 0.38$
H_a: $p < 0.38$

$p =$ proportion of all 2,500 students at your school who would report having engaged in binge drinking in the past month

Method/Conditions: We'll use the one-proportion binomial test.

Conditions:
- SRS: You used a simple random sample.

Note that $np_0 = 30(0.38) = 11.4 < 15$ so we cannot use the z-test.

Mechanics:

$$x = 9 \qquad \hat{p} = \frac{9}{30} = 0.30$$

The data provide evidence for H_a in the form of *low* values of x and $\hat{p}$. Since the data suggest lower values of p, the p-value is the probability of observing a value of X as low or lower than 9 (or a value of $\hat{P}$ *as low or lower* than 0.30) if H_0 were true.

Using the binomial formula with $n = 30$ and $p = p_0 = 0.38$, we get

$$p - value = P(X \le 9) = p(0) + p(1) + \cdots + p(9)$$
$$= \binom{30}{0}(0.38)^0(1 - 0.38)^{30} + \binom{30}{1}(0.38)^1(1 - 0.38)^{29} + \cdots + \binom{30}{9}(0.38)^9(1 - 0.38)^{21}$$
$$= 0.2401$$

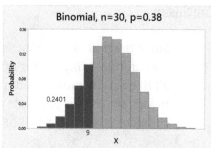

Binomial, n=30, p=0.38

If 38% of all 2,500 students at your school would report having engaged in binge drinking last month, there is still a 24.01% chance that 9 (30%) or fewer in a SRS of 30 students would have done so.

Conclusion:

Since the p-value is not very small, we cannot reject H_0 in favor of H_a. While there is some evidence that the binge drinking problem at your school is not as bad as nationwide, the results are inconclusive. They are not statistically significant, that is, we cannot easily rule out simple random sampling as an explanation for why we're seeing a lower rate in the sample.

Section 6.2 Exercises

1. Determine whether each pair of hypotheses is legitimate. Say what's wrong with the bad ones.

   a. H_0: $p = 0.5$ vs. H_a: $p > 0.5$
   b. H_0: $\hat{p} = 0.6$ vs. H_a: $\hat{p} < 0.6$
   c. H_0: $x = 0.25$ vs. H_a: $x \ne 0.25$
   d. H_0: $z = 0$ vs. H_a: $z > 0$
   e. H_0: $p - value = 0.05$ vs. H_a: $p - value < 0.05$

2. Consider testing the hypotheses H_0: $p = 0.9$ vs. H_a: $p < 0.9$ in a large population. For each sampling scenario, determine whether to use the one-proportion binomial test, one-proportion z-test, or neither. Explain your choice.

   a. STRS of $n = 500$
   b. SRS of $n = 30$
   c. volunteer sample of $n = 2,000$
   d. CRS of 100 clusters each having 50 objects
   e. SRS of $n = 2,000$

3. Consider testing the hypotheses H_0: $p = 0.05$ vs. H_a: $p > 0.05$ in a large population. For each sampling scenario, determine whether to use the one-proportion binomial test, one-proportion z-test, or neither. Explain your choice.

    a.   SRS of $n = 10,000$
    b.   convenience sample of $n = 5,000$
    c.   STRS of $n = 300$
    d.   CRS of 50 clusters each having 10 objects
    e.   SRS of $n = 20$

4.   "Do more than 30% of students at my college have all required course materials within the first two weeks of the semester?" Consider using a hypothesis test to answer this question.

    a.   Identify the objects/subjects, the variable, and type of variable.
    b.   Define the parameter in context.
    c.   Set up the hypotheses.
    d.   For each sampling scenario, determine whether to use the one-proportion binomial test, one-proportion z-test, or neither. Explain your choice.
        i.   sample of 100 students in your dorm
        ii.   SRS of 15 students at your college
        iii.   CRS of 40 classes meeting this semester
        iv.   SRS of 2,000 students at your college

5.   "Is the average range of all electric cars currently more than 150 miles?" Consider using a hypothesis test to answer this question. For each sampling scenario, determine whether to use the one-proportion binomial test, one-proportion z-test, or neither. Explain your choice.

    a.   SRS of 100 electric cars
    b.   all electric cars for a CRS of 5 car manufacturers
    c.   SRS of 10 electric cars

6.   Find the p-value for each hypothesis test using the one-proportion binomial test.

    a. H_0: $p = 0.9$ vs. H_a: $p < 0.9$ using a SRS of $n = 30$ having 25 successes
    b. H_0: $p = 0.9$ vs. H_a: $p < 0.9$ using a SRS of $n = 30$ having 70% successes
    c. H_0: $p = 0.45$ vs. H_a: $p > 0.45$ using a SRS of $n = 8$ having 7 successes
    d. H_0: $p = 0.45$ vs. H_a: $p > 0.45$ using a SRS of $n = 8$ having 62.5% successes

7.   Find the p-value for each hypothesis test using the one-proportion binomial test.

    a. H_0: $p = 0.2$ vs. H_a: $p > 0.2$ using a SRS of $n = 25$ having 32% successes
    b. H_0: $p = 0.2$ vs. H_a: $p > 0.2$ using a SRS of $n = 25$ having 12 successes
    c. H_0: $p = 0.6$ vs. H_a: $p < 0.6$ using a SRS of $n = 8$ having 37.5% successes
    d. H_0: $p = 0.6$ vs. H_a: $p < 0.6$ using a SRS of $n = 8$ having 5 successes

8.   Find the p-value for each hypothesis test using the one-proportion z-test.

    a. H_0: $p = 0.52$ vs. H_a: $p > 0.52$ using a SRS of $n = 100$ having 60 successes
    b. H_0: $p = 0.3$ vs. H_a: $p \neq 0.3$ using a SRS of $n = 500$ having 155 successes
    c. H_0: $p = 0.81$ vs. H_a: $p < 0.81$ using a SRS of $n = 1,500$ having 1,170 successes
    d. H_0: $p = 0.6$ vs. H_a: $p \neq 0.6$ using a SRS of $n = 70$ having 41 successes

9.   Find the p-value for each hypothesis test using the one-proportion z-test.

    a. H_0: $p = 0.25$ vs. H_a: $p < 0.25$ using a SRS of $n = 750$ having 160 successes
    b. H_0: $p = 0.44$ vs. H_a: $p \neq 0.44$ using a SRS of $n = 200$ having 84 successes
    c. H_0: $p = 0.9$ vs. H_a: $p > 0.9$ using a SRS of $n = 3,500$ having 3,193 successes
    d. H_0: $p = 0.2$ vs. H_a: $p \neq 0.2$ using a SRS of $n = 250$ having 52 successes

10. In the case of "directional" alternative hypotheses ("<" or ">"), once in a while the data will surprise the researcher by providing evidence in the opposite direction. In these cases, you don't even need to conduct a hypothesis test because there is no evidence for H_a. For parts a. and b., verify that the one-proportion z-test is appropriate and then verify that the p-value is large.

    a. H_0: $p = 0.25$ vs. H_a: $p < 0.25$ using a SRS of $n = 950$ having 250 successes
    b. H_0: $p = 0.63$ vs. H_a: $p > 0.63$ using a SRS of $n = 75$ having 56% successes
    c.   When testing H_0: $p = p_0$ vs. H_a: $p > p_0$ and a SRS results in $\hat{p} < p_0$, what can you say about the p-value?
    d.   When testing H_0: $p = p_0$ vs. H_a: $p < p_0$ and a SRS results in $\hat{p} > p_0$, what can you say about the p-value?

11. In a SRS of 250 students at a large university, 103 indicated that they had all required course materials within the first two weeks of the semester.

    a.   Identify the objects/subjects, the variable and its type.
    b.   Do these data provide convincing statistical evidence that more than 30% of all students at the university have all required course materials within the first two weeks of the semester?
        i.   Write down the hypotheses and define the parameter in context.
        ii.   Check the conditions.
        iii.   Perform the mechanics.
        iv.   Interpret the p-value.
        v.   Make a conclusion in context.

12. In a SRS of 50 seniors at a particular college, 32 indicated that they had attended a career fair.

    a.   Identify the objects/subjects, the variable and its type.
    b.   Do these data provide convincing statistical evidence that more than 60% of all seniors at this college attended a career fair?
        i.   Write down the hypotheses and define the parameter in context.
        ii.   Check the conditions.
        iii.   Perform the mechanics.
        iv.   Interpret the p-value.
        v.   Make a conclusion in context.

13. Eight hundred twenty in a SRS of 1500 likely voters are in favor of candidate X.

    a.   Identify the objects/subjects, the variable and its type.
    b.   Do these data provide convincing statistical evidence that a majority of all likely voters are in favor of candidate X?
        i.   Write down the hypotheses and define the parameter in context.
        ii.   Check the conditions.
        iii.   Perform the mechanics.
        iv.   Interpret the p-value.
        v.   Make a conclusion in context.

14. A large mail order company is studying its order fulfillment process. An order is shipped "on time" if it is sent out within two business days of when it was received. The company currently claims that "95% of all orders are shipped on time" but is worried that there have been more late orders recently. In a SRS of 300 recently-shipped orders, 279 were shipped on time.

    a.   Identify the objects/subjects, the variable and its type.
    b.   Do these data provide convincing statistical evidence that fewer than 95% of all recent orders were shipped on time?
        i.   Write down the hypotheses and define the parameter in context.
        ii.   Check the conditions.
        iii.   Perform the mechanics.
        iv.   Interpret the p-value.

v.   Make a conclusion in context.

15. Consider exercise 13. How would your conclusion change if a SRS of only 300 gave the same sample proportion in favor of candidate X?

16. Consider exercise 14. How would your conclusion change if a SRS of 1200 gave the same sample proportion of on time orders?

17. According to a Pew Research Center study, 21% in a sample of 5,000 U.S. adults were raised by two parents having different religious views.[5]

a.   Identify the objects/subjects, the variable and its type.
b.   Do these data provide convincing statistical evidence that more than 20% of all U.S. adults were raised in interfaith homes? Conduct a hypothesis test.

18. According to the 2014 National Survey on Drug Use and Health, 28.5% of 18 – 20 year-olds reported having engaged in binge drinking (5 or more alcoholic drinks within a couple hours) in the past month.[6] You think this number is higher at your school and so you take a SRS of 425 18 – 20 year-olds at your school.

a.   Identify the objects/subjects, the variable and its type.
b.   If 160 report having engaged in binge drinking in the past month, would that be convincing statistical evidence that the binge drinking rate among 18 – 20 year-olds at your school is higher than the national rate? Conduct a hypothesis test.

19. Fifty-seven percent of adults in an urban county in my state are white. Selecting people to serve on a jury involves taking a SRS of all adults in a county. Consider a jury selection of 60 adults from your home county.

a.   Identify the objects/subjects, the variable and its type.
b.   If 30 white adults are in your county's sample, would that be convincing statistical evidence that your county is different than the urban county in my state in terms of the proportion of white adults? Conduct a hypothesis test.

20. A magazine is planning to add an online version in addition to its print version but will only do so if more than 15% of current subscribers would be interested. The publisher selects a SRS of 200 current subscribers and finds that 36 are interested in an online version.

a.   Identify the objects/subjects, the variable and its type.
b.   Should the publisher go ahead with the online version? Conduct a hypothesis test and use it to justify your decision.

21. An airline claims that it loses bags for only 5% of all passengers. You've just had several experiences of lost bags in your recent flights with this airline and so you suspect that percentage is higher. In a SRS of 400 recent passengers, how many with lost bags would it take to convince you that the airline's claim is too low? Explain.

22. A car manufacturer learns that some of its 5-year-old cars are experiencing back-up camera malfunctions. The company is hesitant to issue a recall notice but will do so if more than 10% of all cars are currently experiencing the problem. The company selects a SRS of 50 of those cars and finds the back-up cameras on 8 of them are malfunctioning. The p-value for the test is 0.122.

a.   Which of these is the correct interpretation of the p-value?
    i.   There's a 12.2% chance that more than 10% of all cars currently have the problem.
    ii.   There's a 12.2% chance that at most 10% of all cars currently have the problem.
    iii.   There's a 12.2% chance that 16% of all cars currently have the problem.
    iv.   There's a 12.2% chance that at least 16% in a SRS of 50 cars have the problem if only 10% of all cars do.
    v.   There's a 12.2% chance that at least 16% in a SRS of 50 cars have the problem if more than 10% of all cars do.

     vi.   There's a 12.2% chance that at most 16% in a SRS of 50 cars have the problem if only 10% of all cars do.

     vii.   There's a 12.2% chance that at most 16% in a SRS of 50 cars have the problem if more than 10% of all cars do.

     viii. There are about 12.2% of all cars currently experiencing the problem.

  b.   Should the car manufacturer issue a recall notice? Explain.

23. According to the National Center for Education Statistics, the U.S. national high school dropout rate was 5.7% in 2017.[7] To see whether the current rate is lower than that, researchers observe that 5.2% in a SRS of 1000 late-teen to early-20-year-olds in the U.S. have dropped out of high school. The p-value reported is 0.248.

  a.   Which of these is the correct interpretation of the p-value?

     i.   There's a 24.8% chance that the current U.S. high school dropout rate is lower than 5.2%.

     ii.   There's a 24.8% chance that the current U.S. high school dropout rate is at least 5.2%.

     iii.   There's a 24.8% chance that the current U.S. high school dropout rate is 5.2%.

     iv.   There's a 24.8% chance that 5.2% or fewer in a SRS of 1000 late-teen to early-20-year-olds in the U.S. have dropped out of high school if 5.7% of all such young people have.

     v.   There's a 24.8% chance that 5.2% or fewer in a SRS of 1000 late-teen to early-20-year-olds in the U.S. have dropped out of high school if fewer than 5.7% of all such young people have.

     vi.   There's a 24.8% chance that 5.2% or more in a SRS of 1000 late-teen to early-20-year-olds in the U.S. have dropped out of high school if 5.7% of all such young people have.

     vii.   There's a 24.8% chance that 5.2% or more in a SRS of 1000 late-teen to early-20-year-olds in the U.S. have dropped out of high school if fewer than 5.7% of all such young people have.

     viii. There are about 24.8% of all late-teen to early-20-year-olds in the U.S. that have dropped out of high school.

  b.   Can you conclude that the U.S. national high school dropout rate has decreased since 2017? Explain.

24. Consider exercise 22. A hypothesis test is shown below. Identify all the mistakes in this response.

Hypotheses: H_0: $\hat{p} = 0.10$ vs. H_a: $p > 10$ p = proportion of cars experiencing the problem out of 50

Method/Conditions: SRS: We're told we have a SRS.
                      $n = 50 \geq 30$
                      Use the one-proportion z-test.

Mechanics: $p = \frac{8}{50} = 0.16$ $z = \frac{0.16 - 0.10}{\sqrt{\frac{0.16(1-0.16)}{50}}} = 1.16$ $p-value = P(Z \geq 1.16) = 0.8770$

           There's an 87.7% chance that only 10% of all cars currently have the problem.

Conclusion: The sample proportion of cars experiencing the problem is more than 0.10 because the $p - value$ is large.

25. Consider exercise 22. Show the details in the four steps of the hypothesis test.

26. Consider exercise 23. A hypothesis test is shown below. Identify all the mistakes in this response.

Hypotheses: H_0: $\mu < 0.052$ vs. H_a: $\mu = 0.057$ μ = mean national U.S. high school dropout rate in 2017

Method/Conditions: SRS: We're told we have a SRS.
                      $np_0 = 1000(0.057) = 57 \geq 15$
                      Use the one-proportion z-test.

Mechanics: $\hat{p} = 5.2$ $z = \frac{5.7 - 5.2}{\sqrt{\frac{0.057(1-0.057)}{1000}}} = 68.20$ $p-value = 2P(Z \leq 68.20) = 2$

           There's a 2% chance that the sample was accurate.

Conclusion: The sample size was not large enough for the results to be accurate.

27. Consider exercise 23. Show the details in the four steps of the hypothesis test.

6.3 Test for a Population Mean

The previous two sections dealt with research questions about p, either a binomial probability of success or a population proportion of successes, depending on the context. Those questions require the analysis of categorical data. Many research questions are about population averages and require the analysis of numeric data; we'll look at those kinds of questions and analyses in this section.

The idea is to take a random sample of n objects from the population and use the sample data to make a generalization inference about μ, the mean value for all objects in the population.

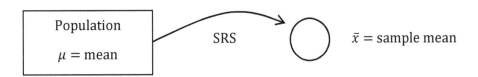

The test is conducted by considering the sampling distribution of $\bar{X}$ = sample mean, under the assumption that the null hypothesis is true.

Shown below are the features of the sampling distribution of sample means. These features are based on SRS's of size n from a population having mean μ and standard deviation σ.

	$\bar{X}$ = sample mean
Center	$E(\bar{X}) = \mu$
Variance*	$Var(\bar{X}) = \dfrac{\sigma^2}{n}$
Standard Deviation*	$SD(\bar{X}) = \dfrac{\sigma}{\sqrt{n}}$
Shape*	Normal if: 1) population data are normal 2) n large enough

* These assume that $\frac{n}{N} < 0.05$, that the sample size is less than 5% of the population size.

The number we use for μ is the one specified by the null hypothesis. If the SRS condition, the independence condition, and one of the shape conditions are satisfied, then the statistic

$$Z = \frac{\bar{X} - \mu}{\sigma/\sqrt{n}}$$

has a standard normal distribution. The problem is that, in generalization inference problems, we're "going the other way." We observe features of the sample and try to generalize to the unknown features of the population. So we don't know the value of σ to use and we rarely specify it in the null hypothesis.

The way out of this "σ unknown" problem is to substitute S for it. That gives us an estimate of the standard deviation of $\bar{X}$ that we call the **standard error** of $\bar{X}$:

$$SE(\bar{X}) = \frac{S}{\sqrt{n}}$$

But now there are two statistics, $\bar{X}$ and S, in the z-score formula instead of just one and that induces more variation in the z-scores than the standard normal distribution can model.

In 1908, William Gosset, a mathematician and data scientist employed at the Guinness Brewing Company, published a paper[8] that showed that a probability model that we now call the **t-distribution** is just the one to use to model the statistic

$$\frac{\bar{X} - \mu}{S/\sqrt{n}}$$

as long as the population data are normal. In practice, this normality condition can be relaxed quite a bit so that we just need to verify that the sample data are nearly-normal. Good thing because we won't have access to the population data anyway!

The t-distribution has many of the same features as the standard normal distribution: it is bell-shaped and symmetric about 0 and probabilities are areas under the curve. The important difference is that it has more variability than the standard normal distribution and the variability is controlled by a number called the **degrees of freedom**. The lower the degrees of freedom, the more variability and the higher the degrees of freedom, the closer the variability gets to the standard normal distribution.

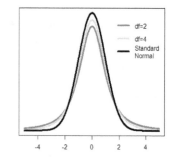

Here are some pictures of two t-distributions, one with 2 degrees of freedom and one with 4 degrees of freedom compared to the standard normal distribution.

T-scores having particular areas to their left are tabulated in Table II. First find the row for the degrees of freedom you want. There are several t-scores in that row and the areas to their left under the curve are the column headings. Note that the t-table is much less detailed than the z-table. The reason is that changing the degrees of freedom changes the variability of the t-distribution and so t-scores having a particular area to their left change.

Just as you did when working normal curve problems, when you work with t-curves, ALWAYS sketch the curve and properly label your picture.

Example: Let t have a t-distribution with 16 degrees of freedom. Find a so that $P(t > a) = 0.95$.

Solution: First sketch the t-curve centered at 0. Then mark the value a on the horizontal axis so that it has an area to its right of 0.95 (and therefore an area of 0.05 to its left). Then use table II with $df = 16$ and read off the t-score in the 0.05 column:

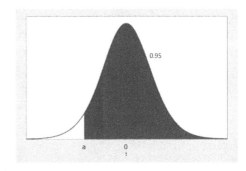

$a = -1.75$

Example: Let t have a t-distribution with 5 degrees of freedom. Find $P(t < 1.48)$.

Solution: First sketch the t-curve centered at 0. Then label the t-score of 1.48 and shade the area under the curve to its left. That's the probability you want. Table II gives this area directly in the $df = 5$ row as the column heading:

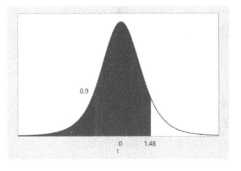

$P(t < 1.48) = 0.9$

Putting all the pieces together results in a hypothesis testing method called the **one-sample t-test**.

One-sample t-test

H_0: $\mu = \mu_0$
H_a: $\mu < \mu_0$ or $\mu > \mu_0$ or $\mu \neq \mu_0$

The p-value is found by comparing the value of $t = \frac{\bar{x} - \mu_0}{s/\sqrt{n}}$ to a t-distribution with $df = n - 1$ under H_0.

Conditions
  - SRS
  - nearly-normal distribution of data*
    * Can be relaxed if n is sufficiently large.

Note that this method is valid only for simple random samples. As you saw in chapter 3, the sampling distribution of a statistic based on data from a fancier random sample like a cluster random sample or a stratified random sample may exhibit very different variability. The analyses of data collected by those sampling methods are beyond the scope of this book.

The research hypothesis can take several forms:

H_a: $\mu < \mu_0$
The researcher believes that the population mean is lower than a certain number. The data provide evidence for this hypothesis in the form of *low* values of $\bar{x}$ and t. The p-value is the probability of observing a value of t *as low or lower* than the one observed in the data under the null hypothesis.

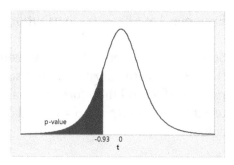

H_a: $\mu > \mu_0$
The researcher believes that the population mean is higher than a certain number. The data provide evidence for this hypothesis in the form of *high* values of $\bar{x}$ and t. The p-value is the probability of observing a value of t *as high or higher* than the one observed in the data under the null hypothesis.

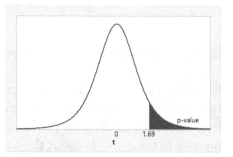

H_a: $\mu \neq \mu_0$
The researcher believes that the population mean differs from a certain number. The data provide evidence for this hypothesis in the form of *low or high* values of $\bar{x}$ and t. If the data suggest lower values of μ, the p-value is *twice* the probability of observing a value of t *as low or lower* than the one observed in the data under the null hypothesis. If the data suggest higher values of μ, the p-value is *twice* the probability of observing a value of t *as high or higher* than the one observed in the data under the null hypothesis.

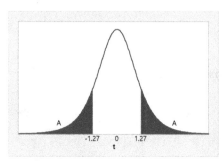

P-value = 2A

Example: Here are the GPAs reported by a simple random sample of 9 students out of 188 students who reported their GPA on a survey at the beginning of a statistics course.

  3.200   3.600   3.800   2.900   2.900   3.350   3.064   3.660   3.100

Suppose you'd like to use these data to determine whether the mean GPA for all respondents to the survey was greater than 3.00. Conduct a test of hypotheses using a significance level of $\alpha = 0.05$.

<u>Hypotheses</u>:

H_0: $\mu = 3.00$
H_a: $\mu > 3.00$

μ = mean GPA for all 188 respondents to the survey

<u>Method/Conditions</u>: We'll use the one-sample t-test.

Conditions:
- SRS: The data are from a simple random sample.
- Nearly-normal distribution of data? A normal probability plot of the data shows a fairly linear pattern, so yes, this condition is satisfied.

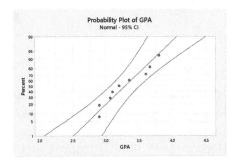

<u>Mechanics</u>:

$$\bar{x} = \frac{3.2 + 3.6 + \cdots + 3.1}{9} = 3.286$$

$$s^2 = \frac{(3.2 - 3.286)^2 + (3.6 - 3.286)^2 + \cdots + (3.1 - 3.286)^2}{9 - 1} = 0.112004$$

$$s = \sqrt{s^2} \approx 0.33467$$

$$t = \frac{\bar{x} - \mu_0}{s/\sqrt{n}} = \frac{3.286 - 3.00}{0.33467/\sqrt{9}} = 2.564 \qquad df = n - 1 = 9 - 1 = 8$$

The data provide evidence for H_a in the form of *high* values of $\bar{x}$ and t. Since the data suggest higher values of μ, the p-value is the probability of observing a value of t as high or higher than 2.564 (or a value of $\bar{X}$ as high or higher than 3.286) if H_0 were true.

Using the t-table and 8 degrees of freedom, we can only approximate the p-value because our t-value is not shown exactly; it's between 2.31 and 2.90:

$$p - value = P(t \geq 2.564)$$
$$= 1 - P(t \leq 2.564)$$
$$= 1 - something\ between\ 0.975\ and\ 0.99$$

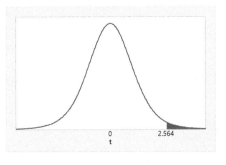

So the p-value is between 0.01 and 0.025.

If the mean GPA for all 188 respondents was 3.00, there is only between a 1% and 2.5% chance of finding an average GPA of 3.286 or higher in a SRS of 9 respondents.

Conclusion:

Since the p-value is quite small (less than the significance level of 0.05), we can reject H_0 in favor of H_a. There is enough evidence at the 5% significance level that the mean GPA for all 188 respondents is higher than 3.00.

If you use Table II to find p-values, you'll almost always need to approximate the p-value since you'll rarely find your t-value in the df row you need. If you want to find the area more precisely, you can use Excel's t-table lookup function, T.DIST, which returns the area to the left of the t-score you input. For example, to find the area below 2.564 in the t-distribution with $df = 8$, enter this in a blank cell in Excel:

$$= T.DIST(2.564, 8, TRUE)$$

The first two arguments are the t-score and degrees of freedom. The last argument is a logical argument that tells Excel what output you want: *TRUE* gives you the area under the curve to the left. *FALSE* gives you the height of the curve.

Here we get 0.9833, so more precisely, the p-value is

$$
\begin{aligned}
p-value &= P(t \geq 2.564) \\
&= 1 - P(t \leq 2.564) \\
&= 1 - 0.9833 = 0.0167.
\end{aligned}
$$

Example: In 2015, 40% of fourth-grade students in the U.S. were classified as proficient in math as assessed by the National Assessment of Educational Progress.[9] We'd like to know whether public fourth-grade schools in Delaware differed from this figure with respect to the mean math proficiency rate in 2015. Shown below are the percent proficient figures in math from a simple random sample of 5 public fourth-grade schools in Delaware from the population of 115 public fourth-grade classes who took the proficiency exam in the state in 2015.[10]

40.09%	22.43%	62.71%	28.33%	43.30%

Hypotheses:

H_0: $\mu = 40\%$
H_a: $\mu \neq 40\%$

μ = mean 2015 math proficiency rate for all public fourth-grade classes in Delaware

Method/Conditions: We'll use the one-sample t-test.

Conditions:

- SRS: The data are from a simple random sample.
- Nearly-normal distribution of data? A normal probability plot of the data shows a fairly linear pattern, so yes, this condition is satisfied.

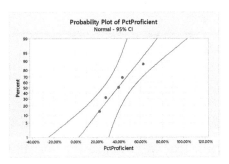

Probability Plot of PctProficient
Normal - 95% CI

Mechanics:

$$\bar{x} = \frac{40.09+22.43+62.71+28.33+43.30}{5} = 39.372$$

$$s^2 = \frac{(40.09-39.372)^2+(22.43-39.372)^2+(62.71-39.372)^2+(28.33-39.372)^2+(43.30-39.372)^2}{5-1} = 242.39102$$

$$s = \sqrt{s^2} \approx 15.5689$$

$$t = \frac{\bar{x}-\mu_0}{s/\sqrt{n}} = \frac{39.372-40}{15.5689/\sqrt{5}} = -0.090 \qquad\qquad df = n-1 = 5-1 = 4$$

The data provide evidence for H_a in the form of *low or high* values of $\bar{x}$ and t. Since the data suggest lower values of μ, the p-value is the probability of observing a value of t as low or lower than -0.090 or as high or higher than +0.090 if H_0 were true. The idea is that we want to consider the *magnitude* of the t-score, not its *sign*. In this problem, a t-score of +0.090 (based on an $\bar{x} = 40.628$) would have been just as much evidence for our H_a, just in the other direction.

Using the t-table and 4 degrees of freedom, we can only approximate the p-value because our t-value is not shown exactly; it's between -1.53 and 0:

$$p - value = 2P(t \le -0.090)$$
$$= 2(something\ between\ 0.1\ and\ 0.5)$$

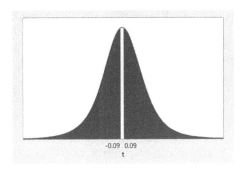

So the p-value is between 0.2 and 1.

If the mean 2015 math proficiency rate for all fourth-grade classes in Delaware was 40%, the chances are not small (between 20% and 100%) of finding a mean 2015 math proficiency rate as far or farther from 40% as 39.372% in a SRS of 5 classes.

Using Excel, we can get the p-value more precisely:

$$T.DIST(-0.090,4,TRUE) = 0.4663$$
$$p - value = 2P(t \le -0.090) = 2(0.4663) = 0.9326$$

Conclusion:

Since the p-value is much greater than any significance level we'd ever use, we cannot reject H_0 in favor of H_a. While there is a tiny bit of evidence that the mean 2015 math proficiency rate for all fourth-grade classes in Delaware was not 40%, the results are inconclusive. They are not at all statistically significant, that is, we cannot easily rule out simple random sampling as an explanation for why we're seeing a different rate in the sample.

Note that we're *not* saying that the data support the hypothesis that the mean rate *is* 40%. All we're saying is that we cannot easily rule out the possibility that the classes where the math proficiency rate was less than 40% were a tiny bit overrepresented in our SRS of 5 classes. When we cannot reject the null hypothesis, it still could be the case that the alternative hypothesis is really true but that we didn't have convincing evidence for it in the data.

Section 6.3 Exercises

1. Determine whether each pair of hypotheses is legitimate. Say what's wrong with the bad ones.

   a. H_0: $x = 40$ vs. H_a: $x < 40$
   b. H_0: $\bar{x} = 0.05$ vs. H_a: $\bar{x} \neq .05$
   c. H_0: $\mu = 100$ vs. H_a: $\mu > 100$
   d. H_0: $t = 0$ vs. H_a: $t < 0$
   e. H_0: $p - value = 0.10$ vs. H_0: $p - value < 0.10$

2. Consider testing the hypotheses H_0: $\mu = 10$ vs. H_a: $\mu > 10$ in a large population. For each sampling scenario, determine whether or not to use the one-sample t-test. Explain your choice.

   a. SRS of $n = 10,000$; data distribution is mildly right skewed
   b. convenience sample of $n = 5,000$; data distribution is nearly-normal
   c. STRS of $n = 300$; data distribution is symmetric, uniform
   d. CRS of 50 clusters each having 10 objects; data distribution is mildly left skewed
   e. SRS of $n = 20$; data distribution is severely skewed

3. Consider testing the hypotheses H_0: $\mu = 515$ vs. H_a: $\mu < 515$ in a large population. For each sampling scenario, determine whether or not to use the one-sample t-test. Explain your choice.

   a. STRS of $n = 500$; data distribution is symmetric
   b. SRS of $n = 30$; data distribution is symmetric
   c. volunteer sample of $n = 2,000$; data distribution is right skewed
   d. CRS of 100 clusters each having 50 objects; data distribution is nearly-normal
   e. SRS of $n = 2,000$; data distribution is symmetric and bimodal

4. "Is the average range of all electric cars currently more than 150 miles?" Consider using a hypothesis test to answer this question.

   a. Identify the objects/subjects, the variable, and type of variable.
   b. Define the parameter in context.
   c. Set up the hypotheses.
   d. For each sampling scenario, determine whether or not to use the one-sample t-test. Explain your choice.
      i. SRS of 100 electric cars; distribution of ranges is right skewed
      ii. all electric cars for a CRS of 5 car manufacturers; distribution of ranges is nearly-normal
      iii. SRS of 10 electric cars; distribution of ranges is right skewed

5. "Do more than 25% of employees at a large company participate in the 401(k) retirement plan?" Consider using a hypothesis test to answer this question. For each sampling scenario, determine whether or not to use the one-sample t-test. Explain your choice.

    a. SRS of 100 employees
    b. STRS of 100 employees, 25 from each of 4 departments
    c. SRS of 5 employees

6. Find the p-value for each hypothesis test. Do it two ways: using the t-table and using technology (e.g. Excel). The t-test conditions are satisfied in each case.

    a. H_0: $\mu = 5$ vs. H_a: $\mu > 5, n = 9, \bar{x} = 14.7, s = 9.1$
    b. H_0: $\mu = 6000$ vs. H_a: $\mu < 6000, n = 14, \bar{x} = 4570.3, s = 3622.8$
    c. H_0: $\mu = 15.6$ vs. H_a: $\mu \neq 15.6, n = 38, \bar{x} = 11.5, s = 8.7$
    d. H_0: $\mu = 430$ vs. H_a: $\mu < 430, n = 20, \bar{x} = 398.45, s = 117.82$

7. Find the p-value for each hypothesis test. Do it two ways: using the t-table and using technology (e.g. Excel). The t-test conditions are satisfied in each case.

    a. H_0: $\mu = 0.338$ vs. H_a: $\mu < 0.338, n = 25, \bar{x} = 0.292, s = 0.047$
    b. H_0: $\mu = 76$ vs. H_a: $\mu > 76, n = 3, \bar{x} = 78, s = 2.65$
    c. H_0: $\mu = 70$ vs. H_a: $\mu \neq 70, n = 501, \bar{x} = 71.5, s = 22.6$
    d. H_0: $\mu = 255$ vs. H_a: $\mu > 255, n = 81, \bar{x} = 260, s = 69.3$

8. In the case of "directional" alternative hypotheses ("<" or ">"), once in a while the data will surprise the researcher by providing evidence in the opposite direction. In these cases, you don't even need to conduct a hypothesis test because there is no evidence for H_a. For parts a. and b., verify that the one-sample t-test is appropriate and then verify that the p-value is large.

    a. H_0: $\mu = 52$ vs. H_a: $\mu > 52$ using a SRS of $n = 101$, moderately skewed data, $\bar{x} = 50, s = 10$
    b. H_0: $\mu = 3$ vs. H_a: $\mu < 3$ using a SRS of $n = 16$, nearly-normal data, $\bar{x} = 3.69, s = 2.44$
    c. When testing H_0: $\mu = \mu_0$ vs. H_a: $\mu > \mu_0$ and a SRS results in $\bar{x} < \mu_0$, what can you say about the p-value?
    d. When testing H_0: $\mu = \mu_0$ vs. H_a: $\mu < \mu_0$ and a SRS results in $\bar{x} > \mu_0$, what can you say about the p-value?

9. Shown here is a normal probability plot of a simple random sample of 50 credit card transaction amounts from the set of all such amounts for a school district in a recent fiscal year. Would a one-sample t-test be appropriate to test hypotheses about the mean transaction amount for all amounts for this school district during the same fiscal year? Explain.

10. Here is a normal probability plot of the sodium amounts in a simple random sample of 10 hot dogs of a certain brand. Would a one-sample t-test be appropriate to test hypotheses about the average sodium content for all hot dogs of this brand? Explain.

11. The time to finish an exam was observed for a simple random sample of 71 students. A normal probability plot of the times is shown here. Would a one-sample t-test be appropriate to test hypotheses about the mean time to finish the exam for all such students? Explain.

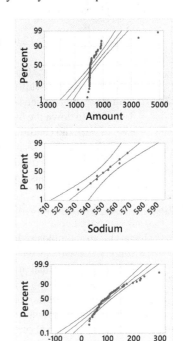

12. A SRS of 29 electric cars is tested to determine each car's 0 – 60mph acceleration time. The mean time was 7.955 seconds and the standard deviation was 3.175 seconds. A histogram of the times is shown here.

    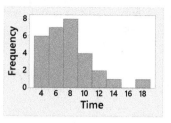

    a. Identify the objects/subjects, the variable and its type.
    b. Do these data provide convincing statistical evidence that the mean 0 – 60mph time for all electric cars is less than 8 seconds?
       i.   Write down the hypotheses and define the parameter in context.
       ii. Check the conditions.
       iii. Perform the mechanics.
       iv. Interpret the p-value.
       v.   Make a conclusion in context.

13. The average number of children under 18 per family in the U.S. is currently 1.8. A simple random sample of German households gave the following data on number of children under 18 per family.

1	3	1	0	2	4	2	0	1	2
4	0	1	4	2	3	2	3	2	0
3	2	1	1	1	1	4	3	2	1
2	1	2	0	0	0	5	1	1	3

    a. Identify the objects/subjects, the variable and its type.
    b. Do these data provide convincing statistical evidence that the average number of children under 18 for all German families is less than 1.8?
       i.   Write down the hypotheses and define the parameter in context.
       ii. Check the conditions.
       iii. Perform the mechanics.
       iv. Interpret the p-value.
       v.   Make a conclusion in context.

14. A small beverage company makes ginger beer and bottles it in 12-oz. bottles. To monitor the filling process, a simple random sample of 5 bottles is selected from the production line every two hours and the amount of ginger beer (ounces) in each bottle is accurately measured. The data for the most recent sample is shown here.

    11.77   11.93   12.09   12.03   11.93

    a. Identify the objects/subjects, the variable and its type.
    b. Do these data provide convincing statistical evidence that the mean amount of ginger beer in all bottles currently being filled is not 12 ounces?
       i.   Write down the hypotheses and define the parameter in context.
       ii. Check the conditions.
       iii. Perform the mechanics.
       iv. Interpret the p-value.
       v.   Make a conclusion in context.

15. An airline has long had a 50-pound limit on the weight of any checked bags. The limit was based on a mean weight of 34 pounds plus a multiple of the standard deviation. In a simple random sample of 101 checked bags from current customers, the average weight was 32.287 pounds and the standard deviation was 5.635 pounds. A normal probability plot of the weights is shown here.

    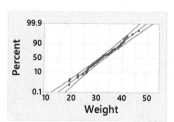

    a. Identify the objects/subjects, the variable and its type.
    b. Do these data provide convincing statistical evidence that the mean weight of all checked bags of current customers is less than 34 pounds?
       i.   Write down the hypotheses and define the parameter in context.

    ii.   Check the conditions.
   iii.   Perform the mechanics.
   iv.   Interpret the p-value.
    v.   Make a conclusion in context.

16. Consider exercise 14.

    a.   How would your conclusion change if the SRS of 5 bottles had given the same standard deviation but a sample mean of 11.75 ounces instead of 11.95 ounces?
    b.   How would your conclusion change if a SRS of 64 bottles gave the same sample mean and standard deviation?

17. Consider exercise 15.

    a.   How would your conclusion change if the SRS of 101 bags had given the same standard deviation but a sample mean of 33.287 pounds instead of 32.287 pounds?
    b.   How would your conclusion change if a SRS of 9 bags gave the same sample mean and standard deviation?

18. As I write this question, the year-to-date (YTD) return of the S&P500 index (an index of 500 of the largest U.S. companies) is 22.4%. The YTD returns of a SRS of 3 mutual funds which invest in the stocks of small U.S. companies (called "small cap funds") are shown here.

    23.39% 22.55% 21.78%

    a.   Identify the objects/subjects, the variable and its type.
    b.   Do these data provide convincing statistical evidence that the average YTD return for all U.S. small cap mutual funds differs from the S&P500 YTD return? Conduct a hypothesis test.

19. Jamie wants to open a juice bar in her town but needs to know whether the incomes of people in the area are high enough to afford such discretionary spending. She will proceed with her venture if the mean annual income of all residents of the town is above $60,000 per year. A SRS of residents yields the following annual incomes (in thousands of dollars:

    60   45   90   81   72   55

    a.   Identify the objects/subjects, the variable and its type.
    b.   Do these data provide convincing statistical evidence for Jamie to open the juice bar? Conduct a hypothesis test.

20. Soil sampling is common in farming because it provides farmers with information about what nutrients are in the soil and which nutrients are lacking. A minimum acceptable nitrate level in corn fields is 30 parts per million (ppm). If the farmer determines that the mean level for all soil specimens in the field is below 30 ppm, she will apply fertilizer. In a SRS of locations in one field, the following nitrate levels (in ppm) were obtained:

    28.0   29.8   26.4   33.1   29.6   25.5   25.6   27.7   31.0   27.3

    a.   Identify the objects/subjects, the variable and its type.
    b.   Should the farmer apply fertilizer to this field? Conduct a hypothesis test.

21. Currently the typical expense ratio for mutual funds which invest in stocks of large-capitalization companies ("large cap equity funds") is 1.25%. You wonder whether the average expense ratio for all mutual funds which invest in stocks of small-capitalization companies ("small cap equity funds") differs from 1.25%. In a SRS of 91 small cap equity funds, you find the mean expense ratio to be 1.4% and the standard deviation to be 0.6%. The p-value for the test is 0.019.

    a.   Which of these is the correct interpretation of the p-value?
        i.   There's a 1.9% chance of getting a mean expense ratio in a SRS of 91 small cap equity funds of at least 0.15% from 1.25% if the mean expense ratio for all small cap equity funds is 1.25%.

    ii.   There's a 1.9% chance of getting a mean expense ratio in a SRS of 91 small cap equity funds of at least 0.15% from 1.25% if the mean expense ratio for all small cap equity funds is not 1.25%.

    iii.   There's a 1.9% chance of getting a mean expense ratio in a SRS of 91 small cap equity funds of at least 1.25% if the mean expense ratio for all small cap equity funds is 1.4%.

    iv.   There's a 1.9% chance that the mean expense ratio for all small cap equity funds is 1.25%.

    v.   There's a 1.9% chance that the mean expense ratio for all small cap equity funds is not 1.25%.

    vi.   There's a 1.9% chance that the mean expense ratio for all small cap equity funds is 1.4%.

  b. Is there convincing statistical evidence that the average expense ratio for all small cap equity funds differs from 1.25%? Explain.

22. Consider the asthma study example from section 2.3. For healthy men, the average forced expiratory volume in one second (FEV1) is 4 liters. You suspect that average FEV1 is lower for mild asthmatics. In a SRS of 22 mild asthmatics, you find the mean and standard deviation of their FEV1 values to be 3.5 and 0.9, respectively. The FEV1 values do not show any outliers and the distribution is fairly symmetric. The p-value for the test is 0.008.

  a. Which of these is the correct interpretation of the p-value?

    i.   There's a 0.8% chance of getting a mean FEV1 in a SRS of 22 mild asthmatics of 3.5 or less if the mean FEV1 for all mild asthmatics is 3.5.

    ii.   There's a 0.8% chance of getting a mean FEV1 in a SRS of 22 mild asthmatics of 3.5 or less if the mean FEV1 for all mild asthmatics is not 4.

    iii.   There's a 0.8% chance of getting a mean FEV1 in a SRS of 22 mild asthmatics of 3.5 or less if the mean FEV1 for all mild asthmatics is 4.

    iv.   There's a 0.8% chance that the mean FEV1 for all mild asthmatics is 4.

    v.   There's a 0.8% chance that the mean FEV1 for all mild asthmatics is less than 4.

    vi.   There's a 0.8% chance that the mean FEV1 for all mild asthmatics is 3.5.

  b. Is there convincing statistical evidence that the average FEV1 for all mild asthmatics is less than 4? Explain.

23. Consider exercise 21. A hypothesis test is shown below. Identify all the mistakes in this response.

Hypotheses: H_0: $\bar{x} = 1.25$ vs. H_a: $\bar{x} > 1.25$ $\bar{x}$ = mean expense ratio for all small cap equity funds

Method/Conditions: SRS: We're told we have a SRS.
                      $n = 91$ is generally sufficiently large enough
                      Use the one-sample z-test.

Mechanics: $z = \frac{1.25-1.4}{0.6} = -0.25$ $p - value = P(Z \le -0.25) = 0.4013$
           There's a 40.13% chance that the mean expense ratio for all small cap equity funds is 1.25%.

Conclusion: The average expense ratio for small cap mutual funds is the same as large cap mutual funds.

24. Consider exercise 21. Show the details in the four steps of the hypothesis test.

25. Consider exercise 22. A hypothesis test is shown below. Identify all the mistakes in this response.

Hypotheses: H_0: $\mu < 3.5$ vs. H_a: $\mu > 4$ μ = mean FEV1 for the 22 mild asthmatics

Method/Conditions: SRS: We're told we have a SRS.
                        $n = 22$ is not large enough, so the t-test is not valid. I'll do it anyway...
                        Use the one-sample t-test.

Mechanics: $t = \frac{3.5-4}{0.9*\sqrt{22}} = -0.12$ $df = 22 - 1 = 21$
           $p - value = P(t \le -0.12) = T.DIST(-0.12, 21, false) = 0.3913$
           There's a 39.13% chance that the mean FEV1 for all mild asthmatics is 3.5.

Conclusion: The sample size is too small for these results to be accurate.

26. Consider exercise 22. Show the details in the four steps of the hypothesis test.

6.4 Type I and Type II Errors

In the hypothesis tests we've considered so far, the research questions could be captured by the alternative hypothesis; we wanted to see if the data provided enough evidence to substantiate the research hypothesis. If so, then we conclude that the research hypothesis is true. If not, we simply say that the data are inconclusive; we did NOT conclude that the null hypothesis is true.

However, some research questions are about making decisions between two competing hypotheses based on data and the goal is to decide which hypothesis is true based on the data. If the data do not convincingly support the alterative hypothesis, the result is that we'd act as if the null hypothesis is true anyway!

Consider the jury trial analogy to hypothesis testing:

Hypothesis Test	Trial by Jury
Researcher	Prosecutor
Question	Is the defendant guilty?
Null Hypothesis	H_0: defendant is innocent
Alternative Hypothesis	H_a: defendant is guilty
Data	Evidence presented at trial
Assume H_0 is true	"Innocent until proven guilty"
Analyze data	Jury deliberations
p-value	Likelihood that the evidence presented came from an innocent defendant
Significance level (α)	"Reasonable doubt"
p-value $< \alpha$	Evidence goes "beyond a reasonable doubt"
Reject H_0; statistically significant results	"Guilty" verdict
Inconclusive results	"Not guilty" verdict

Defendants are presumed "innocent until proven guilty" but if there is enough contradictory evidence of guilt presented at trial, the jury will return a "guilty" verdict; that is, they will reject the null hypothesis. If there is not enough contradictory evidence presented at trial, the jury's verdict is "not guilty," never "innocent." However, we act as if the defendant is innocent! They are immediately free to go as if they were innocent. While the "not guilty" language we use reminds us that the trial has not positively established their innocence, the justice system operates as if the null hypothesis is true. You can think of a trial by jury as, essentially, using data to make a decision between two competing hypotheses.

If we'd never act as if the null hypothesis were true, there would only be one kind of mistake we could make: rejecting a true null hypothesis. Otherwise, if the results of the test were inconclusive, we would not make any decision. However, if an inconclusive result of the test leads to acting as if the null hypothesis were true, then there's a second kind of mistake we could make: accepting a false null hypothesis. The differences are illustrated in the chart below.

	H_0 true in reality	H_a true in reality
decide to accept H_0	correct decision	Type II Error $\beta = P(Type\ II\ Error)$
decide to accept H_a (reject H_0)	Type I Error $\alpha = P(Type\ I\ Error)$	correct decision

We hope the data lead to a correct decision about reality but there are two kinds of mistakes that are possible: rejecting a true H_0 and accepting a false H_0. The significance level of the test, α, is actually the probability of the test resulting in a Type I error. Since we always choose a small significance level on the order of 0.05 or so, we automatically limit the chances of making a Type I error. That's why we're comfortable using language like "enough evidence that H_a is true." We have some protection against Type I errors. The Greek letter *beta* (denoted β) is used for the probability of making a Type II error. You'll notice that β did not appear anywhere in our hypothesis tests so far. If we do not limit the chances of making a Type II error, we will NOT be comfortable using language like "enough evidence that H_0 is true." The problem is that in cases where we'd act as if H_0 is true, if β is fairly large, then we're vulnerable to making a Type II error.

Here's how the chart would look for the jury trial analogy.

H_0: defendant is innocent
H_a: defendant is guilty

	defendant is innocent	defendant is guilty
let defendant go	correct decision	Type II Error (β) free a guilty defendant
punish defendant	Type I Error (α) punish an innocent defendant	correct decision

The protections afforded to the defendant (they're presumed innocent to start, the jury verdict must be unanimous, evidence can be excluded if not obtained correctly, etc.) have the effect of making α small. The justice system wants to avoid making a Type I error and punishing an innocent person!

Using small α's helps us avoid making Type I errors; if the p-value is small enough, we can feel comfortable rejecting the null hypothesis in favor of the alternative hypothesis because we know the test procedure will rarely result in false rejections. So far, however, we have not even considered Type II errors and small β's. Making a decision to accept a null hypothesis without knowing β could be disastrous. Consider a smoke detector and the following hypotheses about the state of a building:

H_0: no fire
H_a: fire

	no fire	fire
behave as if no fire	correct decision	Type II Error (β) potential loss of life
leave the building	Type I Error (α) inconvenience	correct decision

The smoke detector is like a 24/7 hypothesis test, continuously collecting data from the building and signaling whether there is a fire or not. When the smoke detector isn't going off, we act as if H_0 is true (that there is no fire). If it goes off when there's really no fire (like when there's a test), it's just an inconvenience. You have to leave the building, wait for a while, and then re-enter. You'd like α to be low so that false alarms don't happen very often. However, it's much more important that β is as close to 0 as possible! People can die if there really is a fire and the alarm doesn't go off, especially during the night. That's why the system is tested regularly and why you should regularly replace the batteries in your smoke detectors. Accepting the null hypothesis based on the smoke detector without knowing how likely it is to make a Type II error can be disastrous.

Example: Food packagers must periodically sample packaged products to make sure the fill amount (how much product is in the package) is correct. A SRS of 10 2-liter soda bottles is collected and the volume of soda in each is measured. Of course there will be slight variation from one bottle to the next but the company wants to see if there is enough evidence in the sample that the mean fill volume for all bottles at the time is different than 2 liters. If so, adjustments will have to be made. If not, no action will be taken as if the mean fill weight is 2 liters. Set up the hypotheses for this test and describe both a Type I and a Type II error.

Solution: The hypotheses are

H_0: $\mu = 2$
H_a: $\mu \neq 2$

where μ = mean fill volume (in liters) for all soda bottles.

To describe the potential errors, we'll construct the chart:

	H_0: $\mu = 2$ bottles are filled correctly on average	H_a: $\mu \neq 2$ bottles are either over- or under-filled on average
no action	correct decision	Type II Error (β)
adjust the process	Type I Error (α)	correct decision

A Type I error would be to adjust the process when, in fact, the mean fill volume for all soda bottles is 2 liters. That would mean an unnecessary adjustment and downtime, so α should be fairly small.

A Type II error would be to take no action when, in fact, the mean fill volume for all soda bottles is not 2 liters. Bottles would be being consistently over-filled (meaning lost profits) or under-filled (which could mean dissatisfied consumers), so β should also be fairly small.

Example: Market research is an important part of a company's business plan so that products and services will be well-received by the target market. After a fire, the restaurant I used to work in changed much of the booth seating to table seating and replaced much of the carpet with hard flooring. Those changes, considered by many to be upgrades, were not as generally well-received by the restaurant's patrons who were older, more traditional, didn't like change, and preferred the quieter, cozier, carpet-floored, booth-style seating. The restaurant ultimately changed back to the old style but could have saved money by doing a bit of market research first. Consider a SRS of 60 customers where we'd ask each customer whether they prefer booth or table seating. If the data show that, say, more than 50% of all the restaurant's customers prefer table seating, then table seating will be installed. If not, booth seating will be installed. Set up the hypotheses for this test and describe both a Type I and a Type II error.

Solution: The hypotheses are

H_0: $p \leq 0.50$
H_a: $p > 0.50$

where $p =$ proportion of all the restaurant's customers who prefer table seating over booth seating.

To describe the potential errors, we'll construct the chart:

	H_0: $p \leq 0.50$ half or less of all customers prefer tables	H_a: $p > 0.50$ more than half of all customers prefer tables
install booths	correct decision	Type II Error (β)
install tables	Type I Error (α)	correct decision

A Type I error would be to install tables when customers generally prefer booths. This is what happened to the restaurant where I used to work. The result could be dissatisfied customers, loss of business, or extra cost incurred by having to switch back to booths. So α should be quite small.

A Type II error would be to install booths when customers generally prefer tables. The result could be dissatisfied customers, loss of business, or extra cost incurred by having to switch back to tables. So β should also be quite small.

You already know how to avoid making Type I errors by choosing a small significance level. In the next section, you'll learn how to design a study to limit the chances of making a Type II error by using a small value of β.

Section 6.4 Exercises

1.   Which type of error is made when the null hypothesis is falsely accepted?

2. Which type of error is made when the alternative hypothesis is falsely accepted?

3. Identify whether a Type I error, Type II error, or neither is made.

    a. H_0 is rejected when, in reality, H_a is true.
    b. We act as though the alternative hypothesis is true when, in reality, the null hypothesis is true.
    c. The results are declared "statistically significant" when, in reality, the null hypothesis is true.

4. Identify whether a Type I error, Type II error, or neither is made.

    a. H_a is accepted when, in reality, H_0 is true.
    b. We act as though the null hypothesis is true when, in reality, the alternative hypothesis is true.
    c. The results are declared "statistically significant" when, in reality, the alternative hypothesis is true.

5. What symbol is used to represent the probability of a Type I error?

6. What symbol is used to represent the probability of a Type II error?

7. Consider these hypotheses:

    H_0: $\mu = 150$
    H_a: $\mu > 150$

    where μ = average range (in miles) for all electric cars.

    a. If we conclude that $\mu > 150$ when, in fact, $\mu = 150$, what type of error have we made?
    b. If we conclude that $\mu = 150$ when, in fact, $\mu > 150$, what type of error have we made?

8. Consider these hypotheses:

    H_0: $p = 0.60$
    H_a: $p > 0.60$

    where p = proportion of all seniors as a college who have attended a career fair.

    a. If we conclude that $p = 0.60$ when, in fact, $p > 0.60$, what type of error have we made?
    b. If we conclude that $p > 0.60$ when, in fact, $p = 0.60$, what type of error have we made?

9. Consider these hypotheses:

    H_0: $p = 0.95$
    H_a: $p < 0.95$

    where p = proportion of all recent orders that were shipped on time.

    a. If we conclude that 95% of all recent orders were shipped on time when, in fact, fewer than 95% of all recent orders were shipped on time, what type of error have we made?
    b. If we conclude that fewer than 95% of all recent orders were shipped on time when, in fact, 95% of all recent orders were shipped on time, what type of error have we made?

10. Consider these hypotheses:

    H_0: $\mu_1 = \mu_2$
    H_a: $\mu_1 \neq \mu_2$

where μ_1 = average number of children under 18 for all U.S. families and μ_2 = average number of children under 18 for all German families.

   a.   If we conclude that there is no difference between the average number of children under 18 for all U.S. families and all German families when, in fact, there is a difference, what type of error have we made?
   b.   If we conclude that there is a difference between the average number of children under 18 for all U.S. families and all German families when, in fact, there is no difference, what type of error have we made?

11. Consider these hypotheses:

H_0: $p_1 = p_2$
H_a: $p_1 \neq p_2$

where p_1 = proportion of all bags that don't get delivered on time for American Airlines and p_2 = proportion of all bags that don't get delivered on time for Southwest Airlines.

   a.   If we conclude that there is a difference between the proportion of all bags that don't get delivered on time for American Airlines and that for Southwest Airlines when, in fact, there is no difference, what type of error have we made?
   b.   If we conclude that there is no difference between the proportion of all bags that don't get delivered on time for American Airlines and that for Southwest Airlines when, in fact, there is a difference, what type of error have we made?

12. A company's web designer wants to see whether there is a difference between two different versions of the same webpage with respect to the proportion of purchases made from the page. Consider these hypotheses:

H_0: The difference between the two versions does not affect purchase rate.
H_a: The difference between the two versions affects the purchase rate.

   a.   Describe a Type I error.
   b.   Describe a Type II error.
   c.   Which type of error has more serious consequences? Explain.
   d.   Which is more important: to make sure α is small or to make sure β is small? Explain.

13. A pharmaceutical company is testing a new drug to treat insomnia. It must show that the drug is effective using statistical evidence from a clinical trial (testing the drug on people). The drug has some very bad side effects. Consider these hypotheses:

H_0: The drug does not differ from placebo in effectiveness.
H_a: The drug is more effective than placebo.

   a.   Describe a Type I error.
   b.   Describe a Type II error.
   c.   Which type of error do you think has more serious consequences? Explain.
   d.   Which is more important to you: to make sure α is small or to make sure β is small? Explain.

14. A pharmaceutical company is testing a new drug to treat pancreatic cancer (a very aggressive form of cancer). It must show that the drug is effective using statistical evidence from a clinical trial (testing the drug on people). So far, the drug has shown no bad side effects. Consider these hypotheses:

H_0: The drug does not differ from placebo in effectiveness.
H_a: The drug is more effective than placebo.

   a.   Which type of error do you think has more serious consequences? Explain.
   b.   Which is more important to you: to make sure α is small or to make sure β is small? Explain.

15. A car manufacturer learns that some of its 5-year-old cars are experiencing back-up camera malfunctions. The company is hesitant to issue a recall notice but will do so if more than 10% of all cars are currently experiencing the problem. Consider these hypotheses:

    H_0: $p = 0.10$
    H_a: $p > 0.10$

    a. Describe a Type I error.
    b. Describe a Type II error.
    c. Which type of error do you think has more serious consequences? Explain.
    d. Which is more important to you: to make sure α is small or to make sure β is small? Explain.

16. Jamie wants to open a juice bar in her town but needs to know whether the incomes of people in the area are high enough to afford such discretionary spending. She will proceed with her venture if the mean annual income of all residents of the town is above $60,000 per year. Consider these hypotheses:

    H_0: $\mu = 60{,}000$
    H_a: $\mu > 60{,}000$

    a. Describe a Type I error.
    b. Describe a Type II error.
    c. Which type of error do you think has more serious consequences? Explain.
    d. Which is more important to you: to make sure α is small or to make sure β is small? Explain.

17. Soil sampling is common in farming because it provides farmers with information about what nutrients are in the soil and which nutrients are lacking. A minimum acceptable nitrate level in corn fields is 30 parts per million (ppm). If the farmer determines that the mean level for all soil specimens in the field is below 30 ppm, she will apply fertilizer.

    a. Write down the null and alternative hypotheses for the farmer and define the parameter in context.
    b. Describe a Type I error.
    c. Describe a Type II error.
    d. Which type of error do you think has more serious consequences? Explain.

18. A magazine is planning to add an online version in addition to its print version but will only do so if more than 15% of current subscribers would be interested.

    a. Write down the null and alternative hypotheses for the magazine and define the parameter in context.
    b. Describe a Type I error.
    c. Describe a Type II error.
    d. Which type of error do you think has more serious consequences? Explain.

19. A political candidate will campaign in a certain state if she determines that more than 40% of eligible voters prefer her over her closest competitor.

    a. Write down the null and alternative hypotheses for the candidate and define the parameter in context.
    b. Describe a Type I error.
    c. Describe a Type II error.
    d. Which type of error do you think has more serious consequences? Explain.

20. A small beverage company makes ginger beer and bottles it in 12-oz. bottles. To monitor the filling process, a simple random sample of 5 bottles is selected from the production line every two hours and the amount of ginger beer (ounces) in each bottle is accurately measured. The company wants to use the data to see if the mean amount of ginger beer in all bottles currently being filled is not 12 ounces.

    a. Write down the null and alternative hypotheses for the company and define the parameter in context.
    b. Describe a Type I error.

c.   Describe a Type II error.
d.   Which type of error do you think has more serious consequences? Explain.

6.5 Power and Determining the Sample Size

In the last section, you learned about the two kinds of errors that can occur when treating a hypothesis test as a tool to make a decision between two competing hypotheses. In those cases, inconclusive evidence in the data leading us to "fail to reject the null hypothesis" means that we act as if the null hypothesis is true, in effect "accepting the null hypothesis." While choosing to use a small significance level (α) limits the chances of making a Type I error, we haven't yet seen how to do the same for β and Type II errors. Here's that table again showing the two types of errors and their probabilities.

	H_0 true in reality	H_a true in reality
decide to accept H_0	correct decision	Type II Error $\beta = P(Type\ II\ Error)$
decide to accept H_a (reject H_0)	Type I Error $\alpha = P(Type\ I\ Error)$	correct decision

In this section we'll take a deep dive into Type II errors and how both types of errors are related to determining sample size. Let's do it separately for a one-proportion test and a one-sample mean test.

Proportions

Consider the following hypothesis test about a population proportion using a significance level of $\alpha = 0.05$:

H_0: $p = 0.3$
H_a: $p > 0.3$

A SRS of 250 objects from a very large population gives 90 successes. Using the one-proportion z-test (are the conditions satisfied?), we get

$$z = \frac{\hat{p}-p_0}{\sqrt{\frac{p_0(1-p_0)}{n}}} = \frac{\frac{90}{250}-0.3}{\sqrt{\frac{0.3(1-0.3)}{250}}} = 2.07$$

and using Table I,

$$p - value = P(Z \geq 2.07) = 1 - 0.4808 = 0.0192.$$

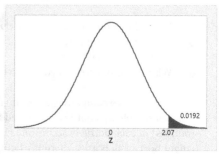

If the population proportion is 0.3, then there's only a 1.92% chance of getting at least 90 successes (resulting in a z-score of at least 2.07) in a SRS of 250. Since this is smaller than the significance level of 0.05, we reject H_0 and conclude that there are more than 30% successes in the population.

In a SRS of 250, how big would the z-score need to be to reject H_0? That's an easy normal curve problem! For some z-score, call it a, the p-value would be exactly 0.05. Any z-score bigger than a would result in a smaller p-value, so we want to find a so that $P(Z \geq a) = 0.05$:

Using Table I, we find that the z-score having an area of 0.05 to its right (an area of 0.95 to its left) under the curve is

$a = 1.645$.

So at a 5% significance level, we'd reject H_0 in favor of H_a in this test if $z > 1.645$. If $z > 1.645$, the p-value would be less than 0.05.

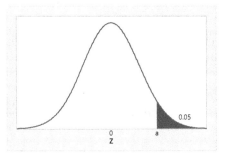

The set of z-scores $z > 1.645$ is called the **rejection region** of the test; if the data produce a z in this region, the null hypothesis is rejected. Let's find out what value of $\hat{p}$ it would take to do that:

We'd need: $z = \dfrac{\hat{p} - p_0}{\sqrt{\dfrac{p_0(1-p_0)}{n}}} > 1.645$

Substitute the value of p_0 from the null hypothesis: $\dfrac{\hat{p} - 0.3}{\sqrt{\dfrac{0.3(1-0.3)}{250}}} > 1.645$

Solve for $\hat{p}$: $\hat{p} > 0.3 + 1.645\sqrt{\dfrac{0.3(1-0.3)}{250}} = 0.3477$

So we'd need about 34.77% or more successes (about 87 or more) in the SRS of 250 to get a p-value small enough to reject H_0.

The probability of a Type II error, β, is the probability of accepting the null hypothesis when, in reality, the alternative hypothesis is true. The problem is that, at a particular significance level, the probability of a Type II error depends on actual value of the parameter in the alternative hypothesis.

In the above example, there is only one value of p specified in the null hypothesis but any value above 0.3 in the alternative hypothesis. To find the probability of a Type II error, we need to select a value in H_a and use it to make the calculation.

Example: Find β, the probability of a Type II error for the test of

H_0: $p = 0.3$
H_a: $p > 0.3$

based on a SRS of 250 using $\alpha = 0.05$ when $p = 0.4$.

Solution: We've already found the rejection region of this test to be $z > 1.645$ which happens when $\hat{p} > 0.3477$. To commit a Type II error, we'd need to accept the null hypothesis when the actual population proportion is 0.4. That means we'd need $\hat{p} \leq 0.3477$. The probability of that happening when $p = 0.4$ is

$$\beta = P(Type\ II\ Error) = P\left(\hat{P} \leq 0.3477\ when\ p = 0.4\right)$$

Then standardize the value of $\hat{p}$ using $p = 0.4$ instead of the null value of $p = 0.3$:

$$\beta = P(Type\ II\ Error) = P\left(Z \leq \frac{0.3477 - 0.4}{\sqrt{\frac{0.4(1-0.4)}{250}}}\right) = P(Z \leq -1.69)$$

and use Table I to get

$$\beta = P(Type\ II\ Error) = 0.0455 \quad \text{when } p = 0.4.$$

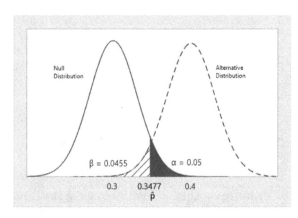

Here's a picture that shows both the null distribution of $\hat{P}$ where $p = 0.3$ and the alternative distribution where $p = 0.4$. This test performs pretty well in this scenario! There's only a 5% chance of rejecting a true null hypothesis (Type I Error) and only a 4.55% chance of accepting a false null hypothesis (Type II error) when p is actually 0.4. The test performs this well because the two distributions are nicely separated.

It turns out that the probability of a Type II error gets larger the closer the null and alternative hypotheses are. That means it's hard to detect when the alternative is true only by a small amount. Here's what happens if you recalculate β for the above example when $p = 0.32$.

$$\beta = P(Type\ II\ Error) = P\left(\hat{P} \leq 0.3477\ when\ p = 0.32\right)$$

Then standardize the value of $\hat{p}$ using $p = 0.32$ instead of the null value of $p = 0.3$:

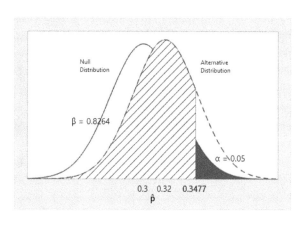

$$\beta = P(Type\ II\ Error) = P\left(Z \leq \frac{0.3477 - 0.32}{\sqrt{\frac{0.32(1-0.32)}{250}}} \right) =$$

$P(Z \leq 0.94)$

and use Table I to get

$\beta = P(Type\ II\ Error) = 0.8264$ when $p = 0.32$.

Since 0.32 is only a bit higher than 0.3, it's quite likely that the "32% success" population will give data that do not look strange under the null hypothesis "30% success" population. While there's still only a 5% chance of rejecting a true null hypothesis (Type I Error), now there's an 82.64% chance of accepting a false null hypothesis! This test can't detect such a subtle difference between H_0 and H_a.

A test having a large Type II error probability is said to lack statistical power. The **power** of a test is the probability of accepting the alternative hypothesis when it's actually true; it's the opposite of a Type II error:

$$power = P(accepting\ H_a\ when\ H_a\ is\ true) = 1 - \beta$$

Just like a Type II error, the power of a test is dependent on a particular value in the alternative hypothesis.

Example: Find the power of the test of

H_0: $p = 0.3$
H_a: $p > 0.3$

based on a SRS of 250 using $\alpha = 0.05$ both when $p = 0.32$ and when $p = 0.4$.

Solution: We've already done the hard work! Just use the Type II error probability to get power.

	$\beta = P(Type\ II\ Error)$	Power
$p = 0.4$	0.0455	$1 - 0.0455 = 0.9545$
$p = 0.32$	0.8264	$1 - 0.8264 = 0.1736$

A test has more power if the null and alternative distributions of the relevant statistic are more separated.

There's a very practical reason why knowing about Type II errors and power is important. It's sample size. How do you determine the sample size for your SRS? The best way to do it is to choose the sample size that results in a test with *both* a low Type I error rate *and* high enough power.

The picture below illustrates (for a ">" alternative) what we'd like the one-proportion z-test to do: sufficiently separate the null distribution of $\hat{P}$ (solid curve centered at p_0) and the alternative distribution of $\hat{P}$ (dashed curve centered at p_a) when $p = p_a$ so that both α and β are small.

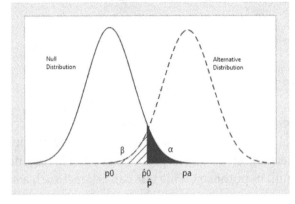

The steps are:

1. Use the null distribution and p_0 to find the value of $\hat{p}$ (call it $\hat{p}_0$) at the threshold of the rejection region for your choice of α. You'll get an expression in terms of n.
2. Find the z-score z_β such that $P(Type\ II\ error) = \beta$.
3. Set $\dfrac{\hat{p}_0 - p_a}{\sqrt{\dfrac{p_a(1-p_a)}{n}}} = z_\beta$ and solve for n.

Example: You want to design a study of the participation rate of a store's loyalty card program. The store owner thinks the participation rate is higher than 60%. How many customers should you randomly select in order to detect a participation rate of at least 65% with 80% power and a significance level of 0.10?

Solution: The test will be of

H_0: $p = 0.60$
H_a: $p > 0.60$

Let's follow the steps.

1. For the significance level of 0.10, the rejection region will be $z > 1.28$. The sample proportion $\hat{p}$ would need to be such that

    $$z = \frac{\hat{p}_0 - 0.60}{\sqrt{\frac{0.60(1-0.60)}{n}}} = 1.28$$

    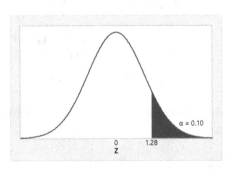

    So $\hat{p}_0 = 0.60 + 1.28\sqrt{\dfrac{0.60(1-0.60)}{n}}$.

2. For 80% power, $\beta = 1 - 0.80 = 0.20$. You want $P(Type\ II\ error) = P(Z \le z_\beta) = 0.20$. From Table I, $z_\beta = -0.84$.

3. You want the test to correctly reject H_0 when $p = 0.65$ or more, so use $p_a = 0.65$ and solve for n:

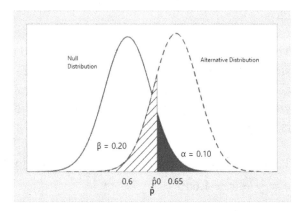

$$\frac{\hat{p}_0 - p_a}{\sqrt{\frac{p_a(1-p_a)}{n}}} = z_\beta$$

$$\frac{0.60 + 1.28\sqrt{\frac{0.60(1-0.60)}{n}} - 0.65}{\sqrt{\frac{0.65(1-0.65)}{n}}} = -0.84$$

Multiply the left-hand side by $\frac{\sqrt{n}}{\sqrt{n}}$ and then simplify:

$$n = \left[\frac{1.28\sqrt{0.60(1-0.60)} + 0.84\sqrt{0.65(1-0.65)}}{0.65 - 0.60}\right]^2 = 422.48$$

So use a SRS of 423 customers.

Means

Calculating Type II error probabilities and finding optimal sample sizes work in a similar way, in theory, for tests about a population mean based on the one-sample t-test procedure. However, the calculations are much more difficult. One workaround is to use the result that the t-distribution approaches the standard normal distribution as n gets large. Instead of using the t-distribution as a model for $t = \frac{\bar{X} - \mu_0}{s/\sqrt{n}}$, use the standard normal distribution as a model for $Z = \frac{\bar{X} - \mu_0}{s/\sqrt{n}}$. The other difficulty is that we may not have collected data yet and would not know the value of s. One solution is to estimate the value of s by dividing the variable's range by 4. If you have reasonable guesses as to the maximum and minimum values you'd ever see, use $s = \frac{range}{4}$. The rationale is that in many numeric distributions, most values will be found within 2 standard deviations of the mean so that the range of the distribution will be about 4 standard deviations wide.

Example: Find β and the power of the test of

H_0: $\mu = 100$
H_a: $\mu < 100$

based on a SRS of 75 using $\alpha = 0.05$ when $\mu = 96$. Assume that the numeric variable has a minimum of 70 and a maximum of 130.

Solution: Since the $range = 130 - 70 = 60$, we can estimate s by using $\frac{range}{4} = \frac{60}{4} = 15$. The next step is to find the rejection region. To do that, we'll use the fact that the test statistic $Z = \frac{\bar{X} - \mu_0}{S/\sqrt{n}}$ has an approximate standard normal distribution under H_0 since $n = 75$ is considered sufficiently large.

To reject H_0 in favor of H_a, the p-value would need to be smaller than $\alpha = 0.05$ which would happen for small values of z. Using Table I we find that any z-score smaller than -1.645 would result in a p-value less than 0.05, so the rejection region is $z < -1.645$.

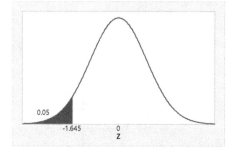

To find out what value of $\bar{x}$ it would take to do that, we'd need $z = \frac{\bar{x} - \mu_0}{s/\sqrt{n}} < -1.645$.

Substitute the value of μ_0 from the null hypothesis and your estimate of s: $\frac{\bar{x} - 100}{15/\sqrt{75}} < -1.645$ and solve for $\bar{x}$:

$$\bar{x} < 100 - 1.645 \frac{15}{\sqrt{75}} = 97.1508$$

So we'd need an average of about 97.1508 in the SRS of 75 (and a sample standard deviation of about 15) to get a p-value small enough to reject H_0.

To commit a Type II error, we'd need to accept the null hypothesis when the actual population mean is 96. That means we'd need $\bar{x} \geq 97.1508$. The probability of that happening when $\mu = 96$ is

$$\beta = P(Type\ II\ Error) = P(\bar{X} \geq 97.1508\ when\ \mu = 96).$$

Then standardize the value of $\bar{x}$ using $\mu = 96$ instead of the null value of $\mu = 100$:

$$\beta = P(Type\ II\ Error) = P\left(Z \geq \frac{97.1508 - 96}{15/\sqrt{75}}\right) = P(Z \geq 0.66)$$

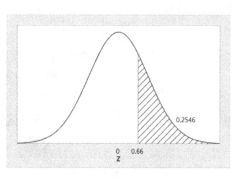

and use Table I to get

$$\beta = P(Type\ II\ Error) = 0.2546\ \ when\ \mu = 96.$$

The power of this test is $1 - \beta = 1 - 0.2546 = 0.7454$ when $\mu = 96$.

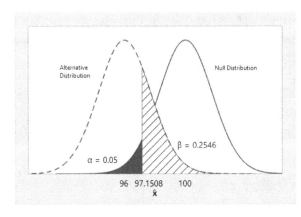

Here's a picture that shows both the null distribution of $\bar{X}$ where $\mu = 100$ and the alternative distribution where $\mu = 96$. This test doesn't perform too badly in this scenario. There's only a 5% chance of rejecting a true null hypothesis (Type I Error) but a larger 25.46% chance of accepting a false null hypothesis (Type II error) when μ is actually 96. There is a 74.54% chance that the test would accept a true alternative hypothesis when the population mean is actually 96. A bit more separation between the two distributions would make the test more powerful against a difference of only 4 between H_0 and H_a.

The picture below illustrates (for a "<" alternative) what we'd like the one-sample z-test to do: sufficiently separate the null distribution of $\bar{X}$ (solid curve centered at μ_0) and the alternative distribution of $\bar{X}$ (dashed curve centered at μ_a) when $\mu = \mu_a$ so that both α and β are small. (Remember the z-test is just "filling in" for the t-test to make our calculations easier!)

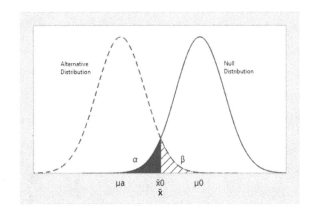

You can set the separation so that both α and β (corresponding to your desired detectable difference between H_0 and H_a) are to your liking and then work backwards to get the required sample size.

The steps are:
1. Estimate the standard deviation of the data, s.
2. Use the null distribution and μ_0 to find the value of $\bar{x}$ (call it $\bar{x}_0$) at the threshold of the rejection region for your choice of α. You'll get an expression in terms of n.
3. Find the z-score z_β such that $P(Type\ II\ error) = \beta$.
4. Set $\frac{\bar{x}_0 - \mu_a}{s/\sqrt{n}} = z_\beta$ and solve for n.

The only catch is that if you get a small n, the standard normal approximation to the t-distribution isn't valid. In that case, ask a statistician to help you with the more difficult t-test approach, especially if you think the data values are skewed.

Example: According to the Environmental Protection Agency (EPA), the 2019 Ford Mustang with a 5.0 liter motor and a 6-speed manual transmission is rated at 18 miles per gallon (mpg).[11] Your 2019 Mustang that is similarly-equipped tends to get around 16 mpg so you think the EPA's figure is too high. How many Mustangs would you need to randomly select in order to detect a population mean gas mileage of 16 mpg for the 2019 5.0L 6-speed Mustang with 85% power and a significance level of 0.02? Suppose actual gas mileages for this vehicle range from 13 mpg to 26 mpg.

Solution: The test will be of

$H_0: \mu = 18$
$H_a: \mu < 18$

Let's follow the steps.

1. Estimate the standard deviation of gas mileages as $s = \frac{range}{4} = \frac{26-13}{4} = 3.25$.

2. For the significance level of 0.02, the rejection region will be $z < -2.05$. The sample mean $\bar{x}$ would need to be such that

    $z = \frac{\bar{x}_0 - 18}{3.25/\sqrt{n}} = -2.05$

    So $\bar{x}_0 = 18 - 2.05 \frac{3.25}{\sqrt{n}}$.

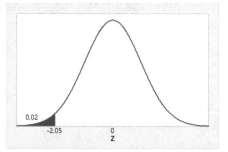

3. For 85% power, $\beta = 1 - 0.85 = 0.15$. You want $P(Type\ II\ error) = P(Z \geq z_\beta) = 0.15$. From Table I, $z_\beta = 1.04$.

4. You want the test to correctly reject H_0 when $\mu = 16$, so use $\mu_a = 16$ and solve for n:

    $\frac{\bar{x}_0 - \mu_a}{s/\sqrt{n}} = z_\beta$

    $\frac{18 - 2.05\frac{3.25}{\sqrt{n}} - 16}{3.25/\sqrt{n}} = 1.04$

    Multiply the left-hand side by $\frac{\sqrt{n}}{\sqrt{n}}$ and then simplify:

    $n = \left[\frac{3.25(2.05+1.04)}{18-16} \right]^2 = 25.2$

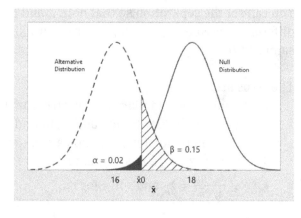

So use a SRS of 26 Mustangs.

If you like formulas, I've provided them below for calculating the required sample size for each of the scenarios in this section. For non-directional alternatives (a "$\neq$" in H_a), use either the upper or lower threshold of the rejection region for z'.

Sample Size for One-Proportion z-test for p	Sample Size for One-Sample t-test* for μ
$$n = \left[\frac{\lvert z' \rvert \sqrt{p_0(1-p_0)} + \lvert z_\beta \rvert \sqrt{p_a(1-p_a)}}{p_a - p_0} \right]^2$$ p_0 = null hypothesized value of p p_a = desired detectable alternative value of p z' = threshold of rejection region z_β = z-score for desired power	$$n = \left[\frac{s(\lvert z' \rvert + \lvert z_\beta \rvert)}{\mu_a - \mu_0} \right]^2$$ s = estimated standard deviation of data μ_0 = null hypothesized value of μ μ_a = desired detectable alternative value of μ z' = threshold of rejection region z_β = z-score for desired power *This gives an approximate sample size for the t-test. If the resulting n is small, seek assistance from a statistician, especially if you think the data values are skewed.

Section 6.5 Exercises

1. How is the power of a hypothesis test related to a Type II error?

2. For each part, fill in either "low" or "high": We'd like it to be _____.

   a. α
   b. β
   c. power

3. For each test, use the one-proportion z-method to find the rejection region.

   a. $H_0: p = 0.5$ vs. $H_a: p > 0.5$ using $\alpha = 0.05$
   b. $H_0: p = 0.5$ vs. $H_a: p < 0.5$ using $\alpha = 0.05$
   c. $H_0: p = 0.5$ vs. $H_a: p \neq 0.5$ using $\alpha = 0.05$

4. For each test, use the one-proportion z-method to find the rejection region.

   a. $H_0: p = 0.2$ vs. $H_a: p > 0.2$ using $\alpha = 0.01$
   b. $H_0: p = 0.2$ vs. $H_a: p < 0.2$ using $\alpha = 0.01$
   c. $H_0: p = 0.2$ vs. $H_a: p \neq 0.2$ using $\alpha = 0.01$

5. For each test, use the z-method (instead of the one-sample t-method) to find the rejection region.

   a. $H_0: \mu = 100$ vs. $H_a: \mu > 100$ using $\alpha = 0.10$
   b. $H_0: \mu = 100$ vs. $H_a: \mu < 100$ using $\alpha = 0.10$
   c. $H_0: \mu = 100$ vs. $H_a: \mu \neq 100$ using $\alpha = 0.10$

6. For each test, use the z-method (instead of the one-sample t-method) to find the rejection region.

   a. $H_0: \mu = 5$ vs. $H_a: \mu < 5$ using $\alpha = 0.05$
   b. $H_0: \mu = 5$ vs. $H_a: \mu < 5$ using $\alpha = 0.05$

c. H_0: $\mu = 5$ vs. H_a: $\mu \neq 5$ using $\alpha = 0.05$

7.   You want to test H_0: $p = 0.3$ vs. H_a: $p > 0.3$ using $\alpha = 0.10$ and a SRS of $n = 100$.

    a.   Find the rejection region.
    b.   Fill in the blank: We would reject H_0 in favor of H_a for $\hat{p}$'s greater than _____.

8.   You want to test H_0: $p = 0.6$ vs. H_a: $p < 0.6$ using $\alpha = 0.03$ and a SRS of $n = 70$.

    a.   Find the rejection region.
    b.   Fill in the blank: We would reject H_0 in favor of H_a for $\hat{p}$'s less than _____.

9.   You want to test H_0: $p = 0.44$ vs. H_a: $p \neq 0.44$ using $\alpha = 0.10$ and a SRS of $n = 200$.

    a.   Find the rejection region using the standard normal distribution.
    b.   Fill in the blanks: We would reject H_0 in favor of H_a for $\bar{x}$'s less than _____ or greater than _____.

10.   You want to test H_0: $\mu = 515$ vs. H_a: $\mu > 515$ using $\alpha = 0.09$ and a SRS of $n = 30$. You expect the smallest and largest data values to be 300 and 700, respectively.

    a.   Find the rejection region using the standard normal distribution.
    b.   Fill in the blank: We would reject H_0 in favor of H_a for $\bar{x}$'s greater than _____.

11.   You want to test H_0: $\mu = 8$ vs. H_a: $\mu < 8$ using $\alpha = 0.04$ and a SRS of $n = 29$. You expect the smallest and largest data values to be 4 and 18, respectively.

    a.   Find the rejection region using the standard normal distribution.
    b.   Fill in the blank: We would reject H_0 in favor of H_a for $\bar{x}$'s less than _____.

12.   You want to test H_0: $\mu = 12$ vs. H_a: $\mu \neq 12$ using $\alpha = 0.08$ and a SRS of $n = 5$. You expect the smallest and largest data values to be 11.7 and 12.3, respectively.

    a.   Find the rejection region using the standard normal distribution.
    b.   Fill in the blanks: We would reject H_0 in favor of H_a for $\bar{x}$'s less than _____ or greater than _____.

13.   Consider the test of H_0: $p = 0.3$ vs. H_a: $p > 0.3$ using $\alpha = 0.10$ and a SRS of $n = 100$.

    a.   Find β and the power of the test when H_a is really true and $p = 0.35$.
    b.   Find β and the power of the test when H_a is really true and $p = 0.40$.

14.   Consider the test of H_0: $p = 0.6$ vs. H_a: $p < 0.6$ using $\alpha = 0.03$ and a SRS of $n = 70$.

    a.   Find β and the power of the test when H_a is really true and $p = 0.50$.
    b.   Find β and the power of the test when H_a is really true and $p = 0.40$.

15.   Consider the test of H_0: $\mu = 1.8$ vs. H_a: $\mu < 1.8$ using $\alpha = 0.02$ and a SRS of $n = 40$. You expect the smallest and largest data values to be 0 and 8, respectively.

    a.   Find β and the power of the test when H_a is really true and $\mu = 1.2$.
    b.   Find β and the power of the test when H_a is really true and $\mu = 0.8$.

16.   Consider the test of H_0: $\mu = 50$ vs. H_a: $\mu > 50$ using $\alpha = 0.05$ and a SRS of $n = 101$. You expect the smallest and largest data values to be 30 and 55, respectively.

    a.   Find β and the power of the test when H_a is really true and $\mu = 51$.
    b.   Find β and the power of the test when H_a is really true and $\mu = 52$.

17. Consider exercise 13. Suppose H_a is really true and $p = p_a$, some value greater than 0.3.

    a.   Describe the relationship between p_a and the probability of a Type II error.
    b.   Describe the relationship between p_a and the power of the test.

18. Consider exercise 14. Suppose H_a is really true and $p = p_a$, some value less than 0.6.

    a.   Describe the relationship between p_a and the probability of a Type II error.
    b.   Describe the relationship between p_a and the power of the test.

19. Consider exercise 15. Suppose H_a is really true and $\mu = \mu_a$, some value less than 1.8.

    a.   Describe the relationship between μ_a and the probability of a Type II error.
    b.   Describe the relationship between μ_a and the power of the test.

20. Consider exercise 16. Suppose H_a is really true and $\mu = \mu_a$, some value greater than 50.

    a.   Describe the relationship between μ_a and the probability of a Type II error.
    b.   Describe the relationship between μ_a and the power of the test.

21. Consider the test of H_0: $p = 0.9$ vs. H_a: $p < 0.9$ using $\alpha = 0.06$ when H_a is really true and $p = 0.89$.

    a.   Find β and the power of the test based on a SRS of $n = 3{,}500$.
    b.   Find β and the power of the test based on a SRS of $n = 5{,}000$.
    c.   Describe the relationship between the power of the test and the sample size.

22. Consider the test of H_0: $\mu = 50$ vs. H_a: $\mu > 50$ using $\alpha = 0.05$ when H_a is really true and $\mu = 51$. You expect the smallest and largest data values to be 30 and 55, respectively.

    a.   Find β and the power of the test based on a SRS of $n = 101$.
    b.   Find β and the power of the test based on a SRS of $n = 500$.
    c.   Describe the relationship between the power of the test and the sample size.

23. Consider the test of H_0: $p = 0.52$ vs. H_a: $p > 0.52$ using a significance level of 0.05. How many objects should be in a SRS so that the test will have

    a.   80% power to detect when p really is 0.55?
    b.   70% power to detect when p really is 0.55?
    c.   80% power to detect when p really is 0.60?
    d.   80% power to detect when p really is 0.525?

24. Consider the test of H_0: $p = 0.25$ vs. H_a: $p < 0.25$ using a significance level of 0.05. How many objects should be in a SRS so that the test will have

    a.   90% power to detect when p really is 0.20?
    b.   85% power to detect when p really is 0.20?
    c.   90% power to detect when p really is 0.10?
    d.   90% power to detect when p really is 0.24?

25. Consider the test of H_0: $\mu = 76$ vs. H_a: $\mu < 76$ using a significance level of 0.05. You expect the smallest and largest data values to be 60 and 80, respectively. How many objects should be in a SRS so that the test will have

    a.   70% power to detect when μ really is 75?
    b.   90% power to detect when μ really is 75?
    c.   70% power to detect when μ really is 75.9?

26. Consider the test of H_0: $\mu = 255$ vs. H_a: $\mu > 255$ using a significance level of 0.05. You expect the smallest and largest data values to be 100 and 400, respectively. How many objects should be in a SRS so that the test will have

    a.   80% power to detect when μ really is 270?
    b.   90% power to detect when μ really is 270?
    c.   80% power to detect when μ really is 256?

27. You think more than 30% of all students at a large university have all required course materials within the first two weeks of the semester. You plan to take collect data on a SRS of 250 students to support your hunch.

    a.   Write down the hypotheses to test your hunch.
    b.   Find the rejection region for a significance level of 0.05.
    c.   If, in fact, 34% of all students at this university have all required course materials within the first two weeks of the semester, find β and the power of the test.
    d.   If you want your test to have 80% power to detect the 34% figure, how many students should you select in your SRS?

28. A magazine is planning to add an online version in addition to its print version but will only do so if more than 15% of current subscribers would be interested. The publisher plans to select a SRS of 200 current subscribers to try to see if enough current subscribers are interested.

    a.   Write down the hypotheses to test the publisher's hunch.
    b.   Find the rejection region for a significance level of 0.05.
    c.   If, in fact, 18% of all current subscribers are interested in an online version, find β and the power of the test.
    d.   If the publisher wants the test to have 90% power to detect the 18% figure, how many current subscribers should the publisher select in its SRS?

29. A large mail order company is studying its order fulfillment process. An order is shipped "on time" if it is sent out within two business days of when it was received. The company currently claims that "95% of all orders are shipped on time" but is worried that there have been more late orders recently. The company plans to collect data on a SRS of 300 recently-shipped orders in an effort to see if its worries are substantiated.

    a.   Write down the hypotheses to test the company's hunch.
    b.   Find the rejection region for a significance level of 0.01.
    c.   If, in fact, only 92% of all recent orders were shipped on time, find β and the power of the test.
    d.   If the company wants the test to have 85% power to detect the 92% figure, how many recent orders should the company select in its SRS?

30. A candidate for public office is considering withdrawing from the race because she suspects that public support isn't there. If that's true, she doesn't want to spend more money on further campaigning. She decides that she'll withdraw if she can gather convincing statistical evidence that fewer than 20% of likely voters would vote for her. She plans to conduct a poll on a SRS of 500 likely voters in her district.

    a.   Write down the hypotheses to test the candidate's suspicion.
    b.   Find the rejection region for a significance level of 0.05.
    c.   If, in fact, only 17% of all likely voters in her district would vote for her, find β and the power of the test.
    d.   If the candidate wants the test to have 80% power to detect the 17% figure, how many likely voters should the candidate select in her SRS?

31. Jamie wants to open a juice bar in her town but needs to know whether the incomes of people in the area are high enough to afford such discretionary spending. She will proceed with her venture if the mean annual income of all residents of the town is above $60,000 per year. She plans to conduct a poll on a SRS of 25 residents in the area. In an earlier sample, she observed that the smallest and largest incomes were $45,000 and $90,000, respectively.

    a.   Write down both the null hypothesis and Jamie's hypothesis.

b. Find the rejection region for a significance level of 0.05. Use the z-method instead of the t-method.
c. If, in fact, the mean annual income of all residents of the town is $62,500, find β and the power of the test.
d. If Jamie wants the test to have 80% power to detect the $62,500 figure, how many residents should she select in her SRS?

32. Soil sampling is common in farming because it provides farmers with information about what nutrients are in the soil and which nutrients are lacking. A minimum acceptable nitrate level in corn fields is 30 parts per million (ppm). A farmer wants to see if the mean level for all soil specimens in the field is below 30 ppm. If so, he will apply fertilizer. He plans to collect a SRS of 20 soil specimens from the field. In a sample of soil specimens earlier in the year, he observed that the smallest and largest nitrate levels were 25.5 ppm and 33.1 ppm, respectively.

   a. Write down both the null hypothesis and the farmer's hypothesis.
   b. Find the rejection region for a significance level of 0.05. Use the z-method instead of the t-method.
   c. If, in fact, the mean nitrate level for all soil specimens in the field is 29.3 ppm, find β and the power of the test.
   d. If the farmer wants the test to have 90% power to detect the 29.3 ppm figure, how many soil specimens should he select in his SRS?

33. For healthy women, the average forced expiratory volume in one second (FEV1) is 3 liters. You suspect that average FEV1 is lower for mild asthmatics. To gather evidence for your hunch, you plan to observe the FEV1 for a SRS of 22 mild asthmatic women. Previous research indicates that a reasonable value for the standard deviation of FEV1 values is 0.9 liters.

   a. Write down the hypotheses to test your hunch.
   b. Find the rejection region for a significance level of 0.05. Use the z-method instead of the t-method.
   c. If, in fact, the average FEV1 for all mild asthmatic women is 2.6 liters, find β and the power of the test.
   d. If you want the test to have 85% power to detect the 2.6 liters figure, how many mild asthmatic women should you select in your SRS?

34. A small beverage company makes ginger beer and bottles it in 12-oz. bottles. To monitor the filling process, a simple random sample of 5 bottles is selected from the production line every two hours and the amount of ginger beer (ounces) in each bottle is accurately measured. The company needs to know if the mean fill amount for all bottles changes so that it can take corrective action. From experience, the company knows the standard deviation of fill amounts is around 0.12 ounces.

   a. Write down the hypotheses for the company.
   b. Find the rejection region for a significance level of 0.05. Use the z-method instead of the t-method.
   c. If, in fact, the average fill amount for all bottles differs is off by 0.1 ounces, find β and the power of the test.
   d. If the company wants the test to have 90% power to detect a difference of 0.1 ounces in either direction, how many bottles should the company select in its SRS's every two hours?

6.6 Common Pitfalls

Before I close this chapter, I wanted to warn you about what can go wrong when conducting a hypothesis test. By that, I don't mean a mistake in a calculation. Even if your "math" is correct, there are important statistical issues to consider.

Random Mechanism

Everything about a hypothesis test is built on the foundation of the random mechanism that is used to generate the data, like simple random sampling from a large population or a process like tossing a coin, rolling a die, or guessing at a multiple choice question. Without the foundation of a random mechanism, the whole hypothesis testing structure is quite shaky.

I came across some data posted on a real estate broker's website on 151 bank-owned properties where the bank foreclosed on the homeowner's mortgage and repossessed the house. Since banks are not in the real estate business, they will often sell these properties for just enough to pay off the balance of the mortgage, which could be well below market value. Buyer beware though...these properties can be in rough shape because if the homeowner couldn't make their mortgage payments, they likely didn't have enough cash to maintain the property. Here is a histogram showing the distribution of the listing prices.

The mean listing price is $135,249 and the median is $115,000. Suppose you're wondering if these data provide enough evidence that the average listing price for all bank-owned properties in this area is above $100,000. Should you carry out a hypothesis test of

H_0: $\mu = \$100{,}000$ vs. H_a: $\mu > \$100{,}000$?

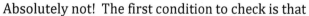

Absolutely not! The first condition to check is that the data are a SRS from the population of all bank-owned properties in the area. That is certainly not the case. Perhaps this particular real estate broker specializes in lower-value bank-owned properties so that these properties are not at all representative of the population of properties. The possibility of selection bias is very real. I would not try to do anything fancier than summarizing the prices using basic data summary tools. Do not try to generalize what you see to any larger population without random sampling!

Interpretation of p-values

In my experience, one of the more difficult things to master is the interpretation of a p-value. Remember that the interpretation of a p-value always has two parts: It's a 1) measure of how likely the data are 2) if the null hypothesis is true. Let's go back to one of the first hypothesis tests in this chapter, the one about football's popularity:

H_0: $p = 0.43$ (no change in popularity since the mid-2000's)
H_a: $p < 0.43$ (a drop in popularity since the mid-2000's)

For the 37% SRS result, the p-value is 0.000042. Here is the correct interpretation along with some common misinterpretations.

Correct: If there's been no change in football's popularity since the mid-2000's, there's only a 0.0042% chance of getting 37% or fewer who prefer football in a SRS of 1,049.

Incorrect: There's a 0.0042% chance that there's been no change in football's popularity since the mid-2000's. (It's NOT the probability that the null hypothesis is true.)

Incorrect: There's a 0.0042% chance that there's been a drop in football's popularity since the mid-2000's. (It's NOT the probability that the alternative hypothesis is true.)

Incorrect: There's a 0.0042% chance of getting 37% or fewer who prefer football in a SRS of 1,049. (This interpretation is incomplete because it fails to include the "if the null hypothesis is true" part.)

The p-value measures the likelihood of the data arising under the null hypothesis *by the random mechanism alone*. If there is no random mechanism like random sampling, then it's more difficult to make sense of what the p-value really means.

Accepting the Null Hypothesis

If you incorrectly think the p-value is the probability that the null hypothesis is true, then you may be tempted to say things like, "The data provide convincing statistical evidence that H_0 is true." After all, if the p-value = 0.99, then betting on H_0 is good, right? I hope you would say "no" because of what you just read above about p-values. A p-value is not a measure of the likelihood of the null hypothesis. It is a measure of how like the *data* are *if* the null hypothesis is true. Accepting H_0 for large p-values can sometimes result in a fairly high chance of making an incorrect decision.

Let's say you're testing these hypotheses:

H_0: all birds are black
H_a: not all birds are black

You observe a SRS of 1 bird and it happens to be a crow. You calculate the p-value (the probability of observing a SRS of 1 black bird if indeed all birds are black) to be exactly 1. If all birds are black, it's a sure thing that the bird you select will be black. Should you therefore conclude that H_0 is true because the p-value is very large? Of course not. There are non-black birds out there; you just didn't see any in your (small) sample. With a sample size of $n = 1$, the chances of making a Type II Error (accepting a false null hypothesis) are fairly large.

Unless you've taken care to quantify β, the probability that your hypothesis test would result in a Type II Error (see section 6.5), avoid accepting the null hypothesis. It is much better to say, "the results are inconclusive" or "there is not convincing statistical evidence for H_a" than to say "H_0 is true."

Peeking at the Data

It is considered statistical cheating to allow the data to determine the research hypothesis and then conduct the hypothesis test *using the same data*. That's not to say that you can't use data to generate research questions; that's common practice. But you should NOT use those same data to test your research hypothesis. Always remember that in the scientific method, the hypotheses come *before* collecting and analyzing the data. Since you may be conducting an analysis on data that's already been collected, avoid peeking at the data before you formulate your hypotheses.

Significance Levels are Arbitrary

The most common significance level is $\alpha = 0.05$ and it can be traced back to at least as far as a comment made by Sir Ronald Fisher in 1925:

> *"The value for which P=0.05, or 1 in 20, is 1.96 or nearly 2; it is convenient to take this point as a limit in judging whether a deviation ought to be considered significant or not."*[12]

Fisher provided no other argument for the 0.05 level; he just as easily could have used 0.06. Therefore, don't be too dogmatic about your conclusion when the p-value is fairly close to the significance level. I like to use language like "marginal" for p-values close to α. The evidence from a high-quality study showing a p-value of 0.001 is much stronger than from a similar study showing a p-value of 0.04, even though both would be considered statistically significant results at the 0.05 level. If the statistical significance of your result depends on the α you choose (commonly 0.05, 0.10, or 0.01), your results are "marginal."

Multiple Testing

You're a basketball player who has only a 5% chance of missing a foul shot. Yep, you're that good! Now, what's the chance you'll miss at least one foul shot if you attempt 20 in a row? If you said "more than 5%," you're right. In fact, it's around 64%. It's pretty likely that you'll miss at least 1 out of 20.

The same idea holds for multiple hypothesis tests. If we use a significance level of $\alpha = 0.05$, that means that we'll commit a Type I error (reject a true null hypothesis) for every 1 test in 20 in the long run. Conduct enough tests and it's bound to happen eventually. If there are many tests conducted about the same research question, and the null hypothesis is really true, the chances are good that at least one of the tests will falsely conclude that the research hypothesis is true.

The following xkcd.com cartoon nicely illustrates this issue. The research question is about whether there's a link between jelly beans and acne. The hypotheses are

H_0: eating jelly beans is not related to acne
H_a: eating jelly beans is related to acne

The scientists conducted 20 tests using data on jelly beans of 20 different colors and found only 1 test (for green jelly beans) where the p-value was less than 0.05. Now even if there is no link between eating jelly beans and acne, we'd expect to find one Type I error in 20 tests at the 5% significance level; that's exactly what $\alpha = 0.05$ means. Had the green jelly bean test been the only test conducted, then we may take notice. Knowing that there were 20 tests conducted, the 1 significant result is perfectly consistent with the null hypothesis and the effects of simple random sampling. Even if, in general, folks that eat jelly beans of any color get acne at the same rate as those who don't eat jelly beans, we'd eventually get a SRS where we'd just happen to randomly choose more of the jelly bean eaters who got acne than the non-jelly bean eaters.

Often results are reported in the news without disclosure as to the research process. It is a critical consumer of statistical information who asks, "Was that the only test conducted or were there others?" Beware of the multiple testing problem!

Statistical Significance vs. Practical Significance

A statistically significant result is one where the data are not likely to have arisen due to the random mechanism alone in a universe where the null hypothesis is true. It does not necessarily mean that the result is earth-shatteringly important. It does not necessarily mean you should rush to tell everyone you know about it. Don't confuse a statistically significant result with a practically significant result, that is, one that indicates a big difference from the null hypothesis.

Suppose you and your friend want to know whether the mean annual household income in your metropolitan area exceeds $60,000. You set up your hypotheses:

H_0: $\mu = \$60,000$
H_a: $\mu > \$60,000$

"Significant" © Randall Munroe, xkcd.com. Used by permission.

μ = mean annual household income for all households in the area

Both you and your friend go out and collect data on a SRS of households. Lo and behold the only difference between your two samples is the sample size!

Your sample	Your friend's sample
SRS of $n = 15$	SRS of $n = 1000$
$\bar{x} = \$61{,}000$	$\bar{x} = \$61{,}000$
$s = \$10{,}000$	$s = \$10{,}000$
$t = \dfrac{61{,}000 - 60{,}000}{10{,}000/\sqrt{15}} = 0.387 \qquad df = 14$	$t = \dfrac{61{,}000 - 60{,}000}{10{,}000/\sqrt{1000}} = 3.162 \qquad df = 999$
$p - value = 0.352$	$p - value = 0.001$

While there is very little evidence in your sample that the mean annual household income for all households in the area exceeds $60,000, your friend's evidence is very strong, even though both sample means are the same at just a bit over $60,000. That difference of $1,000 is quite likely in a SRS of only 15 but very unlikely in a SRS of 1,000. Your friend's result is highly *statistically* significant (unlikely due to random sampling alone), but is not very *practically* significant. The $1,000 difference in terms of annual salary is fairly small. A $10,000 or $20,000 difference would be more practically significant. Your friend's test provides a large amount of evidence *that* the population mean exceeds $60,000; it does not provide evidence that the population mean income exceeds $60,000 by a large amount. Did you catch the difference between those last two statements? The hypothesis test highlights evidence *of* a difference but doesn't highlight the *magnitude* of the difference. Don't confuse the two. In fact, with a large enough sample size, any difference, no matter how small between the sample and the null hypothesis can be made to be statistically significant.

Section 6.6 Exercises

1. "Do recent stock market losses worry you?" was the question asked in an online poll. There were 115 "yes" responses and 219 "no" responses.[13] The 95% score interval is (0.295, 0.397).

    a. Can we be 95% confident that the proportion of all U.S. adults who were worried (at the time) about the stock market losses was between 0.295 and 0.397? Explain.
    b. Can we be 95% confident that the proportion of all Huntington, WV adults who were worried (at the time) about the stock market losses was between 0.295 and 0.397? Explain.
    c. Can we be 95% confident that the proportion of all readers of the Herald-Dispatch who were worried (at the time) about the stock market losses was between 0.295 and 0.397? Explain.

2. As of April 20, 2020, Worldometer reported that, in the U.S., there were 781,368 Coronavirus cases and 41,575 deaths.[14] The 95% score interval is (0.053, 0.054).

    a. Why is the confidence interval so narrow (precise)? Is this a good thing or a bad thing? Explain.
    b. Can we be 95% confident that the proportion of all U.S. coronavirus cases that end in death is between 5.3% and 5.4%? Explain.
    c. Several studies using more representative samples indicated that the actual number of coronavirus cases was far higher than those reported. At the time, many testing centers tested only those with a doctor's order. What kind of bias is present here? Explain how such a testing system will underreport the actual number of cases.

d. One study found that there could be at least 50 times as many coronavirus infections as reported cases because a very large fraction of cases are asymptomatic.[15] Multiply the number of cases reported by Worldometer by 50 and, using the same number of deaths, recompute the 95% confidence interval.

3. In a study designed to evaluate the effectiveness of Teen Court (TC), an alternative to the criminal justice system (CJS) for first-time juvenile offenders, it was found that 16 of 50 juveniles whose cases were randomly assigned to TC recidivated (returned to their bad behavior) while 12 of the 50 individuals whose cases were randomly assigned to the CJS recidivated.[16] Let p_1 = proportion of TC cases like these who recidivated and p_2 = proportion of CJS cases like these who recidivated. The p-value for testing H_0: $p_1 = p_2$ vs. H_a: $p_1 \neq p_2$ is 0.504. Which of these is the best interpretation of the p-value?

a. There's a 50.4% chance that a case like the ones used in the study will recidivate.
b. There's a 50.4% chance there is no difference between the proportions of TC cases like these who recidivated and the proportion of CJS cases like these who recidivated.
c. There's a 50.4% chance there is a difference between the proportions of TC cases like these who recidivated and the proportion of CJS cases like these who recidivated.
d. There's a 50.4% chance of getting two sample proportions at least 8 percentage points apart in an experiment like this.
e. There's a 50.4% chance of getting two sample proportions at least 8 percentage points apart in an experiment like this, if there is no difference between the proportions of TC cases like these who recidivated and the proportion of CJS cases like these who recidivated.
f. There's a 50.4% difference in the proportion of TC cases who recidivated and the proportion of CJS cases who recidivated.

4. Why should you avoid concluding that the null hypothesis is true for large p-values?

5. When would it be acceptable to conclude that the null hypothesis is true for large p-values?

6. Consider exercise 3. Say what's wrong with these conclusions.

a. There is no evidence of a difference between the proportions of TC cases like these who recidivated and the proportion of CJS cases like these who recidivated.
b. We have convincing statistical evidence that there is no difference between the proportions of TC cases like these who recidivated and the proportion of CJS cases like these who recidivated.
c. We have convincing statistical evidence that there is a difference between the proportions of TC cases like these who recidivated and the proportion of CJS cases like these who recidivated.

7. Consider exercise 3.

a. Calculate the sample proportion who recidivated in each group.
b. Why shouldn't you set up the alternative hypothesis as H_a: $p_1 > p_2$ based on your answer to part a.?
c. Give an argument for using a non-directional alternative H_a: $p_1 \neq p_2$ instead of H_a: $p_1 < p_2$, an alternative that reflects "TC is more effective than CJS."

8. Consider testing H_0: $p = 0.5$ vs. either H_a: $p < 0.5$ or H_a: $p > 0.5$, where the choice of "<" or ">" is made based on the data, i.e. whether the sample proportion $\hat{p}$ is less than or greater than 0.5. A simulation is conducted where 1,000 SRS's are drawn from a population that is split 50/50, success/failure (i.e. H_0 is true). For each sample, the p-value is calculated based on the choice of direction in H_a. A histogram of the 1,000 p-values is shown below.

a. At the 5% significance level, what proportion of the tests resulted in an incorrect rejection of the null hypothesis?
b. What proportion of tests conducted at a 5% significance level *should* reject a true null hypothesis?
c. At the 10% significance level, what proportion of the tests resulted in an incorrect rejection of the null hypothesis?
d. What proportion of tests conducted at a 10% significance level *should* reject a true null hypothesis?
e. What effect does "peeking at the data" have on the performance of a hypothesis test?
f. Where did the 24 p-values in the interval [0.5, 0.55) come from?

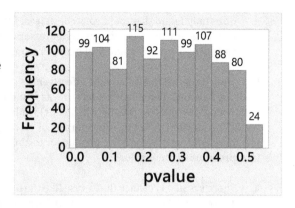

9. A hypothesis test results in a p-value of 0.049.

a. For what significance levels would you reject H_0?
b. Does this p-value indicate that there is strong statistical evidence for H_a? Explain.

10. A hypothesis test results in a p-value of 0.051.

a. For what significance levels would you reject H_0?
b. Does this p-value indicate that there is no statistical evidence for H_a? Explain.
c. Does this p-value indicate that there is very weak statistical evidence for H_a? Explain.

11. Instead of reporting the actual p-value for a test, many researchers report only that the p-value is either less than or greater than a certain significance level.

a. If you see "p-value < 0.05," does that necessarily mean there is strong statistical evidence for H_a? Explain.
b. If you see "p-value > 0.05," does that necessarily mean there is very little statistical evidence for H_a? Explain.

12. An airline has long had a 50-pound limit on the weight of any checked bags. The limit was based on a mean weight of 34 pounds plus a multiple of the standard deviation. In a simple random sample of 101 checked bags from current customers, the average weight was 32.287 pounds and the standard deviation was 5.635 pounds. The weights are nearly-normal and a t-test of H_0: $\mu = 34$ vs. H_a: $\mu < 34$ gives a p-value = 0.001.

a. Keeping all else the same, change the sample size so that evidence for H_a is only marginally significant.
b. Keeping all else the same, change the sample mean so that evidence for H_a is only marginally significant.

13. Soil sampling is common in farming because it provides farmers with information about what nutrients are in the soil and which nutrients are lacking. A minimum acceptable nitrate level in corn fields is 30 parts per million (ppm). If the farmer determines that the mean level for all soil specimens in the field is below 30 ppm, she will apply fertilizer. In a SRS of locations in one field, the following nitrate levels (in ppm) were obtained:

28.0   29.8   26.4   33.1   29.6   25.5   25.6   27.7   31.0   27.3

The t-test of H_0: $\mu = 30$ vs. H_a: $\mu < 30$ gives a p-value = 0.035.

a. Keeping all other summaries the same, change the sample size so that evidence for H_a is highly significant.
b. Keeping all other summaries the same, change the sample mean so that evidence for H_a is highly significant.

14. Consider the soil sampling problem in exercise 13. In practice, a farmer will perform tests on several fields and make several fertilizer application decisions. Suppose the farmer performs a soil sample test on each of 20 fields. for each field, she decides to apply fertilizer if the p-value is less than 0.05. She finds that one field out of the 20 gave a p-value less than 0.05. Should she apply fertilizer to that field? Explain.

15. Dave decides to test all the coins in his collection to see if any are biased. He has 100 coins and tosses each one 50 times. He uses the data on each coin to test H_0: $p = 0.5$ vs. H_a: $p \neq 0.5$ at the 5% significance level.

    a.   If, in reality, all 100 coins are fair, how many coins will Dave expect to declare "not fair?"
    b.   After all 100 tests, Dave finds that 5 coins gave p-values less than 0.05. Should Dave conclude that those 5 coins are not fair? Explain.

16. In the 1970's the makers of Trident gum ran a TV ad that said, "4 out of 5 dentists surveyed recommend sugarless gum for patients that chew gum." Of course we don't know how those data were collected, but let's assume they were from an SRS of dentists. Consider using the data to test H_0: $p = 0.79$ vs. H_a: $p > 0.79$ where p = proportion of all dentists who recommend sugarless gum for patients that chew gum.

    a.   If the SRS was of 500 dentists and 400 of them recommend sugarless gum for patients that chew gum, would those results be statistically significant?
    b.   If the SRS was of 50,000 dentists and 40,000 of them recommend sugarless gum for patients that chew gum, would those results be statistically significant?
    c.   Would you consider the result of part b practically significant? Explain.

17. In a SRS of 100,000 boys, the average growth in one month after their first birthday was 0.47 inches and the standard deviation was 0.22 inches. Consider using the data to test H_0: $\mu = 0.46"$ vs. H_a: $\mu > 0.46"$ where μ = mean growth in one month for all one-year-old boys.

    a.   Would these results be statistically significant?
    b.   You are a manufacturer of children's clothing. You currently make 12-month and 18-month sizes for boys. Would these results convince you to offer 13-month sizes as well? Explain.

Notes

[1] Gallup, January 4, 2018, "Football Still Americans' Favorite Sport to Watch."

[2] Ibid.

[3] Pew Research Center, January, 2017, "The World Facing Trump: Public Sees ISIS, Cyberattacks, North Korea as Top Threats."

[4] SAMHSA. 2014 National Survey on Drug Use and Health (NSDUH). Table 6.89B—Binge Alcohol Use in the Past Month among Persons Aged 18 to 22, by College Enrollment Status and Demographic Characteristics: Percentages, 2013 and 2014. Available at: http://www.samhsa.gov/data/sites/default/files/NSDUH-DetTabs2014/NSDUH-DetTabs2014.htm#tab6-89b.

[5] Pew Research Center, Oct. 26, 2016, "One-in-Five U.S. Adults Were Raised in Interfaith Homes."

[6] SAMHSA. 2014 National Survey on Drug Use and Health (NSDUH).

[7] nces.ed.gov.

[8] Gosset, W. (1908). "The probable error of a mean," Biometrika, 6, 302 – 310. At the time, the paper was published under Gosset's penname "Student" and the distribution he discovered is often called "Student's t-distribution" today.

[9] www.nationsreportcard.gov.

[10] data.delaware.gov.

[11] Environmental Protection Agency (2019), "Fuel Economy Guide."

[12] Fisher, R. A. (1925), "Statistical Methods for Research Workers."

[13] The Herald-Dispatch, Huntington, West Virginia, March 4, 2020.

[14] www.worldometers.info.

[15] Bendavid, et al. (Apr 11, 2020), "COVID-19 Antibody Seroprevalence in Santa Clara County, California," https://doi.org/10.1101/2020.04.14.20062463.

[16] Based on Stickle, et al. (2008), "An experimental evaluation of teen courts," *Journal of Experimental Criminology*, 4: 137 – 163.

Chapter 7: Confidence Intervals

7.1 Introduction to Parameter Estimation

At the end of the last chapter, we considered an example where both you and your friend collect data to test these hypotheses:

H_0: $\mu = \$60,000$
H_a: $\mu > \$60,000$

μ = mean annual household income for all households in the area

Suppose the only difference between your two samples is the sample size:

Your sample	Your friend's sample
SRS of $n = 15$	SRS of $n = 1000$
$\bar{x} = \$61,000$	$\bar{x} = \$61,000$
$s = \$10,000$	$s = \$10,000$
$t = \frac{61,000-60,000}{10,000/\sqrt{15}} = 0.387 \quad df = 14$	$t = \frac{61,000-60,000}{10,000/\sqrt{1000}} = 3.162 \quad df = 999$
$p - value = 0.352$	$p - value = 0.001$

We saw that, while there is very little evidence in your sample that the mean annual household income for all households in the area exceeds $60,000, your friend's evidence is very strong, even though both sample means are the same at just a bit over $60,000. Your friend has collected enough evidence to detect a population mean income that is just a little bit above $60,000. The problem with hypothesis tests and p-values is that they do not help us with questions like, "How much above $60,000 is the mean annual household income for all households?"

Hypothesis tests naturally apply to research questions that can be answered with a "yes" or "no" like these:

- Is the mean annual household income for all households in the area above $61,000?
- Do more than half of a restaurant's customers prefer table seating over booth seating?

They do not explicitly answer research questions about the actual value of a parameter, like these:

- What is the mean annual household income for all households in the area?
- What proportion of a restaurant's customers prefer table seating over booth seating?

Since the entire population data set is rarely available, we must rely on sampling. Random sampling is best because it eliminates selection bias and results in a representative sample. However, we need to account for the random mechanism in the analysis of the data. To do that, we can use the hypothesis testing methodology in a different way.

Consider the hypothesis test about the mean 2015 math proficiency rate for all public fourth-grade classes in Delaware that we did in the last chapter:

$H_0: \mu = 40\%$
$H_a: \mu \neq 40\%$

$\mu =$ mean 2015 math proficiency rate for all public fourth-grade classes in Delaware

The SRS of 5 public fourth-grade classes in Delaware gave

$\bar{x} = 39.372$
$s = 15.5689$

and the results of the one-sample t-test were

$t = -0.090$
$p - value = 2P(t \leq -0.090) = 2(0.4663) = 0.9326.$

Since the research question was specifically about whether the mean math proficiency rate for all public fourth-grade classes in Delaware was not 40%, that statement was naturally the research hypothesis. If convincing evidence is found (if the p-value $< \alpha$), the null hypothesis $H_0: \mu = 40\%$ is rejected and we conclude that the data convincingly support the research hypothesis. The SRS of 5 classes provided some, but not convincing, support for the research hypothesis and we would not reject H_0 based on the sample. If the population of classes had a mean proficiency rate of 40%, a SRS of 5 classes giving the results above is reasonably likely. Another way to think about that conclusion is that, based on your sample data, a population mean of 40% is a reasonable value. You can think of a hypothesis test as using sample data to evaluate the reasonableness of a proposed value of a parameter (in this case, μ). However, just because 40% is reasonable based on the data doesn't mean that it's the *only* reasonable value. There are others!

The goal of this chapter is to find reasonable values of a parameter based on a sample of data. That set of reasonable values is called a **confidence interval**. Here is a straight-forward way to find a confidence interval for any population parameter using the hypothesis testing methodology. I call it the "guess-and-check" method:

1. Propose a value for the parameter as the null hypothesis.
2. Compare the data to this value using a non-directional alternative hypothesis (e.g. $\neq$).
3. If the p-value $< \alpha$, reject H_0 and the proposed value. Otherwise, keep the proposed value in the list; it is reasonable based on your data.
4. Propose another value for the parameter.
5. Repeat steps 1 – 4 until you have found all of the reasonable parameter values.

The result of this process is a set of reasonable values called a $100(1 - \alpha)\%$ confidence interval for the parameter. If $\alpha = 0.05$, you would end up with a 95% confidence interval.

In the fourth-grade math proficiency rate example, we already know that 40% is a reasonable value for μ. Let's propose another: $\mu = 50\%$. Using the same data from the SRS of 5 fourth-grade classes in Delaware ($n = 5$, $\bar{x} = 39.372$, and $s = 15.5689$) we test these hypotheses

H_0: $\mu = 50\%$
H_a: $\mu \neq 50\%$

and get

$$t = \frac{\bar{x}-\mu}{s/\sqrt{n}} = \frac{39.372-50}{15.5689/\sqrt{5}} = -1.526 \text{ and } p - value = 2P(t \leq -1.526) = 2(0.1008) = 0.2016.$$

Using $\alpha = 0.05$ we do not reject H_0 and so 50% is also a reasonable value of μ.

Let's repeat using $\mu = 60\%$:

H_0: $\mu = 60\%$
H_a: $\mu \neq 60\%$

and get

$$t = \frac{\bar{x}-\mu}{s/\sqrt{n}} = \frac{39.372-60}{15.5689/\sqrt{5}} = -2.963 \text{ and } p - value = 2P(t \leq -2.963) = 2(0.0207) = 0.0414.$$

Since the p-value is smaller than 0.05 we reject H_0; 60% is *not* a reasonable value of μ because a SRS of 5 from a population where the mean math proficiency rate is 60% is unlikely. So we know that the highest reasonable value of μ is somewhere between 50% and 60%. We can keep "guessing and checking" to find more precisely where it is and we can do the same to find the lowest reasonable value.

Here is a table that summarizes the results of several proposed values for μ.

μ	Hypotheses		Test Statistic	p-value	Is μ Reasonable?
20%	H_0: $\mu = 20\%$	H_a: $\mu \neq 20\%$	$t = 2.782$	0.0497	no
20.1%	H_0: $\mu = 20.1\%$	H_a: $\mu \neq 20.1\%$	$t = 2.768$	0.0504	yes
30%	H_0: $\mu = 30\%$	H_a: $\mu \neq 30\%$	$t = 1.346$	0.2495	yes
39.372%	H_0: $\mu = 39.372\%$	H_a: $\mu \neq 39.372\%$	$t = 0$	1.0000	yes
40%	H_0: $\mu = 40\%$	H_a: $\mu \neq 40\%$	$t = -0.090$	0.9326	yes
50%	H_0: $\mu = 50\%$	H_a: $\mu \neq 50\%$	$t = -1.526$	0.2016	yes
58.7%	H_0: $\mu = 58.7\%$	H_a: $\mu \neq 58.7\%$	$t = -2.776$	0.0500	yes
58.8%	H_0: $\mu = 58.8\%$	H_a: $\mu \neq 58.8\%$	$t = -2.790$	0.0493	no
60%	H_0: $\mu = 60\%$	H_a: $\mu \neq 60\%$	$t = -2.963$	0.0414	no

Values between 20.1% and 58.7% would be considered reasonable and so we would say that (20.1%, 58.7%) is a 95% confidence interval for μ. A SRS like the one we obtained would not be considered too unusual if the population mean is somewhere in that interval.

Note that, according to this methodology, the sample mean will always be a reasonable value because it will give a test statistic value of $t = 0$ and a p-value = 1. That makes perfect sense. A SRS gives a representative sample and so the sample mean, in that sense, should be a good estimate of the population mean. The sample mean is our best single guess. We call it the *point estimate* (single value estimate) of the population mean. In general, a **point estimate** is a good single value estimate of a population parameter based on sample data. Here are some common parameters along with their point estimates.

Parameter	Point Estimate
μ: population mean	$\bar{x}$: sample mean
p: population proportion of successes	$\hat{p}$: sample proportion of successes
σ^2: population variance	s^2: sample variance
σ: population standard deviation	s: sample standard deviation

Interpretation of a Confidence Interval

Every hypothesis test should end with a conclusion that is in layman's terms in context. Likewise, every confidence interval should be accompanied by an interpretation of the endpoints in context. Here is a model interpretation of the 95% confidence interval for μ in the math proficiency example:

We are 95% confident that the mean 2015 math proficiency rate for all public fourth-grade classes in Delaware was between 20.1% and 58.7%.

The interpretation of any confidence interval should have these components:

- <u>Use the language of "confidence."</u> Do not use language like "95% chance" or "95% probability." Instead, use language like "95% confident;" it's a loaded term! We'll look at exactly what we mean by "confidence" shortly.

- <u>Explicitly refer to the population parameter.</u> In this example, for instance, make sure that your reader knows you are talking about the *mean* rate for *all* public fourth-grade classes in Delaware.

- <u>Include as much context about the population and the variable as you can.</u> Tell the reader about the population and the variable you're studying.

Here are some common mistakes I've seen over the years from folks trying to interpret a confidence interval like (20.1, 58.7).

Wrong Interpretation	Mistake
We are 95% ~~sure~~ that the mean 2015 math proficiency rate for all public fourth-grade classes in Delaware was between 20.1% and 58.7%.	Doesn't use the language of "confidence." The word "confident" has important statistical meaning and must accompany any interpretation.

confident

Wrong Interpretation	Mistake
We are 95% confident that the mean 2015 math proficiency rate for the 5 public fourth-grade classes in Delaware was between 20.1% and 58.7%.	Refers to the sample, not the population. We do not need to make an inference about the sample; we know with certainty that the sample mean is in the interval!
We are 95% confident that the 2015 math proficiency rate for all public fourth-grade classes in Delaware was between 20.1% and 58.7%.	Refers to the population but not the population *mean*. We are *not* making a generalization about individual classes, only about the mean rate for all classes.
About 95% of all public fourth-grade classes in Delaware had a math proficiency rate between 20.1% and 58.7% in 2015.	No language of confidence and does not refer to the population *mean*.
We are 95% confident that μ is between 20.1 and 58.7.	While technically correct, this interpretation doesn't give the reader any context about the population or the variable the researcher is studying.

Meaning of Confidence

Why don't we use probability language like "95% chance" or "95% sure" in an interpretation of a confidence interval? The answer is because the word "confidence" is a statistical term with a special meaning. We call 95% (or whatever confidence level you use) the **confidence level**. Don't confuse the terms *confidence level* and *confidence interval*. The confidence *level* is the criteria used to determine the set of reasonable values that we call the confidence *interval*. Since $\alpha = 0.05$ is often used as the default significance level for a hypothesis test, 95% confidence is often used as the default confidence level.

The reason we don't use probability language is that probability is used to describe how uncertain a particular event is *before* we observe the outcome of the experiment. We say things like "there's a 50% chance that a coin will come up heads" *before* we toss it. To say $P(head) = 0.50$ is to describe the long-run nature of tossing the coin. In the long run, over many tosses, the fraction that come up heads will get very close to 0.50. It doesn't make much sense to look at a coin *after* it has been tossed and *has come up* tails and say, "there's a 50% chance the coin will come up heads." On any one toss, the result is either heads or it is not heads; there are only two possibilities.

So how then do we talk about the uncertainty of an outcome *after* we observe it? A confidence interval is a set of reasonable values of a parameter derived from data collected *after* performing a random sampling experiment. It doesn't make much sense to say, "There's a 95% chance that μ is in the confidence interval." For any one random sample, the confidence interval either contains the value of the parameter or it does not; there are only two possibilities. The solution is to use the language of confidence to talk about data observed *after* a data collection experiment has been done while at the same time describing the long-run behavior of the method used to construct the confidence interval. By using the language of confidence, we're really describing the long-run behavior of the method used (the "guess-and-check" repeated hypothesis testing described above) to construct the interval.

Look again at the interpretation of the 95% confidence interval for the mean math proficiency rate:

We are 95% confident that the mean 2015 math proficiency rate for all public fourth-grade classes in Delaware was between 20.1% and 58.7%.

The phrase "95% confident" means that if we'd collect many, many SRS's of 5 fourth-grade classes and construct a 95% confidence interval for each, about 95% of those intervals would include the value of μ, the mean 2015 math proficiency rate for all public fourth-grade classes in Delaware, and about 5% would not. Think of the phrase "95% confident" as describing the "long-run capture rate" of the method.

Here's a picture that may help you understand what I'm saying. It just so happens that I have the data on all 115 public fourth-grade classes in Delaware for 2015 and the mean math proficiency rate for the population is $\mu = 43.9\%$. The vertical line in the picture is drawn at this value. The horizontal lines represent 95% confidence intervals constructed for each of 100 SRS's of 5 taken from this population. The one at the bottom is the one we constructed: (20.1%, 58.7%), which of course includes 43.9%. The other horizontal lines show the 95% confidence intervals for the other 99 SRS's. Different samples give different results, which is why our confidence is not in any one particular interval but in the *method* we use. You'll notice a few confidence intervals that do not include 43.9%. In fact, if we use 95% confidence, we'd expect 5 intervals out of 100 to miss the mark, which is what we see here.

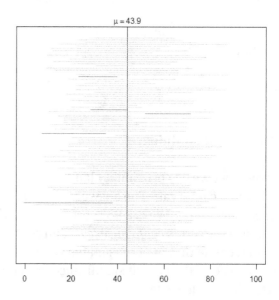

Suppose you're looking for a good surgeon for your upcoming procedure. It's a serious procedure and so you want a good surgeon with a high success rate. Her performance on any one procedure isn't all that telling; what you want to know is how she has performed over many procedures. What is her overall success rate? In the same way, we use the language of confidence to describe the success rate of the method over many random samples.

After reading this section, you have all the tools you need to construct confidence intervals for population parameters, as long as you know how to conduct the corresponding hypothesis tests and use the guess-and-check method. While straight-forward, the guess-and-check approach is terribly inefficient and time-consuming. It may take a while to "home in" on the endpoints of the interval. Fortunately there are much more efficient ways to construct confidence intervals. Turn to the next two sections to learn about them.

Section 7.1 Exercises

1. Consider these symbols: $\mu, s, \hat{p}, \sigma^2, s^2, p, \bar{x}, \sigma$

    a. Clearly define each symbol in words.
    b. Pair each parameter with its point estimate.

2. Find the confidence level that goes with each of the following significance levels.

    a. $\alpha = 0.2$
    b. $\alpha = 0.05$
    c. $\alpha = 0.001$
    d. $\alpha = 0.7$

3. Find the confidence level that goes with each of the following significance levels.

    a. $\alpha = 0.1$
    b. $\alpha = 0.6$
    c. $\alpha = 0.03$
    d. $\alpha = 0.01$

4. Find the significance level that goes with each of the following confidence levels.

    a. 82%
    b. 93%
    c. 95%
    d. 88%

5. Find the significance level that goes with each of the following confidence levels.

    a. 40%
    b. 90%
    c. 99%
    d. 80%

6. A SRS of 29 electric cars is tested to determine each car's 0 – 60mph acceleration time. The mean time was 7.955 seconds and the standard deviation was 3.175 seconds. A histogram of the times is shown here. You'd like to use the data to estimate the mean 0 – 60mph acceleration time for all electric cars.

    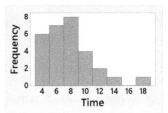

    a. Define the parameter in context. Use the correct notation.
    b. What symbol would you use for the point estimate? What is its value?
    c. Consider testing $H_0: \mu = \mu_0$ vs. $H_a: \mu \neq \mu_0$. Find the p-values for the tests when μ_0 is 6, 7, 8, 9, and 10.
    d. If your significance level is $\alpha = 0.05$, which of these values of μ (6, 7, 8, 9, 10) would be considered plausible?

7. The average number of children under 18 per family in the U.S. is currently 1.8. A simple random sample of German households gave the following data on number of children under 18 per family. You'd like to use these data to estimate the average number of children under 18 per family for all German families.

1	3	1	0	2	4	2	0	1	2
4	0	1	4	2	3	2	3	2	0
3	2	1	1	1	1	4	3	2	1
2	1	2	0	0	0	5	1	1	3

    a.   Define the parameter in context. Use the correct notation.
    b.   What symbol would you use for the point estimate? What is its value?
    c.   Consider testing H_0: $\mu = \mu_0$ vs. H_a: $\mu \neq \mu_0$. Find the p-values for the tests when μ_0 is 1, 1.2, 1.5, 1.8, 2, and 2.5.
    d.   If your significance level is $\alpha = 0.05$, which of these values of μ (1, 1.2, 1.5, 1.8, 2, 2.5) would be considered plausible?

8.   Eight hundred twenty in a SRS of 1500 likely voters are in favor of candidate X. You'd like to use these data to estimate the proportion of all likely voters who favor candidate X.

    a.   Define the parameter in context. Use the correct notation.
    b.   What symbol would you use for the point estimate? What is its value?
    c.   Consider testing H_0: $p = p_0$ vs. H_a: $p \neq p_0$. Find the p-values for the tests when p_0 is 0.5, 0.52, 0.54, 0.56, and 0.58.
    d.   If your significance level is $\alpha = 0.05$, which of these values of p (0.5, 0.52, 0.54, 0.56, 0.58) would be considered plausible?

9.   A magazine is planning to add an online version in addition to its print version but will only do so if enough current subscribers would be interested. The publisher selects a SRS of 200 current subscribers and finds that 36 are interested in an online version. You'd like to use these data to estimate the proportion of all current subscribers who would be interested in an online version.

    a.   Define the parameter in context. Use the correct notation.
    b.   What symbol would you use for the point estimate? What is its value?
    c.   Consider testing H_0: $p = p_0$ vs. H_a: $p \neq p_0$. Find the p-values for the tests when p_0 is 0.1, 0.15, 0.2, and 0.25.
    d.   If your significance level is $\alpha = 0.05$, which of these values of p (0.1, 0.15, 0.2, 0.25) would be considered plausible?

10. Let's say you test H_0: $\mu = 10$ vs. H_a: $\mu \neq 10$ and get a p-value = 0.487. Why should you not conclude that H_0 is true, i.e. that $\mu = 10$?

11. Let's say you test H_0: $\mu = 45$ vs. H_a: $\mu \neq 45$ and get a p-value = 0.390. Why should you not conclude that H_0 is true, i.e. that $\mu = 45$?

12. Let's say you test H_0: $p = 0.5$ vs. H_a: $p \neq 0.5$ and get a p-value = 0.622. Why should you not conclude that H_0 is true, i.e. that $p = 0.5$?

13. Let's say you test H_0: $p = 0.8$ vs. H_a: $p \neq 0.8$ and get a p-value = 0.622. Why should you not conclude that H_0 is true, i.e. that $p = 0.8$?

14. Consider exercise 6. A 95% confidence interval for μ is (6.75, 9.16). Interpret this interval.

15. Consider exercise 7. A 95% confidence interval for μ is (1.35, 2.20). Interpret this interval.

16. Consider exercise 8. A 95% confidence interval for p is (0.521, 0.572). Interpret this interval.

17. Consider exercise 9. A 95% confidence interval for p is (0.133, 0.239). Interpret this interval.

18. In a simple random sample of 101 checked bags from an airline's current customers, the average weight was 32.287 pounds and the standard deviation was 5.635 pounds. A 95% confidence interval for μ based on these data is (31.17, 33.40). Identify the best interpretation and say what's wrong with the others.

    a.   There's a 95% chance that a customer's bag weighs between 31.17 pounds and 33.40 pounds.
    b.   About 95% of customers' bags weigh between 31.17 pounds and 33.40 pounds.
    c.   We are 95% confident that the mean weight for the 101 bags is between 31.17 pounds and 33.40 pounds.
    d.   We are 95% confident that the mean weight for all bags of current customers is between 31.17 pounds and 33.40 pounds.

e.   We are 95% confident that μ is between 31.17 pounds and 33.40 pounds.

f.   We are 95% confident that a customer's bag weighs between 31.17 pounds and 33.40 pounds.

g.   There's a 95% chance that the mean weight for all bags of current customers is between 31.17 pounds and 33.40 pounds.

19. In a SRS of 91 small cap equity funds, you find the mean expense ratio to be 1.4% and the standard deviation to be 0.6%. A 95% confidence interval for μ based on these data is (1.275%, 1.525%). Identify the best interpretation and say what's wrong with the others.

a.   We are 95% confident that μ is between 1.275% and 1.525%.

b.   We are 95% confident that a small cap equity fund has an expense ratio between 1.275% and 1.525%.

c.   There's a 95% chance that the mean expense ratio for all small cap equity funds is between 1.275% and 1.525%.

d.   There's a 95% chance that the expense ratio for a small cap equity fund is between 1.275% and 1.525%.

e.   About 95% of small cap equity funds have expense ratios between 1.275% and 1.525%.

f.   We are 95% confident that the mean expense ratio for the 91 funds is between 1.275% and 1.525%.

g.   We are 95% confident that the mean expense ratio for all small cap equity funds is between 1.275% and 1.525%.

20. Suppose that in a SRS of 425 18 – 20 year-olds at your school, 160 report having engaged in binge drinking in the past month. A 95% confidence interval for p based on these data is (0.332, 0.423). Identify the best interpretation and say what's wrong with the others.

a.   About 95% of 18 – 20 year-olds at your school have a binge drinking rate between 0.332 and 0.423.

b.   We are 95% confident that the mean binge drinking rate for all 18 – 20 year-olds at your school is between 0.332 and 0.423.

c.   We are 95% confident that between 33.2% and 42.3% of all 18 – 20 year-olds at your school have engaged in binge drinking in the past month.

d.   We are 95% confident that p is between 0.332 and 0.423.

e.   We are 95% confident that between 33.2% and 42.3% of the 425 18 – 20 year-olds have engaged in binge drinking in the past month.

f.   There's a 95% chance that between 33.2% and 42.3% of all 18 – 20 year-olds at your school have engaged in binge drinking in the past month.

21. In a SRS of 60 adults from a particular county to serve on jury duty, 30 are white. A 95% confidence interval for p based on these data is (0.377, 0.623). Identify the best interpretation and say what's wrong with the others.

a.   We are 95% confident that between 37.7% and 62.3% of all adults in this county are white.

b.   We are 95% confident that p is between 0.377 and 0.623.

c.   We are 95% confident that between 37.7% and 62.3% of the 60 adults are white.

d.   There's a 95% chance that between 37.7% and 62.3% of all adults in this county are white.

e.   We are 95% confident that the mean proportion of white adults in this county is between 0.377 and 0.623.

22. Consider exercise 18. What is meant by the phrase "95% confidence?"

23. Consider exercise 19. What is meant by the phrase "95% confidence?"

24. Consider exercise 20. What is meant by the phrase "95% confidence?"

25. Consider exercise 21. What is meant by the phrase "95% confidence?"

7.2 Estimating a Population Mean

In the last section I illustrated the "guess-and-check" method for constructing a confidence interval for the mean 2015 math proficiency rate for all fourth-grade classes in Delaware:

1. Propose a value for μ as the null hypothesis.
2. Compare the data to this value using a non-directional alternative hypothesis (e.g. $\neq$).
3. If the p-value $< \alpha$, reject H_0 and the proposed value. Otherwise, keep the proposed value in the list; it is reasonable based on your data.
4. Propose another value for μ.
5. Repeat steps 1 – 4 until you have found all of the reasonable parameter values.

There is a much easier way to construct the confidence interval by locating the extremes. The key is to use the criterion in step #3 of the guess-and-check method. A value of μ is considered reasonable if the p-value for the non-directional test using that value in the null hypothesis is not too small ($\geq \alpha$). For that to happen, the value of the t-test statistic must be within certain bounds; let's call them $-t^*$ and t^* (see the picture). If the t-test statistic is one of those, then the p-value is exactly equal to α. Any value between them will produce a p-value $> \alpha$.

Setting the t-test statistic equal to these two values of t will give us the two values of μ that are at the endpoints of the confidence interval:

$$\frac{\bar{x}-\mu}{s/\sqrt{n}} = t^* \quad \Rightarrow \quad \mu = \bar{x} - t^*\frac{s}{\sqrt{n}}$$

$$\frac{\bar{x}-\mu}{s/\sqrt{n}} = -t^* \quad \Rightarrow \quad \mu = \bar{x} + t^*\frac{s}{\sqrt{n}}$$

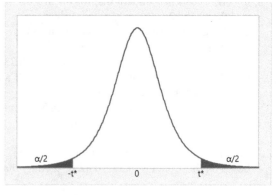

This interval has a fairly simple structure: the center is the sample mean which is the point estimate of μ. To get the endpoints, the quantity $t^*\frac{s}{\sqrt{n}}$ is added and subtracted. We call the quantity $t^*\frac{s}{\sqrt{n}}$ the **margin of error** because it measures how far off we might be from the actual value of μ if our guess was the sample mean, $\bar{x}$.

The one-sample t-test is the basis for the confidence interval so the conditions required for the validity of the t-test also apply here. Putting all the pieces together results in a method called the **one-sample t-interval**.

One-sample t-interval

A $100(1 - \alpha)\%$ confidence interval for μ is

$$\bar{x} \pm t^* \frac{s}{\sqrt{n}}$$

where t^* is the value having an area of $\alpha/2$ to its right under the t-distribution with $df = n - 1$.

Conditions
- SRS
- nearly-normal distribution of data*
  * Can be relaxed if n is sufficiently large.

Note that this method is valid only for simple random samples. As you saw in chapter 3, the sampling distribution of a statistic based on data from a fancier random sample like a cluster random sample or a stratified random sample may exhibit very different variability. The analyses of data collected by those sampling methods are beyond the scope of this book.

It is also important to note that the margin of error does not account for error due to selection bias, non-response bias, or response bias. That's why good studies strive to eliminate these sources of bias when collecting data.

Example: Let's look again at the data from the SRS of 5 fourth-grade classes in Delaware.

40.09%   22.43%   62.71%   28.33%   43.30%

$n = 5$
$\bar{x} = 39.372$
$s = 15.5689$

We want to estimate μ = mean 2015 math proficiency rate for all public fourth-grade classes in Delaware.

Conditions:
- SRS: The data are from a simple random sample.
- Nearly-normal distribution of data? A normal probability plot of the data shows a fairly linear pattern, so yes, this condition is satisfied.

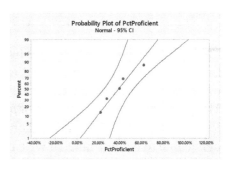

The conditions to use the one-sample t-interval are satisfied so we can calculate the endpoints of the 95% confidence interval:

For 95% confidence, $\alpha = 1 - 0.95 = 0.05$ so t^* is the value having an area of 0.025 to its right (0.975 to its left) in the t-distribution with $df = n - 1 = 5 - 1 = 4$. Using Table II we get $t^* = 2.78$.

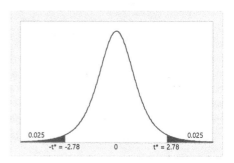

Now just substitute into the formula to get the endpoints:

$\bar{x} \pm t^* \frac{s}{\sqrt{n}}$

$39.372 \pm 2.78 \frac{15.5689}{\sqrt{5}}$

39.372 ± 19.356

$(20.02, 58.73)$

The margin of error is 19.356 meaning that we are 95% confident that our observed sample mean of 39.372% is within 19.356% of the mean 2015 math proficiency rate for all public fourth-grade classes in Delaware. So we can be 95% confident that the mean 2015 math proficiency rate for all public fourth-grade classes in Delaware was between 20.02% and 58.73%.

Notice that this matches the results of the guess-and-check method but it took a lot less time!

The level of confidence we use affects the precision of our estimation. All else equal, using a higher confidence level results in lower precision (larger margin of error and a wider confidence interval). I like to use a fishing analogy: the population parameter is the fish we're trying to catch and the confidence interval is the net we use. All else equal, in order to be more confident of catching the fish, we must use a bigger net.

Example: Using the Delaware math proficiency data, we can see the effect of the confidence level on the interval. The following table shows the margin of error and the endpoints of the interval for various levels of confidence.

Confidence Level	t^*	Margin of Error	Confidence Interval
80%	1.53	10.653%	(28.72%, 50.02%)
90%	2.13	14.830%	(24.54%, 54.20%)
95%	2.78	19.356%	(20.02%, 58.73%)
99%	4.60	32.028%	(7.34%, 71.40%)

This brings us to a dilemma. We like higher confidence levels. High confidence levels assure people that our results can be trusted. We also like narrow confidence intervals. Wide confidence intervals aren't that helpful. (Notice how wide the 99% confidence interval is in the table above! The range of reasonable values of μ is too wide.) So, all else equal, we can't have our cake and eat it too; we can't have both narrow confidence intervals and high confidence levels. But there is hope of being able to eat our cake! If we can change something else within our control to counter the effect of increasing the confidence level, we can get a narrow interval at the same time. That "something else" is the sample size and section 7.4 addresses that issue.

Section 7.2 Exercises

1. What is the symbol we use for the point estimate of a population mean?

2. What is the symbol we use for the parameter that is estimated by $\bar{x}$?

3. What are the conditions required for the one-sample t-interval to be valid?

4. Find t^* for the one-sample t-interval in these situations.

    a. 95% confidence using 10 degrees of freedom
    b. 80% confidence and a SRS of $n = 22$
    c. 99% confidence using 14 degrees of freedom
    d. 90% confidence and a SRS of $n = 7$

5. Find t^* for the one-sample t-interval in these situations.

    a. 99.9% confidence using 16 degrees of freedom
    b. 90% confidence and a SRS of $n = 10$
    c. 98% confidence using 4 degrees of freedom
    d. 95% confidence and a SRS of $n = 18$

6. For a particular value of degrees of freedom, what happens to t^* as the confidence level increases?

7. For a particular value of degrees of freedom, what happens to t^* as the confidence level decreases?

8. What is the highest level of confidence you can get by using Table II?

9. What is the lowest level of confidence you can get by using Table II?

10. Shown here is a normal probability plot of a simple random sample of 50 credit card transaction amounts from the set of all such amounts for a school district in a recent fiscal year.[1] Would a one-sample t-interval be appropriate to estimate the mean transaction amount for all amounts for this school district during the same fiscal year? Explain.

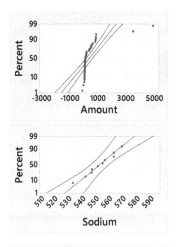

11. Here is a normal probability plot of the sodium amounts in a simple random sample of 10 hot dogs of a certain brand. Would a one-sample t-interval be appropriate to estimate the average sodium content for all hot dogs of this brand? Explain.

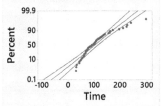

12. The time to finish an exam was observed for a simple random sample of 71 students. A normal probability plot of the times is shown here. Would a one-sample t-interval be appropriate to estimate the mean time to finish the exam for all such students? Explain.

13. How do the conditions of the one-sample t-interval and the one-sample t-test compare?

14. How do you calculate the margin of error in a one-sample t-interval? What does it tell you?

15. Numeric data observed on a SRS of 16 objects from a large population gave $\bar{x} = 22.8$ and $s = 5.1$.

    a. What is the point estimate of μ?
    b. Calculate the margin of error using 95% confidence.
    c. Find the 95% confidence interval for μ.
    d. What condition are you not able to check? Why not?
    e. How would the 90% confidence interval compare to the 95% confidence interval?

16. Numeric data observed on a SRS of 28 objects from a large population gave $\bar{x} = 1.2$ and $s = 0.04$.

    a. What is the point estimate of μ?
    b. Calculate the margin of error using 95% confidence.
    c. Find the 95% confidence interval for μ.
    d. What condition are you not able to check? Why not?
    e. How would the 99% confidence interval compare to the 95% confidence interval?

17. Numeric data observed on a SRS of 38 objects from a large population gave $\bar{x} = 96.8$ and $s = 13.9$.

    a. What is the point estimate of μ?
    b. Calculate the margin of error using 95% confidence.
    c. Find the 95% confidence interval for μ.
    d. Under what condition might the one-sample t-interval not be appropriate?
    e. How would the 80% confidence interval compare to the 95% confidence interval?

18. Numeric data observed on a SRS of 61 objects from a large population gave $\bar{x} = -27.0$ and $s = 11.3$.

    a. What is the point estimate of μ?
    b. Calculate the margin of error using 95% confidence.
    c. Find the 95% confidence interval for μ.
    d. Under what condition might the one-sample t-interval not be appropriate?
    e. How would the 99.9% confidence interval compare to the 95% confidence interval?

19. A SRS of 29 electric cars is tested to determine each car's 0 – 60mph acceleration time. The mean time was 7.955 seconds and the standard deviation was 3.175 seconds. A histogram of the times is shown here.

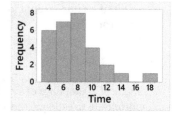

    a. Identify the objects/subjects, the variable and its type.
    b. Estimate the mean 0 – 60mph time for all electric cars using a 95% confidence interval.
        i. Define the parameter in context.
        ii. Check the conditions.
        iii. Perform the mechanics.
        iv. Interpret the confidence interval.

20. The average number of children under 18 per family in the U.S. is currently 1.8. A simple random sample of German households gave the following data on number of children under 18 per family.

1	3	1	0	2	4	2	0	1	2
4	0	1	4	2	3	2	3	2	0
3	2	1	1	1	1	4	3	2	1
2	1	2	0	0	0	5	1	1	3

    a.   Identify the objects/subjects, the variable and its type.
    b.   Estimate the average number of children under 18 for all German families using a 95% confidence interval.
        i.   Define the parameter in context.
        ii.   Check the conditions.
        iii.   Perform the mechanics.
        iv.   Interpret the confidence interval.
    c.   Do these data provide convincing statistical evidence that the average number of children under 18 for all German families differs from 1.8? Use your confidence interval to answer.

21. A small beverage company makes ginger beer and bottles it in 12-oz. bottles. To monitor the filling process, a simple random sample of 5 bottles is selected from the production line every two hours and the amount of ginger beer (ounces) in each bottle is accurately measured. The data for the most recent sample is shown here.

        11.77   11.93   12.09   12.03   11.93

    a.   Identify the objects/subjects, the variable and its type.
    b.   Estimate the mean amount of ginger beer in all bottles currently being filled using a 95% confidence interval.
        i.   Define the parameter in context.
        ii.   Check the conditions.
        iii.   Perform the mechanics.
        iv.   Interpret the confidence interval.
    c.   Do these data provide convincing statistical evidence of a problem in the filling process? Explain using your confidence interval.

22. An airline has long had a 50-pound limit on the weight of any checked bags. The limit was based on a mean weight of 34 pounds plus a multiple of the standard deviation. In a simple random sample of 101 checked bags from current customers, the average weight was 32.287 pounds and the standard deviation was 5.635 pounds. A normal probability plot of the weights is shown here.

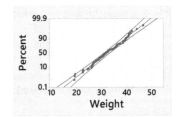

    a.   Identify the objects/subjects, the variable and its type.
    b.   Estimate the mean weight of all checked bags of current customers using a 95% confidence interval.
        i.   Define the parameter in context.
        ii.   Check the conditions.
        iii.   Perform the mechanics.
        iv.   Interpret the confidence interval.
    c.   Do these data provide convincing statistical evidence that the mean weight of all checked bags of current customers is less than 50 pounds? Explain using your confidence interval.

23. The YTD returns of a SRS of 3 mutual funds which invest in the stocks of small U.S. companies (called "small cap funds") are shown here.

        23.39% 22.55% 21.78%

    a.   Identify the objects/subjects, the variable and its type.
    b.   Calculate and interpret the 95% confidence interval for the mean return for all U.S. small cap mutual funds. Be sure to check the appropriate conditions.

24. In a SRS of locations in one field, the following nitrate levels (in ppm) were obtained:

        28.0   29.8   26.4   33.1   29.6   25.5   25.6   27.7   31.0   27.3

    a.   Identify the objects/subjects, the variable and its type.
    b.   Calculate and interpret the 95% confidence interval for the mean nitrate level for all locations in the field. Be sure to check the appropriate conditions.

25. In a SRS of 22 mild asthmatics, you find the mean and standard deviation of their FEV1 values to be 3.5 and 0.9, respectively. The FEV1 values do not show any outliers and the distribution is fairly symmetric.

    a.   Identify the objects/subjects, the variable and its type.
    b.   Calculate and interpret the 95% confidence interval for the mean FEV1 for all mild asthmatics. Be sure to check the appropriate conditions.

26. In a SRS of 91 small cap equity funds, you find the mean expense ratio to be 1.4% and the standard deviation to be 0.6%.

    a.   Identify the objects/subjects, the variable and its type.
    b.   Calculate and interpret the 95% confidence interval for the mean expense ratio for all small cap equity funds. Be sure to check the appropriate conditions.

27. Consider the medal count data for countries having at least one medal from exercise 1 in section 2.2. Comment on the use of the one-sample t-interval to estimate the mean number of gold medals among all countries having at least one medal in the 2018 Winter Olympics.

28. Consider the data on the top 50 Major League Baseball players ranked by the number of career home runs from exercise 8 in section 2.2. The mean number of games played in that data set is 2494.2 and the standard deviation is 375.2. Comment on the use of the one-sample t-interval to estimate the mean number of games played for all 50 players.

29. Consider the data on basketball team from exercise 16 in section 2.2. The mean number of fouls committed is 10.09 and the standard deviation is 8.88. Comment on the use of the one-sample t-interval to estimate the mean number of fouls committed for the players.

30. Consider the data on median student loan debt of students entering loan repayment on a simple random sample of institutions from exercise 14 in section 2.3.

    a.   Calculate and interpret a 95% confidence interval for $\mu_{Associate's}$ and for $\mu_{Bachelor's}$.
    b.   Do these data provide convincing statistical evidence that the average median student loan debt of all students graduating from Associate's degree-granting institutions differs from that for Bachelor's degree-granting institutions? Explain using your confidence intervals.

31. Consider the house appraisal data from exercise 15 in section 2.3.

    a.   Calculate and interpret a 95% confidence interval for μ_{Jill} and μ_{Anya}.
    b.   Do these data provide convincing statistical evidence of a difference between Jill's and Anya's mean appraisal value for all houses for sale? Explain using your intervals from part a.
    c.   For each property, calculate the difference in appraised value (Jill – Anya).
    d.   Calculate and interpret a 95% confidence interval for $\mu_{Difference}$ using the differences from part c. as the data.
    e.   Do these data provide convincing statistical evidence of an average difference between Jill's and Anya's appraisal value for all houses for sale? Explain using your interval from part d.
    f.   Which analysis method (part a. or part d.) should you use? Why?

32. A simple random sample of 10 8th grade classes in a certain state was selected and the percent of students classified as proficient in an English language arts (ELA) test was recorded for each class. The mean percent proficient was 53.73% and the standard deviation was 14.79%. A 95% confidence interval for μ is (43.1, 64.3). Identify the best interpretation and say what's wrong with the others.

    a.   We are 95% confident that μ is between 43.1% and 64.3%.
    b.   We are 95% confident that between 43.1% and 64.3% of students in an 8th grade class in this state are proficient in ELA.

    c. There's a 95% chance that the mean proficiency in ELA for all 8th grade classes in this state is between and 64.3%.

    d. We are 95% confident that the mean proficiency in ELA for all 8th grade classes in this state is between 43.1 and 64.3%.

    e. We are 95% confident that the mean proficiency in ELA for the 10 8th grade classes was between 43.1% and 64.3%.

    f. About 95% of all 8th grade classes in this state have an ELA proficiency between 43.1% and 64.3%.

    g. About 95% of the sample 8th grade classes had an ELA proficiency between 43.1% and 64.3%.

33. A simple random sample of 9 bags of chips resulted in an average product weight of 9.4 ounces and a standard deviation of 0.8 ounces. A 95% confidence interval for μ is (8.8, 10.0). Identify the best interpretation and say what's wrong with the others.

    a. About 95% of all bags of chips have a product weight between 8.8 ounces and 10.0 ounces.

    b. About 95% of the sample bags of chips had a product weight between 8.8 ounces and 10.0 ounces.

    c. We are 95% confident that μ is between 8.8 ounces and 10.0 ounces.

    d. We are 95% confident that a bag of chips will have a product weight between 8.8 ounces and 10.0 ounces.

    e. There's a 95% chance that the mean product weight for all bags of chips is between 8.8 ounces and 10.0 ounces.

    f. We are 95% confident that the mean product weight for all bags of chips is between 8.8 ounces and 10.0 ounces.

    g. We are 95% confident that the mean product weight for the 9 bags of chips is between 8.8 ounces and 10.0 ounces.

34. Consider exercise 23. What is meant by the phrase "95% confident?"

35. Consider exercise 24. What is meant by the phrase "95% confident?"

36. Consider exercise 25. What is meant by the phrase "95% confident?"

37. Consider exercise 32. What is meant by the phrase "95% confident?"

7.3 Estimating a Population Proportion

The results of public opinion polls are all around us, from social media polls to election polling during an election cycle. If you know anything about the 2016 U.S. Presidential election and the polling that occurred leading up to it, you know that a good understanding of the fundamentals of polling is important for the general public. The issues of non-response bias and response bias plague public opinion polls and despite the efforts of very smart people to compensate for them, their effects can impact the results of polls. Trying to accurately estimate p, the population proportion who hold a certain opinion is a difficult task.

In this section, we'll consider the problem of estimating a population proportion in a best-case scenario: a SRS from a large population where there is no non-response to worry about and no response bias. It turns out we can use the guess-and-check method applied to a hypothesis test for p in the same way we applied the guess-and-check method to a hypothesis test for μ in the last section:

1. Propose a value for p as the null hypothesis.
2. Compare the data to this value using a non-directional alternative hypothesis (e.g. $\neq$).

, reject H_0 and the proposed value. Otherwise, keep the proposed value in
ɔnable based on your data.
value for p.
until you have found all of the reasonable parameter values.

th the one-proportion binomial test and the one proportion z-test.
way to eliminate the guessing and checking if you use the binomial test so
this section.

Remember that the statistic

$$z = \frac{\hat{p} - p}{\sqrt{\dfrac{p(1-p)}{n}}}$$

has approximately a standard normal distribution
under certain conditions: 1) a SRS from the
population and 2) $np \geq 15$ and $n(1-p) \geq 15$. A
value of p is considered reasonable if the p-value for
the non-directional test using that value in the null
hypothesis is not too small ($\geq \alpha$). For that to
happen, the value of the z-test statistic must be
within certain bounds, let's call them $-z^*$ and z^* (see
the picture).   If the z-test statistic is one of those,
then the p-value is exactly equal to α. Any value
between them will produce a p-value $> \alpha$.

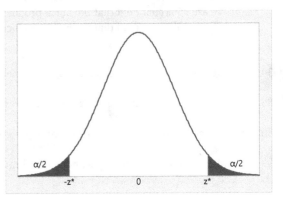

Setting the z-test statistic equal to either of these two values of z will give us the two values of p that
are at the endpoints of the confidence interval. Consider z^* first:

$$\frac{\hat{p}-p}{\sqrt{\frac{p(1-p)}{n}}} = z^*$$

Square both sides, cross multiply to get

$$(\hat{p} - p)^2 = \frac{z^{*2} p(1-p)}{n},$$

and then expand to get

$$\hat{p}^2 - 2\hat{p}p + p^2 = \frac{z^{*2}}{n} p - \frac{z^{*2}}{n} p^2 \,.$$

This is a quadratic equation which we can solve using the quadratic formula. To use the formula
the equation must be in the general form $ap^2 + bp + c = 0$ so rearrange the terms to get

$$\left(1 + \frac{z^{*2}}{n}\right)p^2 + \left(-2\hat{p} - \frac{z^{*2}}{n}\right)p + \hat{p}^2 = 0.$$

Using the quadratic formula with

$$a = 1 + \frac{z^{*2}}{n}$$

$$b = -2\hat{p} - \frac{z^{*2}}{n}$$

and

$$c = \hat{p}^2$$

results in the following solutions:

$$p = \frac{-b \pm \sqrt{b^2 - 4ac}}{2a}.$$

After substitution and a lot of simplification, we get the confidence interval endpoints as

$$\frac{\hat{p} + \frac{z^{*2}}{2n}}{1 + \frac{z^{*2}}{n}} \pm z^* \frac{\sqrt{\frac{\hat{p}(1 - \hat{p})}{n} + \frac{z^{*2}}{4n^2}}}{1 + \frac{z^{*2}}{n}}.$$

If you set the z-test statistic equal to $-z^*$,

$$\frac{\hat{p} - p}{\sqrt{\frac{p(1-p)}{n}}} = -z^*$$

and then square both sides and cross multiply, you get

$$(\hat{p} - p)^2 = \frac{z^{*2}p(1-p)}{n}.$$

From there the process is identical to what we just did, resulting in the same confidence interval endpoints. The interval is often called the **score interval**.

One-proportion score interval

A $100(1 - \alpha)\%$ confidence interval for p is

$$\frac{\hat{p} + \dfrac{z^{*2}}{2n}}{1 + \dfrac{z^{*2}}{n}} \pm z^* \frac{\sqrt{\dfrac{\hat{p}(1 - \hat{p})}{n} + \dfrac{z^{*2}}{4n^2}}}{1 + \dfrac{z^{*2}}{n}}$$

where z^* is the value having an area of $\alpha/2$ to its right under the standard normal distribution.

Condition
- SRS

Note that this method is valid only for simple random samples. As you saw in chapter 3, the sampling distribution of a statistic based on data from a fancier random sample like a cluster random sample or a stratified random sample may exhibit very different variability. The analyses of data collected by those sampling methods are beyond the scope of this book.

It is also important to note that the margin of error does not account for error due to selection bias, non-response bias, or response bias. That's why good studies strive to eliminate these sources of bias when collecting data.

I know what you're thinking: "Are you kidding?! That's a crazy complicated formula!" You're right; compared to the t-interval for μ, this formula is much more involved. But before you get too mad at me, consider these points:

- I didn't discover the formula; Edwin Wilson did in the 1920's.[2] So don't shoot the messenger!
- I did all the algebra for you. It took me over a page. Seriously.
- The guess-and-check method is much more tedious.
- It turns out this interval performs quite well even in cases where the $np \geq 15$ and $n(1 - p) \geq 15$ conditions don't hold![3] That's a good thing because the population proportion p is unknown, so we can't check that condition anyway. All that is necessary in almost any practical problem is a SRS from a fairly large population.

Example: In a nationwide simple random sample of 1501 U.S. adults conducted in April 2017 by the Pew Research Center, only 300 (about 20%) said they can trust the Federal government to do what is right "just about always" or "most of the time."[4] Clearly public trust in Washington has eroded since the late 1950s when some sources put the figure at over 70%...or has it? The 20% figure is almost surely not correct; it's based on a sample. If we could survey all 240ish million U.S. adults, what would the actual population percentage be? Construct a 95% confidence interval for the proportion of all U.S. adults who would say they can trust the Federal government to do what is right "just about always" or "most of the time."

We want to estimate p = proportion of all U.S. adults who would say they can trust the Federal government to do what is right "just about always" or "most of the time."

Condition:

- SRS: The data are from a simple random sample.

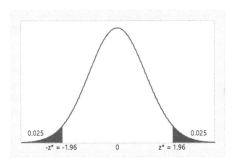

The condition to use the one-proportion score interval is satisfied so we can calculate the endpoints of the 95% confidence interval:

For 95% confidence, $\alpha = 1 - 0.95 = 0.05$ so z^* is the value having an area of 0.025 to its right (0.975 to its left) in the standard normal distribution. Using Table I we get $z^* = 1.96$.

Now just substitute into the formula to get the endpoints:

$$\frac{\hat{p} + \frac{z^{*2}}{2n}}{1 + \frac{z^{*2}}{n}} \pm z^* \frac{\sqrt{\frac{\hat{p}(1-\hat{p})}{n} + \frac{z^{*2}}{4n^2}}}{1 + \frac{z^{*2}}{n}}$$

$$\frac{\frac{300}{1501} + \frac{1.96^2}{2 \cdot 1501}}{1 + \frac{1.96^2}{1501}} \pm 1.96 \frac{\sqrt{\frac{\frac{300}{1501}\left(1 - \frac{300}{1501}\right)}{1501} + \frac{1.96^2}{4(1501)^2}}}{1 + \frac{1.96^2}{1501}}$$

$$0.2006 \pm 0.0202$$
$$(0.180, .221)$$

We can be 95% confident that the proportion of all U.S. adults who would say they can trust the Federal government to do what is right "just about always" or "most of the time" is between 0.180 and 0.221 (between 18.0% and 22.1%). Even the highest plausible proportion is far below 70% so we can be confident that this proportion has decreased since the 1950s.

This analysis assumes the best, that all of the U.S. adults that were randomly selected actually respond to the survey. In practice, that is rarely the case and there could be significant non-response. In addition, folks with both landlines and cell phones are more likely to be randomly selected by the random digit dialing method used. To account for these complexities, the Pew Research Center reports a larger margin of error (2.9%) based on a fancier method that goes beyond the scope of this book.

Beware of non-random samples, especially from surveys. There are lots of survey reports out there like that and you would not want to apply the score method to construct a confidence interval based on those data. In the case of estimating a population proportion, there is a conservative way to get the lowest and highest possible values for p even when the data are from a non-random sample.

Example: A survey was sent to all 1,187 faculty members at a large university. One of the questions asked faculty about their satisfaction with their experience of being a faculty member. Of the 433 faculty who responded to the question, 220 (50.8%) were either "satisfied" or "very satisfied."[5] Construct an interval of possible values for the proportion of all faculty at this university who would have reported they were "satisfied" or "very satisfied" with their experience.

We want to estimate p = proportion of all faculty at this university who would have reported they were "satisfied" or "very satisfied" with their experience.

If you're tempted to use the score interval, don't! The method is based on a SRS from the population. The 433 responses are from those faculty who chose to respond; their responses may or may not represent the faculty as a whole. They certainly are not a SRS.

Worst-case scenario: None of the 1,187 − 433 = 754 non-respondents would have responded "satisfied" or "very satisfied." Then we'd have

$$p = \frac{number\ of\ "satisfied"\ or\ "very\ satisfied"\ responses}{1187} = \frac{220}{1187} = 0.185.$$

Best-case scenario: All of the 1,187 − 433 = 754 non-respondents would have also responded "satisfied" or "very satisfied." Then we'd have

$$p = \frac{number\ of\ "satisfied"\ or\ "very\ satisfied"\ responses}{1187} = \frac{220+754}{1187} = \frac{974}{1187} = 0.821.$$

We know that between 18.5% and 82.1% of all faculty at this university who would have reported they were "satisfied" or "very satisfied" with their experience. That's a wide range, but it's difficult to do any better without a representative, random sample. If the satisfied faculty were more likely to respond to the survey, that would leave fewer satisfied non-respondents, and the actual figure would be closer to the low end of this range. On the other hand, if the dissatisfied faculty were more likely to respond, that would leave more satisfied non-respondents, and the actual figure would be closer to the high end.

Check out how much more precise the estimation could have been if the 433 were a SRS. Here's the one-proportion score interval based on the 220 "satisfied" and "very satisfied" responses assuming the 433 respondents were from a SRS: $(0.461, 0.555)$. That's the power of random sampling without non-response!

For non-random samples, here's a summary of what I call the **lowest/highest method** for estimating a population proportion.

One-proportion lowest/highest interval

The following give the lowest and highest possible values of p:

Lowest: $p = \frac{x}{N}$ Highest: $p = \frac{x+(N-n)}{N}$

x = number of successes in the sample
n = sample size
N = population size

Section 7.3 Exercises

1. What is the symbol we use for the point estimate of a population proportion?

2. What is the symbol we use for the parameter that is estimated by $\hat{p}$?

3. What are the conditions required for the one-proportion score interval to be valid?

4. Find z^* for these confidence levels.

    a. 94%
    b. 80%
    c. 99%
    d. 87%

5. Find z^* for these confidence levels.

    a. 82%
    b. 96%
    c. 99.99%
    d. 85%

6. What happens to z^* as the confidence level increases?

7. What happens to z^* as the confidence level decreases?

8. What is the highest level of confidence you can get by using Table I?

9. In a simple random sample of 20 potential customers, 15 said they would use your service over your closest competitor's service. Would the one-proportion score interval be appropriate to estimate the proportion of all potential customers who would use your service over your closest competitor's service? Explain.

10. In a simple random sample of 1.5 million U.S. undergraduates, 298,583 responded. Of the respondents 225,684 said they have, at least once, not used an information source because of its questionable quality.[6] Would the one-proportion score interval be appropriate to estimate the proportion of all U.S. undergraduates who have, at least once, not used an information source because of its questionable quality? Explain.

11. You create a poll about belief in the supernatural using your favorite social media tool. Of the 1,744 responses, 1,221 believe in the supernatural. Would the one-proportion score interval be appropriate to estimate the proportion of all U.S. adults who believe in the supernatural? Explain.

12. A county takes a cluster random sample of households in the county and records the ethnicity of each person in each household. Twenty-two percent of the 849 people in the sample were minorities. Would the one-proportion score interval be appropriate to estimate the minority proportion in the county? Explain.

13. How do the conditions of the one-proportion score interval and the one-proportion z-test compare?

14. How do the conditions of the one-proportion score interval and the one-proportion binomial test compare?

15. What is a good thing about the one-proportion score interval that outweighs its complex formula?

16. Success/failure data observed on a SRS of 160 objects from a large population gave $\hat{p} = 0.65$.

    a.   What is the point estimate of p?
    b.   Calculate the margin of error using 95% confidence.
    c.   Find the 95% confidence interval for p.
    d.   How would the 90% confidence interval compare to the 95% confidence interval?

17. Success/failure data observed on a SRS of 2000 objects from a large population gave $\hat{p} = 0.21$.

    a.   What is the point estimate of p?
    b.   Find the 92% confidence interval for p.
    c.   How would the 99% confidence interval compare to the 92% confidence interval?

18. Success/failure data observed on a SRS of 500 objects from a large population gave 49 successes.

    a.   What is the point estimate of p?
    b.   Find the 82% confidence interval for p.
    c.   How would the 95% confidence interval compare to the 82% confidence interval?

19. Success/failure data observed on a SRS of 90 objects from a large population gave 68 successes.

    a.   What is the point estimate of p?
    b.   Find the 98% confidence interval for p.
    c.   How would the 95% confidence interval compare to the 98% confidence interval?

20. In a SRS of 250 students at a large university, 103 indicated that they had all required course materials within the first two weeks of the semester.

    a.   Identify the objects/subjects, the variable and its type.
    b.   Estimate the proportion of all students at this university who have all required course materials within the first two weeks of the semester using a 95% confidence interval.
        i.   Define the parameter in context.
        ii.   Check the conditions.
        iii.   Perform the mechanics.
        iv.   Interpret the confidence interval.

21. In a SRS of 50 seniors at a particular college, 32 indicated that they had attended a career fair.

    a.   Identify the objects/subjects, the variable and its type.
    b.   Estimate the proportion of all seniors at this college who have attended a career fair using a 95% confidence interval.
        i.   Define the parameter in context.
        ii.   Check the conditions.
        iii.   Perform the mechanics.
        iv.   Interpret the confidence interval.

    c.   Do these data provide convincing statistical evidence that more than 60% of all seniors at this college have attended a career fair? Use your confidence interval to answer.

22. Eight hundred twenty in a SRS of 1500 likely voters are in favor of candidate X.

    a.   Identify the objects/subjects, the variable and its type.
    b.   Estimate the proportion of all likely voters who are in favor of candidate X using a 95% confidence interval.
         i.   Define the parameter in context.
        ii.   Check the conditions.
       iii.   Perform the mechanics.
       iv.   Interpret the confidence interval.
    c.   Do these data provide convincing statistical evidence that a majority of all likely voters are in favor of candidate X? Explain using your confidence interval.

23. A large mail order company is studying its order fulfillment process. An order is shipped "on time" if it is sent out within two business days of when it was received. In a SRS of 300 recently-shipped orders, 279 were shipped on time.

    a.   Identify the objects/subjects, the variable and its type.
    b.   Estimate the proportion of all recently-shipped orders that were shipped on time using a 90% confidence interval.
         i.   Define the parameter in context.
        ii.   Check the conditions.
       iii.   Perform the mechanics.
       iv.   Interpret the confidence interval.
    c.   The company currently claims that "95% of all orders are shipped on time" but is worried that there have been more late orders recently. Do these data provide convincing statistical evidence that fewer than 95% of all recently-shipped orders were shipped on time? Explain using your confidence interval.

24. Consider exercise 22. Calculate a 99.99% confidence interval. Would using the 99.99% interval give you a different answer to part c.? Explain.

25. Consider exercise 23. Calculate an 85% confidence interval. Would using the 85% interval give you a different answer to part c.? Explain.

26. According to a Pew Research Center study, 21% in a sample of 5,000 U.S. adults were raised by two parents having different religious views.[7]

    a.   Identify the objects/subjects, the variable and its type.
    b.   Calculate and interpret the 95% confidence interval for the proportion of all U.S. adults who were raised in interfaith homes. Be sure to check the appropriate conditions.

27. In a SRS of 425 18 – 20 year-olds at your school, 160 reported having engaged in binge drinking in the past month.

    a.   Identify the objects/subjects, the variable and its type.
    b.   Calculate and interpret the 95% confidence interval for the proportion of all 18 – 20 year-olds at your school who have engaged in binge drinking in the past month. Be sure to check the appropriate conditions.

28. There were 30 white adults in a SRS of 60 adults from a particular county.

    a.   Identify the objects/subjects, the variable and its type.
    b.   Calculate and interpret the 95% confidence interval for the proportion of all adults in this county who are white. Be sure to check the appropriate conditions.

29. A magazine publisher selects a SRS of 200 current subscribers and finds that 36 are interested in an online version.

a. Identify the objects/subjects, the variable and its type.
b. Calculate and interpret the 95% confidence interval for the proportion of all current subscribers who are interested in an online version.

30. Consider the cause of death data from exercise 3 in section 2.1. Comment on the use of the one-proportion score interval to estimate the proportion of all deaths by cancer in the U.S. in 2016.

31. Consider the mutual fund asset class data from exercise 6 in section 2.1. Comment on the use of the one-proportion score interval to estimate the proportion of all mutual funds available at Fidelity that are U.S. equity funds.

32. Consider the arrival status data from exercise 7 in section 2.1. Comment on the use of the one-proportion score interval to estimate the proportion of all February 2019 flights that were on time.

33. Consider the survey about time spent with children from exercise 7 in section 2.3. Consider the data a SRS of U.S. parents of children under 19.

    a. Calculate and interpret a 95% confidence interval for $p_{mothers}$, the proportion of all U.S. mothers in 2015 who thought they spend too little time with their child/children.
    b. Calculate and interpret a 95% confidence interval for $p_{fathers}$, the proportion of all U.S. fathers in 2015 who thought they spend too little time with their child/children.
    c. Do these data provide convincing statistical evidence of a difference between all U.S. mothers and fathers in 2015 with respect to the proportion who think they spend too little time with their child/children? Explain using your confidence intervals.

34. Consider the online dating survey data from the example in section 2.3. Consider the data a SRS of U.S. adults.

    a. Calculate and interpret a 95% confidence interval for p_{18-24}, the proportion of all U.S. adults aged 18 – 24 in 2015 who had used internet dating.
    b. Calculate and interpret a 95% confidence interval for p_{25-34}, the proportion of all U.S. adults aged 25 – 34 in 2015 who had used internet dating.
    c. Do these data provide convincing statistical evidence of a difference between all U.S. 18 – 24 year-olds and all U.S. 25 – 34 year-olds in 2015 with respect to the proportion who had used internet dating? Explain using your confidence intervals.

35. In August of 2017, a simple random sample of U.S. adults was asked about their TV-watching habits. Of the 1,893 adults in the sample, 530 reported using a streaming service to watch TV.[8] A 90% confidence interval for p is (0.263, 0.297). Identify the best interpretation and say what's wrong with the others.

    a. We are 90% confident that p is between 0.263 and 0.297.
    b. There's a 90% chance that between 26.3% and 29.7% of the 1,893 adults used a streaming service to watch TV.
    c. We are 90% confident that between 26.3% and 29.7% of the 1,893 adults used a streaming service to watch TV.
    d. We are 90% confident that between 26.3% and 29.7% of all U.S. adults in August of 2017 used a streaming service to watch TV.
    e. The probability that more U.S. adults in August of 2017 used a streaming service to watch TV than other methods is between 0.263 and 0.297.
    f. About 90% of all U.S. adults in August of 2017 used a streaming service to watch TV between 26.3% and 29.7% of the time.
    g. We are 90% confident that $\hat{p}$ is between 0.263 and 0.297.

36. In a survey of a SRS of 1,350 U.S. adults, 1,202 have definitely heard about the Holocaust.[9] A 95% confidence interval for p is (0.876, 0.904). Identify the best interpretation and say what's wrong with the others.

    a. The probability that more U.S. adults have definitely heard about the Holocaust than have not is between 0.876 and 0.904.
    b. About 95% of all U.S. adults definitely hear about the Holocaust between 87.6% and 90.4% of the time.
    c. We are 95% confident that $\hat{p}$ is between 0.876 and 0.904.

d. We are 95% confident that p is between 0.876 and 0.904.

e. There's a 95% chance that between 87.6% and 90.4% of the 1,350 adults have definitely heard about the Holocaust.

f. We are 95% confident that between 87.6% and 90.4% of the 1,350 adults have definitely heard about the Holocaust.

g. We are 95% confident that between 87.6% and 90.4% of all U.S. adults have definitely heard about the Holocaust.

37. Consider exercise 26. What is meant by the phrase "95% confident?"

38. Consider exercise 27. What is meant by the phrase "95% confident?"

39. Consider exercise 28. What is meant by the phrase "95% confident?"

40. Consider exercise 29. What is meant by the phrase "95% confident?"

41. A survey was sent to the parents of all 4,809 elementary children in a school district asking whether or not the child was diagnosed with autism. Of the 517 responses, 21 indicated an autism diagnosis.

    a. Use the one-proportion lowest/highest interval to give an interval for p, the proportion of all 4,809 elementary children in the district with autism.
    b. How confident are you that p is in the interval in part a.?
    c. Suppose the data would have come from a SRS. Calculate a 95% confidence interval for p using the score method.
    d. How confident are you that p is in the interval in part c.?
    e. What is the value of the SRS compared to the volunteer sample?

42. You want to estimate the proportion of all people in the U.S. who use alternative medical treatments that improve in their condition. You ask 25 of your friends who use alternative medical treatments about whether their condition has improved and 20 say that it has. Suppose there are 100,000,000 people in the U.S. who use alternative medical treatments.

    a. Use the one-proportion lowest/highest interval to give an interval for p, the proportion of all 100,000,000 people in the U.S. who use alternative medical treatments that improve in their condition.
    b. How confident are you that p is in the interval in part a.?
    c. Suppose the data would have come from a SRS. Calculate a 95% confidence interval for p using the score method.
    d. How confident are you that p is in the interval in part c.?
    e. What is the value of the SRS compared to the sample of your friends?

43. A bank wants to estimate the proportion of the week's 22,850 transactions that were processed correctly. The bank selects the last 100 transactions at the end of the day on Friday and finds that 99 were processed correctly.

    a. Use the one-proportion lowest/highest interval to give an interval for p, the proportion of the week's 22,850 transactions that were processed correctly.
    b. How confident are you that p is in the interval in part a.?
    c. Suppose the data would have come from a SRS. Calculate a 95% confidence interval for p using the score method.
    d. How confident are you that p is in the interval in part c.?
    e. What is the value of the SRS compared to the sample at the end of the day on Friday?

44. An online poll asked viewers, "Are you more of a morning person or a night person?" Of the 50,353 respondents, 54% said "morning person." Consider trying to estimate the proportion of all Americans who are "morning persons."

    a. Use the one-proportion lowest/highest interval to give an interval for p, the proportion of all Americans who are "morning persons." Use the current U.S. population.

    b.   How confident are you that p is in the interval in part a.?

    c.   Suppose the data would have come from a SRS. Calculate a 95% confidence interval for p using the score method.

    d.   How confident are you that p is in the interval in part c.?

    e.   What is the value of the SRS compared to the online sample?

7.4 Determining the Sample Size

I left you with a dilemma at the end of section 7.2. It went something like this:

- We like both high confidence levels and narrow confidence intervals.
- All else equal, increasing the confidence level makes the confidence interval wider.

So what do you do if your desired level of confidence results in an interval that is too wide? Or what if you need to use an unacceptably low level of confidence to get the interval to have the width you want? Unfortunately there's not much that can be done to remedy such situations *after* the data have been collected. However, with careful planning *before* collecting the data, you can assure that you'll be able to report a narrow confidence interval with high confidence. To have your cake and eat it too, you can counter the effect that the confidence level has on the width of the interval by adjusting the sample size. In this section, you'll learn how to do that.

Estimating a Population Mean

Suppose you plan to collect a SRS from a large population in hopes of estimating the population mean and you plan to use the one-sample t-interval to do it. It turns out that the margin of error is very important in determining what sample size to use since the margin of error both controls the width of the interval and captures the confidence level. For the one-sample t-interval the margin of error is

$$MOE = t^* \frac{s}{\sqrt{n}}.$$

It's tempting to think of this as only a formula that gives you the margin of error if you plug in t^*, s, and n. But if you think of it as an equation that specifies a relationship between four things, it becomes much more useful. You can determine what any one of those four things are by plugging in for the other three! Here we want to know what n should be if we want to specify the margin of error and the confidence level. Solving the above equation for n gives

$$n = \left(\frac{t^* s}{MOE}\right)^2.$$

While that looks nice and tidy, there are two problems.

Problem 1: You won't know the sample standard deviation, s. Remember, at this stage, you haven't collected data yet! One solution is to estimate the value of s by dividing the variable's range by 4. If you have reasonable guesses as to the maximum and minimum values you'd ever see, use $s = \frac{range}{4}$. The rationale is that in many numeric distributions, most values will be found within 2 standard deviations of the mean so that the range of the distribution will be about 4 standard deviations wide.

Problem 2: To get t^*, you need to know the degrees of freedom, $n - 1$, which also involves n. That means you would have to use guess-and-check to find n: guess a value of n, plug it into the equation along with s, your desired MOE, and the t^* for your confidence level and degrees of freedom, and see if both sides are the same. If not, guess another value of n. Repeat the process until both sides are as close as possible. To avoid that process, I suggest substituting z^* for t^* in the formula. If you find that your required n is somewhat small, use that n as your first guess in the formula with t^* and use the guess-and-check process.

Sample Size Required for Estimating a Population Mean

$$n = \left(\frac{z^* s}{MOE}\right)^2$$

z^* = value having an area of $\alpha/2$ to its right under the standard normal distribution for a desired $100(1 - \alpha)\%$ confidence level

s = estimate of the standard deviation of the variable of interest; can use $\frac{range}{4}$ as an approximation

MOE = desired margin of error; half the desired width of the confidence interval

Note: The resulting sample size is appropriate for only a SRS.

Example: One of the most important outcomes in treating a migraine headache is the time it takes to start feeling better. Suppose you want to design a study to estimate the mean time to the onset of relief for all migraine sufferers in a population. You plan to collect a SRS of migraine sufferers and you would like to use the data to construct a 90% confidence interval having a width of 7 minutes. You anticipate that patients may start to feel better in as little as 20 minutes up to as long as 2 hours. How large should the sample size be?

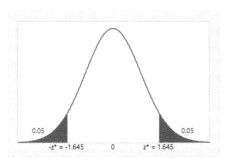

For 90% confidence, $\alpha = 1 - 0.90 = 0.10$ so z^* is the value having an area of 0.05 to its right (0.95 to its left) in the standard normal distribution. Using Table I we get $z^* = 1.645$.

To get an interval with a width of 7 minutes requires a margin of error that is half as long, so $MOE = 3.5$ minutes.

To estimate s we can use the $\frac{range}{4}$ approximation, so $s = \frac{120-20}{4} = 25$ minutes. Then substitute these values into the sample size formula:

$$n = \left(\frac{z^* s}{MOE}\right)^2 = \left(\frac{1.645 \cdot 25}{3.5}\right)^2 = 138.1.$$

As a matter of practice, you would round up and use a SRS of 139 migraine sufferers.

Estimating a Population Proportion

Now suppose you plan to collect categorical data for a SRS from a large population in hopes of estimating the population proportion of successes and you plan to use the one-proportion score interval to do it. Once again the action is in the margin of error:

$$MOE = z^* \frac{\sqrt{\dfrac{\hat{p}(1 - \hat{p})}{n} + \dfrac{z^{*2}}{4n^2}}}{1 + \dfrac{z^{*2}}{n}}.$$

Once again, there are a couple of problems:

Problem 1: Solving for n is a bit messy; think quadratic formula again! One solution is to simplify the margin of error to this:

$$MOE = z^* \sqrt{\frac{\hat{p}(1 - \hat{p})}{n}}.$$

Solving this simplified equation for n gives

$$n = \frac{z^{*2} \hat{p}(1 - \hat{p})}{MOE^2}.$$

Problem 2: You won't know the sample proportion, $\hat{p}$. Remember, at this stage, you haven't collected data yet! One nice solution is to use $\hat{p} = 0.5$ as this results in the largest possible value of $\hat{p}(1 - \hat{p})$ and hence the largest possible sample size. That way, you're covered regardless of the sample results.

Sample Size Required for Estimating a Population Proportion

$$n = \left(\frac{z^*}{2 \cdot MOE}\right)^2$$

z^* = value having an area of $\alpha/2$ to its right under the standard normal distribution for a desired $100(1 - \alpha)\%$ confidence level

MOE = desired margin of error; half the desired width of the confidence interval

Note: The resulting sample size is appropriate for only a SRS.

Example: Many employers today provide 401(k) plans or similar kinds of defined contribution retirement benefits for their employees. Under these plans, employees must contribute a certain portion of their pay in order to participate in the plan. The employer often contributes a sum of money to the plan for employees that participate. Suppose you wanted to estimate the proportion of all employees at your company that participate in the plan. Your company employs a large number of people and doesn't have easy access to 401(k) participation data, so sampling is the only way to do this. How many employees should you select in a SRS to be able to estimate the proportion of all employees who participate to within 3% with 88% confidence?

For 88% confidence, $\alpha = 1 - 0.88 = 0.12$ so z^* is the value having an area of 0.06 to its right (0.94 to its left) in the standard normal distribution. Using Table I we get $z^* = 1.555$.

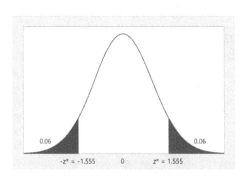

To get a sample proportion within 3% of the population proportion means we need a margin of error small enough to do that. Since we can be confident that p will be in the resulting confidence interval, taking $MOE = 0.03$ means we can be confident that $\hat{p}$ will be at most 0.03 away from p.

Then substitute these values into the sample size formula:

$$n = \left(\frac{z^*}{2 \cdot MOE}\right)^2 = \left(\frac{1.555}{2 \cdot 0.03}\right)^2 = 671.7$$

You would want to use a SRS of 672 employees.

Section 7.4 Exercises

1. All else equal, will higher or lower levels of confidence result in a narrower confidence interval?

2. All else equal, will larger or smaller sample sizes result in a narrower confidence interval?

3. All else equal, what kind of confidence intervals are preferred: narrow ones or wide ones?

4. You want to estimate a population mean μ with a 95% confidence interval having a margin of error of 25. You expect the smallest and largest data values to be 300 and 700, respectively. How many objects do you need in a SRS?

5. You want to estimate a population mean μ with a 90% confidence interval having a margin of error of 0.5. You expect the smallest and largest data values to be 4 and 18, respectively. How many objects do you need in a SRS?

6. You want to estimate a population mean μ with a 99% confidence interval having a width of 0.01. You expect the smallest and largest data values to be 11.7 and 12.3, respectively. How many objects do you need in a SRS?

7. You want to estimate a population mean μ with a 80% confidence interval having a width of 10. You expect the smallest and largest data values to be 55 and 145, respectively. How many objects do you need in a SRS?

8. Calculate the value of $\hat{p}(1 - \hat{p})$ for these values of $\hat{p}$: 0, 0.05, 0.1, 0.2, 0.3, 0.4, 0.5, 0.6, 0.7, 0.8, 0.9, 0.95, 1. Then plot the values of $\hat{p}(1 - \hat{p})$ vs. $\hat{p}$. Is it pretty clear that the maximum occurs at $\hat{p} = 0.5$?

9. You want to estimate a population proportion p with a 99% confidence interval having a margin of error of 0.03. How many objects do you need in a SRS?

10. You want to estimate a population proportion p with a 95% confidence interval having a margin of error of 0.06. How many objects do you need in a SRS?

11. You want to estimate a population proportion p with a 90% confidence interval having width of 0.1. How many objects do you need in a SRS?

12. You want to estimate a population proportion p with a 99% confidence interval having a width of 0.01. How many objects do you need in a SRS?

13. You want to estimate the proportion of all students at a large university who have all required course materials within the first two weeks of the semester. You plan to take collect data on a SRS of students and use the data to construct a 95% confidence interval. If you want a margin of error of no larger than 0.02, how many students should you select in your SRS?

14. A magazine wants to estimate the proportion of all its current subscribers that would be interested in an online version. How many subscribers should be selected as part of a SRS in order to estimate this parameter with 99% confidence and a margin of error of ±3%?

15. Jamie wants to open a juice bar in her town but needs to know whether the incomes of people in the area are high enough to afford such discretionary spending. She will collect a SRS of all residents in the town in order to estimate the mean annual income of all residents. She comes to you with her plan and adds that she'd like to end up with a 95% confidence interval having a margin of error of $5,000. In an earlier sample, she observed that the smallest and largest incomes were $45,000 and $90,000, respectively. How many residents would you advise her to select?

16. Soil sampling is common in farming because it provides farmers with information about what nutrients are in the soil and which nutrients are lacking. A farmer wants to select a SRS of soil specimens from a field in order to estimate the average nitrate level of all such specimens in the field. If a confidence interval based on the data is to have 85% confidence and a margin of error of 0.5 parts per million, how many soil specimens should be selected? In a sample of soil specimens earlier in the year, the farmer observed that the smallest and largest nitrate levels were 25.5 ppm and 33.1 ppm, respectively.

17. The coronavirus of 2019 and 2020 (COVID-19) did not produce any observable symptoms in many people who were infected by it. Many governments pushed for more extensive testing even though tens of thousands of people had already been tested.

    a. How many people would be needed in a SRS in order to estimate the proportion of all COVID-19 cases that are asymptomatic using a 95% confidence interval having a width of 3% points?
    b. How does your answer compare to the numbers of people who were tested in various countries?
    c. What does your comparison between parts a. and b. tell you about the value of random sampling?
    d. Do you think it would be practical/ethical to require testing of a SRS of people in the U.S.?

18. A mail order company wants to estimate the proportion of all orders that are shipped "on time" (within 2 days of when the order was received). A 90% confidence interval with a width of 0.05 is to be constructed based on the data from a SRS.

19. A candidate for public office is considering withdrawing from the race because he suspects that public support isn't there. In order to estimate the percentage of all likely voters that currently plan to vote for him using a 95% confidence interval that is 4 percentage points wide, how many likely voters should be selected in a SRS?

20. For healthy women, the average forced expiratory volume in one second (FEV1) is 3 liters. You suspect that average FEV1 is lower for mild asthmatics. To gather evidence for your hunch, you plan to observe the FEV1 for a SRS of mild asthmatic women and then construct a 90% confidence interval based on the data. Previous research indicates that a reasonable value for the standard deviation of FEV1 values is 0.9 liters. How large should your SRS be?

21. Consider exercise 13. You want to consider several margins of error: 0.005, 0.01, 0.02, 0.05, 0.10.

    a. Calculate the required sample sizes (still using 95% confidence) and present your results in a table.
    b. Calculate the required sample sizes for the margins of error shown above AND for 80%, 90%, 95%, and 99% confidence. Present your results in a 2-way table where the rows are the margins of error and the columns are the confidence levels.

22. Consider exercise 15. Instead of giving Jamie only one recommend sample size, you decide to show her several possibilities based on several margins of error: $1,000, $2,500, $5,000, $10,000, $15,000.

    a. Calculate the required sample sizes (still using 95% confidence) and present your results in a table.
    b. Calculate the required sample sizes for the margins of error shown above AND for 80%, 90%, 95%, and 99% confidence. Present your results in a 2-way table where the rows are the margins of error and the columns are the confidence levels.

Notes

[1] data.delaware.gov.

[2] Wilson, E. (1927). "Probable inference, the law of succession, and statistical inference." Journal of the American Statistical Association, 22(158), 209 – 212.

[3] Agresti, A. and Coull, B. A. (1998). "Approximate is better than 'exact' for interval estimation of binomial proportions," *The American Statistician*, 52 (2), 119 – 126.

[4] Pew Research Center, May, 2017, "Public Trust in Government Remains Near Historic Lows as Partisan Attitudes Shift."

[5] University of Delaware Faculty Climate Survey (2018).

[6] Based on National Survey of Student Engagement (2019): Experiences with Information Literacy, The Trustees of Indiana University.

[7] Pew Research Center, Oct. 26, 2016, "One-in-Five U.S. Adults Were Raised in Interfaith Homes."

[8] Pew Research Center (2017). "Summer 2017 Political Landscape Re-interview Survey," conducted Aug. 15 – 17, 2017.

[9] Schoen Consulting (2018). "Holocaust Knowledge and Awareness Study," commissioned by The Conference on Jewish Material Claims against Germany.

Chapter 8: Inference About Two Groups

I grew up on a dairy farm and spent a good bit of time around corn fields. Have you ever noticed fields of corn having signs posted along the road with different numbers on them? Those are test plots. The farmer is trying to determine whether one variety of corn is superior to another.

Have you ever wondered whether brand-name batteries are worth the higher price compared to generic batteries?

Do you snore? Have you ever been woken up by someone snoring? Have you ever woken someone else up by your snoring? Is there a difference between men and women with respect to the fraction of snorers?

Answers to these questions and many others can be found by careful data collection and the use of the methods in this chapter. I often tell my students at this point in the course that, "This is where it gets interesting!" It's not that what we've done so far isn't important. What we've learned about the fundamentals of collecting and summarizing data and how probability distributions are used in making inferences is foundational. A house cannot stand very long if the foundation is shaky underneath it. On the other hand, when you look at a house, you rarely look at the foundation. You see the structure built on top. That's the interesting part: the architecture, entrance design, window layout, siding and roofing materials and colors, and so on. In the same way, one of the more visible tool sets in data science is the set of tools used to make comparisons between two groups.

8.1 Introduction to Inference About Two Groups

When making an inference (a statement that goes beyond the data at hand) about two groups, there are two key concepts: 1) The type of inference (generalization inference or causal inference) and 2) the data structure (independent samples or paired samples). In this section, we'll explore these concepts.

You're already familiar with the difference between generalization inferences and causal inferences. Here we'll focus specifically on comparing only two groups.

Generalization Inference

The term **generalization inference** refers to making a statement about one or several large groups of objects (called **populations**) based on data collected from one or several smaller groups of objects (called **samples**). The variable used to compare the groups is called the **response**. Comparing the populations can be done using appropriate summaries of the population data sets (called **parameters**). As you saw in an earlier chapter, **random sampling** is the best way to obtain the samples because it guarantees representative samples, on average. We like random sampling but it introduces uncertainty. Is a difference you see in the samples reflecting a difference in the populations or is it just an artifact of the randomness? We have to be able to distinguish between the two.

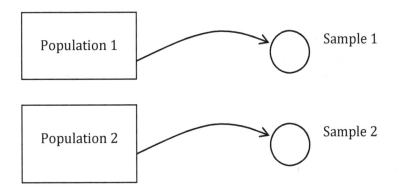

Example: Suppose you wanted to compare a certain brand-name battery to a generic battery sold in your favorite store. You'd like to know if the brand-name batteries sold in the store last longer, on average, than the generic batteries. You take two simple random samples of 10 batteries of each type and test each battery to determine how long it lasted (in hours).

Populations: all brand-name batteries and all generic batteries currently in stock at the store
Samples: the set of 10 brand-name batteries and the set of 10 generic batteries you randomly selected
Response: time-to-failure (hours)
Parameters: μ_1 = mean time-to-failure for all brand-name batteries currently in stock at the store
μ_2 = mean time-to-failure for all generic batteries currently in stock at the store

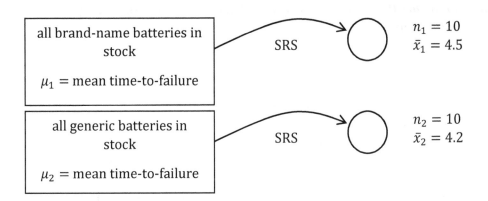

What would you conclude if, in the samples, the average time-to-failure for the brand-name batteries was 4.5 hours vs. 4.2 hours for the generic batteries? Clearly the brand-name batteries are better, right? Not so fast. Even batteries of the same brand vary in their times-to-failure. Is it possible that you just happened to randomly select more of the stronger brand-name batteries and more of the weaker generic batteries? Of course! A better question is, *even if the average times-to-failure for all brand-name and generic batteries in stock at the store are the same, how likely is it to*

get two random samples of 10 batteries where the brand-name sample mean is at least 0.3 hours higher?

Example: Suppose you wanted to compare males and females with respect to the fraction of snorers. You'd like to know if there is a difference between the fraction of all male adults in your country who snore and the fraction of all female adults who snore. You take a simple random sample of 1000 adults and observe the gender and snoring status of each. Let's say you get 550 females and 450 males.

> *Populations*: all male adults and all female adults in your country
> *Samples*: the set of 550 female adults and the set of 450 male adults you randomly selected
> *Response*: snoring status (yes or no)
> *Parameters*: p_1 = proportion of all males in your country who snore
> p_2 = proportion of all females in your country who snore

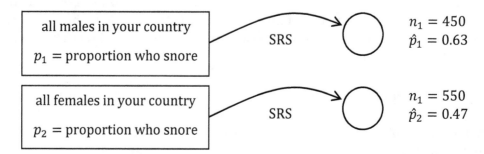

What would you conclude if, in the samples, 63% of the males snored and only 47% of the females snored? Obviously males snore more than females, right? Whoa, back up the train. Is it possible that you just happened to randomly select more of male snorers and fewer of the female snorers in your country? Of course! A better question is, *even if the fractions of snorers for all male and female adults in your country are the same, how likely is it to get two random samples of 550 and 450 where the sample percentages are at least 16 percentage points apart?*

Causal Inference

The term **causal inference** refers to drawing a conclusion about the *efficacy* (the capacity for producing a desired result or effect) of some intervention. It answers the question, "Does it work?" about the intervention. Recall from chapter 3 that good causal inferences are made in the context of an **experiment** designed to isolate the effect of the intervention. The experimenter manipulates the settings (called **levels**) of one or several variables (called **factors**) in the hopes of seeing a change in some outcome (called the **response**). As you saw in chapter 3, **random assignment** of objects to treatments is the best way to assign the treatments because it balances the treatment groups with respect to any potential confounding variable. We like random assignment but it introduces uncertainty. Does the difference you see between the treatment groups reflect the effect of the intervention or is it just an artifact of the randomness? We have to be able to distinguish between the two.

Example: You are an analyst for a company who wants to know whether putting the "Buy Now" button at the top or bottom of the web page for a product works better. Does the location of the button affect sales? You design an experiment where you randomly assign (using a SRA) all visits to the product page to one of two versions of the page: one where the "Buy Now" button is at the top and one where the "Buy Now" button is at the bottom. For each visit, you determine whether the customer purchased the product or not.

Factor: button location
Levels: top, bottom
Treatments: top, bottom
Response: purchase status (yes or no)

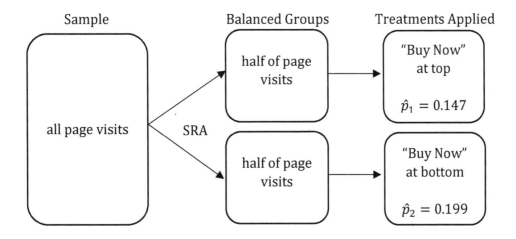

What would you conclude if 14.7% of customers who saw the "top" version purchased the product vs. 19.9% in the "bottom" group? The company should put the button at the bottom of the page, right? Hold your horses. Is it possible that you just happened to randomly assign more of the customers who planned to purchase anyway to the "bottom" group and fewer of the customers who were not going to purchase to the "top" group? Of course! A better question is, *even if the location of the button does not affect purchase status, how likely is it to randomly assign customers to the two locations and get purchase percentages at least 5.2 percentage points apart?*

Example: A farmer comes to you to help her decide which variety of corn (A or B) she should use to maximize yield. You design an experiment where you randomly assign each of her 20 fields to either variety A or variety B, 10 fields each. At harvest time, you determine the corn yield (in bushels) for each field.

Factor: variety
Levels: A, B
Treatments: A, B
Response: yield (bushels)

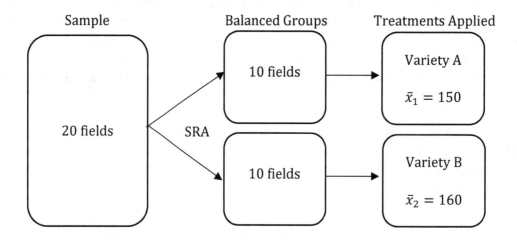

What would you conclude if the average yield for the 10 variety A fields was 150 bushels compared to 160 bushels for the variety B fields? The farmer should plant variety B, right? Wait a minute. Is it possible that you just happened to randomly assign the higher-yielding fields to variety B and the lower-yielding fields to variety A? Of course! A better question is, *even if the variety of the corn does not affect yield, how likely is it to randomly assign fields to the two varieties and get average yields at least 10 bushels apart?*

Notice how in each of these four examples there is a need to quantify the effect of the random mechanism used (random sampling or random assignment). We like random mechanisms. The next two sections of this chapter are devoted to answering questions about the effect of the random mechanism.

Generalization inference and causal inference are two very different kinds of inferences. They are designed to answer two very different kinds of questions. You must be very careful to limit the scope of your conclusions based on the kind of inference you do. In particular, generalization inference (and random sampling) allows statements about differences between populations but not about the *causes* of those differences. Causal inference (and random assignment) allows statements about the cause of an observed difference but *generalizing* those effects to larger groups is often limited to objects like those in the sample.

Example: In the button location experiment, let's say you determine that the 5.2 percentage point observed difference is statistically significant, unlikely to have been due to random assignment alone. In addition to random assignment, customers were blinded as to the treatment group. In fact, customers would not have known that they were part of an experiment.

Justified by the data: The bottom location for the "Buy Now" button *caused* higher sales compared to the top location *for customers like those in the study*.

Not justified by the data: The bottom location for the "Buy Now" button *caused* higher sales compared to the top location *for all customers*.

Example: A reporter for the New York Times, in an article titled, "Cataract Surgery May Prolong Your Life,"[1] writes about a woman who had her cataracts surgically removed, "And I was thrilled to be able to tell her that the surgery very likely did more than improve her poor vision. According to the results of a huge new study it may also prolong her life." The article cited results from a study of over 74,000 women with cataracts aged 65 and older that found a 60% lower mortality rate among the women who had their cataracts removed. While this result is likely statistically significant, statements about the cause of the difference are not justified by the data.

Justified by the data: Among all women with cataracts, those who elect to have their cataracts removed experience a lower mortality rate than those who do not (assuming the sample of women was representative of all women).

Not justified by the data: The surgery *caused* the lower mortality rate.

About a year after the results of this study were published, the article was retracted because the authors found an error in the analysis. After correcting it, the difference in mortality rates was in the opposite direction; women who had their cataracts removed had a *higher* mortality rate than those who did not! While I have no doubt that it was a completely honest mistake, this is a perfect example of the fallibility of people who use the scientific method. Those who place their faith in science and the scientific method are building on a foundation that is imperfect not only because there is a chance for error when basing decisions on incomplete information but because the method is implemented by imperfect human beings.

Independent Samples vs. Paired Samples

There is another complexity in a comparison of two groups. It is the difference between independent samples and paired samples. We say two samples are **paired** if there is a natural, one-to-one way to match the response data values across the two samples. Otherwise, the samples are **independent**. I think the best way to understand the difference is to look at examples.

Example: In order to determine which variety of corn (A or B) is better for yield, a farmer divides each of her 10 fields into 2 plots. She tosses a coin for each field to assign a plot to either variety A or variety B. At harvest time she observes the yield (in bushels) for each plot. A sketch of the data is shown here.

Notice that the yield figures are naturally matched by field. They are also matched one-to-one so that it would make sense to match the first yield value for A with only the first yield value for B (and not with any other yield in that column). These samples of yields are *paired* by field.

Field	Yield (variety A)	Yield (variety B)
1	- ⟵————⟶ -
2	- ⟵————⟶ -
3	- ⟵————⟶ -
4	- ⟵————⟶ -
5	- ⟵————⟶ -
6	- ⟵————⟶ -
7	- ⟵————⟶ -
8	- ⟵————⟶ -
9	- ⟵————⟶ -
10	- ⟵————⟶ -

The yield numbers themselves aren't important in making the determination of independent or paired. I didn't even show them here so you don't focus on the values themselves. The important thing is how they were collected.

Example: Consider the same question about determining which variety of corn (A or B) is better for yield. Instead of dividing each field into 2 plots, another farmer randomly assigned each of his 20 fields to either variety A or variety B, ten to each. At harvest time he observes the yield (in bushels) for each plot. A sketch of the data is shown here.

Now there is no natural way to match the yield figures across the two samples. These samples of yields are *independent*.

Yield (variety A)	Yield (variety B)
-	-
-	-
-	-
-	-
-	-
-	-
-	-
-	-
-	-
-	-

The difference between independent and paired samples is important, not only for analyzing the data but also for designing studies. Both structures have their advantages and disadvantages and we'll explore those later in the chapter. For now, ask yourself how you would design the study to compare variety A to variety B. Would you follow the paired samples design or the independent samples design? Why? Keep reading for the answer.

Section 8.1 Exercises

Exercises 1 – 10: Determine whether each problem requires a generalization inference or a causal inference.

If the problem requires a generalization inference: Identify
a) the objects/subjects,
b) the response variable and type,
c) the populations,
d) the samples, and
e) the parameters.
Also f) say whether a generalization inference is permitted and why/why not.

If the problem requires a causal inference: Identify
a) the objects/subjects,
b) the response variable and type,
c) the factor(s), and
d) the treatments.
Also e) say whether a causal inference is permitted and why/why not.

1. Shown here are the distributions of the percent of faculty that are full-time for schools whose top degree is the Bachelor's and those whose top degree is a graduate degree. The data come from the set of top 20 schools ranked by tuition that you'll find in chapter 1.

	$\bar{x}$	s
Bachelor's	92.2	5.34
Graduate	73.4	21.01

We'd like to use these data to compare all Bachelor's schools and all schools granting graduate degrees with respect to the percent of all faculty that are full-time.

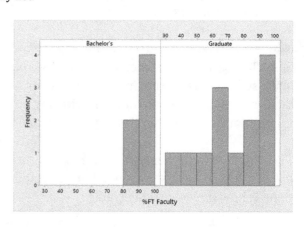

2. Refer to the asthma study example from section 2.3. Here are summarized data for two of the treatment groups at the end of the treatment period using data on peak expiratory flow in the morning (AMPEF) in L/min. AMPEF measures the speed at which a person can force air out of their lungs. The higher the number, the healthier their lung function.

	Salmeterol	Triamcinolone
n	54	54
$\bar{x}$	467.7	474.6
s	106.6	115.4

We'd like to use these data to see if triamcinolone would improve AMPEF compared to salmeterol, for asthma patients like these.

3. In a survey of U.S. parents of children under 18 years old, the Pew Research Center asked a sample of parents, "Do you think you spend too much time with your child/children, too little time, or about the right amount of time?" Suppose the responses summarized below are from the 1,807 people who responded out of a random sample of 12,047 U.S. parents of children under 18.[2]

	too much	right amount	too little
Mothers	70	579	221
Fathers	19	472	446

We'd like to use these data to see if there's a difference in the proportions of all U.S. mothers and all U.S. fathers who think they spend too little time with their child(ren).

4. Consider exercise 3 and suppose the summarized responses are from 1,807 parents in a SRS. Would that change any of your answers? If so, which ones?

5. Consider the accounting homework study from exercise 13 in section 2.3. The improvement in test scores (posttest – pretest) was recorded for each student. Some summaries of the score differences are shown here.

	no assistance	complete solution
n	30	20
$\bar{x}$	2.43	1.95
m	3	2
min	-3	-5
max	7	8

Is there a difference in how no assistance vs. complete solutions affect test scores for students like these?

6. Consider exercise 5 and suppose that the data were actually from a study where an instructor used a different homework assistance type in each of two different sections of an accounting course one semester. Would that change any of your answers? If so, which ones?

7. Consider the data on median student loan debt for the samples of institutions from exercise 14 in section 2.3. The data are grouped by the predominant degree granted.

Associate's:	9.5	5.8	6.3	9.6	4.0	9.9	6.8	9.0	4.0	15.7
	3.5									

Bachelor's:	15.6	17.0	12.6	12.0	17.3	20.1	15.0	16.0	19.8

Is there a difference between Associate's and Bachelor's institutions in the U.S. with respect to the average median student loan debt?

8. Consider exercise 7 and suppose that the data were from a SRS of institutions in a single state. Would that change any of your answers? If so, which ones?

9. A rural ice cream shop wants to see if changing the name of one of their flavors from "Chocolate Mud" to "Chocolate Dream" would change sales of that flavor. They keep track of sales (in dollars) of Chocolate Mud for each day during May. The name is then changed to Chocolate Dream and sales are observed each day during June. The data show that sales were higher on average for June.

10. Consider exercise 9. How should the ice cream shop collect data in order to justify a causal inference? Be specific in your explanation.

Exercises 11 – 20: For each problem, determine
a) the two groups to be compared,
b) the response variable and type, and
c) whether the samples would be independent or paired.

11. A real estate firm wants to compare two appraisers (Helen and Samara) that it tends to use for its real estate transactions. To do so, it randomly selects 10 properties currently for sale and has both Helen and Samara appraise each house.

12. A real estate firm wants to compare two appraisers (Juan and Mark) that it tends to use for its real estate transactions. To do so, it randomly selects 20 properties currently for sale, has Juan appraise 10 of them and Mark appraise the other 10. Who gets which house is determined at random.

13. An energy company claims that cars will get better gas mileage on its brand of regular, 87 octane, unleaded gasoline. A test is conducted where the 40 cars in one rental car company's fleet are used. Twenty are randomly chosen and driven through a test course using a tank of the company's gas and the other 20 use a tank of a competitor's gas.

14. An energy company claims that vehicles will get better gas mileage on its brand of regular, 87 octane, unleaded gasoline. A test is conducted where 20 vehicles in one rental car company's fleet are used. Each car is driven through a test course using both a tank of the company's gas and a tank of a competitor's gas. Which brand of gas is used first is determined at random for each vehicle.

15. A company that advertises on social media wants to compare the effectiveness of its ad. It's considering two different color palettes (blue/green and orange/yellow) for the ad graphic. On one social media platform, it displays the ad to its target market but randomly determines which color palette to use for each user. Whether or not the user clicks on the ad is recorded. Data were collected for 2.8 million blue/green presentations and 2.75 million orange/yellow presentations.

16. In a random sample of 280 married couples, each spouse is asked about their political leanings (liberal, middle, conservative). The data are to be used to see if either husbands or wives tend to be more conservative.

17. A group of 100 volunteer Alzheimer patients is randomly assigned to one of two experimental treatments. One is a new drug and the other is a restricted diet. Each patient is given a memory test at the end of the study period that measures cognitive ability. Higher numbers indicate better cognitive function.

18. Are 12-oz or 16-oz beverage cans stronger? Random samples of 30 of each kind were subjected to strength tests and the downward force applied to the can top before the can began to deform was recorded.

19. In a random sample of 1500 likely voters, each voter is asked whether or not they favor candidate X, both before and after a major debate. The data are to be used to determine whether there was a change in the proportion of all likely voters who favored candidate X after the debate.

20. A group of youth offender cases in Maryland was randomly assigned to be processed through either the traditional justice system or through a teen court, an alternative system thought to curb re-offence. Out of 56 teen court cases, 18 were found to re-offend later. Out of 51 traditional justice system cases, 12 were found to re-offend later.[3]

8.2 Generalization Inference About Two Populations

In this section, we'll explore how to use sample data to draw generalizations about two large populations. Always keep in mind that the key to doing this well is random sampling. You would have a tough time justifying generalization inferences based on data from non-random samples!

This section is divided into three parts based on the kinds of data used to compare the two populations:

- Numeric Data (Independent Samples)
- Numeric Data (Paired Samples)
- Categorical Data (Independent Samples)

Numeric Data (Independent Samples)

Let's consider the battery example from the last section:

Example: Suppose you wanted to compare a certain brand-name battery to a generic battery sold in your favorite store. You'd like to know if the brand-name batteries sold in the store last longer, on average, than the generic batteries. You take two simple random samples of 10 batteries of each type and test each battery to determine how long it lasted (in hours).

Populations: all brand-name batteries and all generic batteries currently in stock at the store
Samples: the set of 10 brand-name batteries and the set of 10 generic batteries you randomly selected
Response: time-to-failure (hours)
Parameters: μ_1 = mean time-to-failure for all brand-name batteries currently in stock at the store
μ_2 = mean time-to-failure for all generic batteries currently in stock at the store

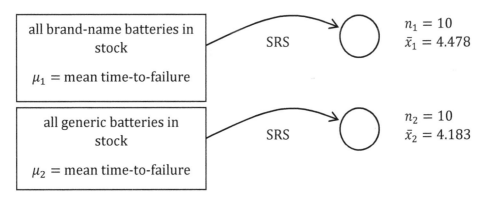

Shown below are the times-to-failure data (in hours) and some summaries.

Time-to-Failure (hours) Brand-Name	Time-to-Failure (hours) Generic
4.55	4.31
4.74	4.93
4.78	4.14
4.53	4.51
4.65	3.78
4.24	3.95
4.25	4.95
4.13	4.15
4.57	3.52
4.34	3.59
$n_1 = 10$	$n_2 = 10$
$\bar{x}_1 = 4.478$	$\bar{x}_2 = 4.183$
$s_1 = 0.225$	$s_2 = 0.503$

In this sample, the brand-name batteries lasted 0.295 hours longer (4.478 – 4.183), on average, than the generic batteries. What would you conclude? Is it possible that you just happened to randomly select more of the stronger brand-name batteries and more of the weaker generic batteries? Of course! A better question is, *even if the average times-to-failure for all brand-name and generic batteries in stock at the store are the same, how likely is it to get two random samples of 10 batteries where the brand-name sample mean is at least 0.295 hours higher?*

You can answer this question by conducting a hypothesis test about the difference $\mu_1 - \mu_2$ by considering the sampling distribution of $\bar{X}_1 - \bar{X}_2$, the difference between the sample means, under the assumption that the null hypothesis is true.

Shown below are the features of the sampling distribution of $\bar{X}_1 - \bar{X}_2$. These features are based on two <u>independent</u> SRS's from populations having means μ_1 and μ_2 and standard deviations σ_1 and σ_2.

	$\bar{X}_1 - \bar{X}_2$
Center	$E(\bar{X}_1 - \bar{X}_2) = \mu_1 - \mu_2$
Variance	$Var(\bar{X}_1 - \bar{X}_2) = \dfrac{\sigma_1^2}{n_1} + \dfrac{\sigma_2^2}{n_2}$
Standard Deviation	$SD(\bar{X}_1 - \bar{X}_2) = \sqrt{\dfrac{\sigma_1^2}{n_1} + \dfrac{\sigma_2^2}{n_2}}$

Shape	Normal if:
	    1) data from both populations are normal
	    2) n_1 and n_2 are both sufficiently large

The number we use for $\mu_1 - \mu_2$ is 0, the one usually specified by the null (no difference) hypothesis. If the independent SRS's condition and one of the shape conditions are satisfied, then the statistic

$$Z = \frac{X_1 - \bar{X}_2}{\sqrt{\frac{\sigma_1^2}{n_1} + \frac{\sigma_2^2}{n_2}}}$$

has a standard normal distribution under the null hypothesis. Just as we did in an earlier chapter, we'll substitute the sample variances for the unknown population variances and use

$$t = \frac{X_1 - \bar{X}_2}{\sqrt{\frac{s_1^2}{n_1} + \frac{s_2^2}{n_2}}}.$$

You'll notice I used "t" for this statistic so you might suspect this statistic has a t-distribution. For a technical reason (that we'll not discuss here), this statistic does not have a t-distribution. However, we can approximate using a t-distribution as long as we use the right degrees of freedom. The following t-test is often attributed to Bernard Welch[4] (for proposing the test) and Franklin Satterthwaite[5] (for discovering the optimal degrees of freedom to use).

Welch's t-method

$H_0: \mu_1 = \mu_2$
$H_a: \mu_1 < \mu_2$ or $\mu_1 > \mu_2$ or $\mu_1 \neq \mu_2$

Test Statistic: $t = \dfrac{\bar{x}_1 - \bar{x}_2}{\sqrt{\frac{s_1^2}{n_1} + \frac{s_2^2}{n_2}}}$ $\qquad df = \dfrac{\left(\frac{s_1^2}{n_1} + \frac{s_2^2}{n_2}\right)^2}{\frac{\left(\frac{s_1^2}{n_1}\right)^2}{n_1 - 1} + \frac{\left(\frac{s_2^2}{n_2}\right)^2}{n_2 - 1}}$ (round down)

$100(1 - \alpha)\%$ CI for $\mu_1 - \mu_2$: $\quad \bar{x}_1 - \bar{x}_2 \pm t^* \sqrt{\frac{s_1^2}{n_1} + \frac{s_2^2}{n_2}}$

where t^* is the value having an area of $\alpha/2$ to its right under the t-distribution with df shown above.

Conditions
- independent SRS's
- two nearly-normal data distributions (can be relaxed for large sample sizes or roughly symmetric distributions with no outliers)

Let's apply Welch's method to the batteries data.

Hypotheses

$H_0: \mu_1 = \mu_2$
$H_a: \mu_1 > \mu_2$

μ_1 = mean time-to-failure for all brand-name batteries currently in stock at the store
μ_2 = mean time-to-failure for all generic batteries currently in stock at the store

Method/Conditions: We'll use Welch's t-method.

Conditions:
- independent SRS's: Both samples were SRS's as given in the problem. There is no natural, one-to-one way to match the times-to-failure across the two samples, so they are independent samples.
- two nearly-normal data distributions: Normal probability plots of the data in both samples are both fairly linear, so the two samples of times are nearly-normal.

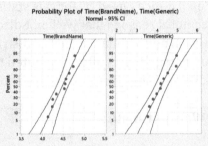

Mechanics (test statistic / p-value approach)

$$t = \frac{\bar{x}_1 - \bar{x}_2}{\sqrt{\frac{s_1^2}{n_1} + \frac{s_2^2}{n_2}}} = \frac{4.478 - 4.183}{\sqrt{\frac{0.225^2}{10} + \frac{0.503^2}{10}}} = 1.692 \qquad df = \frac{\left(\frac{s_1^2}{n_1} + \frac{s_2^2}{n_2}\right)^2}{\frac{\left(\frac{s_1^2}{n_1}\right)^2}{n_1-1} + \frac{\left(\frac{s_2^2}{n_2}\right)^2}{n_2-1}} = \frac{\left(\frac{0.225^2}{10} + \frac{0.503^2}{10}\right)^2}{\frac{\left(\frac{0.225^2}{10}\right)^2}{10-1} + \frac{\left(\frac{0.503^2}{10}\right)^2}{10-1}} = 12.46$$

We'll round down and use $df = 12$.

The data provide evidence for H_a in the form of *high* values of t so the p-value is the probability of observing a value of t as high or higher than 1.692 if H_0 were true.

Using the t-table and 12 degrees of freedom, we can only approximate the p-value because our t-value is not shown exactly; it's between 1.36 and 1.78:

$$p - value = P(t \geq 1.692)$$
$$= 1 - P(t \leq 1.692)$$
$$= 1 - something\ between\ 0.9\ and\ 0.95$$

So the p-value is between 0.05 and 0.10. Using some kind of technology (e.g. Minitab or Excel) you can get the $p - value$ more precisely:

$$p - value = 0.0582$$

If there is no difference between the mean time-to-failure for all brand-name batteries currently in stock at the store and that for all generic batteries, there is a 5.8% chance of finding a difference at least as high as 0.295 hours in the means of two random samples of 10 batteries from this store.

Mechanics (confidence interval approach)

You can also calculate a 95% confidence interval for $\mu_1 - \mu_2$ as a way to assess the evidence for H_a:

For 95% confidence, $\alpha = 1 - 0.95 = 0.05$ so t^* is the value having an area of 0.025 to its right (0.975 to its left) in the t-distribution with $df = 12$. Using Table II we get $t^* = 2.18$.

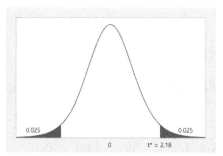

Now just substitute into the formula to get the endpoints:

$$\bar{x}_1 - \bar{x}_2 \pm t^* \sqrt{\frac{s_1^2}{n_1} + \frac{s_2^2}{n_2}}$$

$$4.478 - 4.183 \pm 2.18 \sqrt{\frac{0.225^2}{10} + \frac{0.503^2}{10}}$$

$$0.295 \pm 0.380$$

$$(-0.085, 0.675)$$

We are 95% confident that the difference between the mean time-to-failure for all brand-name batteries currently in stock at the store and that for all generic batteries is between – 0.085 hours and 0.675 hours.

Conclusion

The p-value is marginally insignificant (using the typical 5% significance level), so we cannot reject H_0 at that level. The 95% confidence interval for $\mu_1 - \mu_2$ includes 0. Since H_0: $\mu_1 = \mu_2$ ($\mu_1 - \mu_2 = 0$) is plausible based on the data, we cannot reject H_0 using the confidence interval approach, which tells the same story as the test statistic / p-value approach. We have somewhat marginally insignificant evidence at the 5% significance level that the mean time-to-failure for all brand-name batteries currently in stock at the store is higher than that for all generic batteries.

Because the evidence is so marginal, don't dismiss it. Definitely do not conclude there is no difference! It may be that the difference is quite small and we would need more data to discover it.

Simulation

I've proposed that Welch's t-method is a good method under the stated conditions. Let's do a simulation study to see how. In chapter 5 you saw how I used simulations to motivate what happens when taking random samples from large populations. What would a simulation look like here? Well, suppose the two populations of batteries have the same mean time-to-failure. To simulate our SRS process, do the following:

- Take two SRS's of 10 batteries each and observe their times-to-failure.
- Calculate the sample means ($\bar{x}_1$ and $\bar{x}_2$) and sample standard deviations (s_1 and s_2) for each sample.
- Calculate the t-test statistic using Welch's method.
- Repeat this process 500 times.

The results are shown here.

Rep #	$\bar{x}_1$	$\bar{x}_2$	s_1	s_2	t
1	4.36	4.45	0.13	0.42	-0.638
2	4.27	4.14	0.16	.075	0.547
3	4.39	4.60	0.15	0.43	-1.464
⋮	⋮	⋮	⋮	⋮	⋮
500	4.27	4.28	0.20	0.56	-0.045

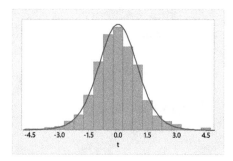

The graph shows a histogram of the distribution of the 500 values of t along with the t-distribution having 12 degrees of freedom. Notice that this t-distribution is a fairly good model. It adequately describes the distribution of t-values we can expect for independent SRS's of 10 from two populations having the same mean. You can think of the t-distribution as a model for simple random sampling under the null hypothesis. You would reject the null hypothesis if you observe a value of t that would look unusual when compared to this distribution.

Numeric Data (Paired Samples)

Comparison of means using paired samples can be more efficient than using independent samples. If you're the analyst, knowing this can help you choose a more efficient analysis technique than Welch's t-method. If you're the one designing the study and responsible for collecting data, knowing this can help you design a more efficient study. Let's see how all that works.

Example: A real estate firm wants to compare two appraisers (Jill and Anya) that it tends to use for its real estate transactions. To do so, it randomly selects 5 properties currently for sale and has both Jill and Anya appraise each house. The appraised values are shown below.

Property	Value (Jill)	Value (Anya)
1	$175,600	$181,650
2	$94,800	$96,800
3	$227,300	$236,500
4	$419,650	$423,850
5	$275,200	$279,900

Note that the two samples of values are naturally paired one-to-one by property because Jill and Anya are appraising the *same properties*. Also notice how much variability there is in the appraised values across properties: the values range from under $100,000 to over $400,000! Of course you know that there is considerable variation across different properties; that's not news and it's not the focus of the analysis. The goal is to compare Jill and Anya. Notice that Anya appraised every property in the sample a bit higher than did Jill. Does that mean Anya's average appraised value would be higher than Jill's for all properties currently for sale? Not necessarily. What if we just happened to randomly select five properties that Anya appraised higher? Perhaps there are properties in the population that Jill would have appraised higher and we just didn't select them. Again, we need to account for the randomness in the data collection process.

How is that done? Actually, fairly simply! Since the goal is to compare Jill and Anya, and they appraised the same properties, let's consider the **paired differences** in values:

Property	Value (Jill)	Value (Anya)	Diff (Jill – Anya)
1	$175,600	$181,650	-$6,050
2	$94,800	$96,800	-$2,000
3	$227,300	$236,500	-$9,200
4	$419,650	$423,850	-$4,200
5	$275,200	$279,900	-$4,700

The differences capture exactly what we're after: the difference between Jill's and Anya's valuations. Since Anya's appraisals were all higher than Jill's in this sample, the differences (Jill – Anya) are all negative. Now just *treat the differences as the data and use a one-sample t-method*! Calculating the differences reduces the analysis from a two-sample problem to a one-sample problem. The resulting one-sample t-method applied to the paired differences is often referred to as the **paired t-method**. Here are the details:

Hypotheses

$H_0: \mu = 0$
$H_a: \mu \neq 0$

μ = *mean difference* in appraised values (Jill – Anya) for all properties currently for sale (if we'd have them both appraise them all!)

Method/Conditions: We'll use the paired t-method.

- SRS: The data are from a simple random sample of properties.
- Nearly-normal distribution of data? A normal probability plot of the differences shows a fairly linear pattern, so yes, this condition is satisfied.

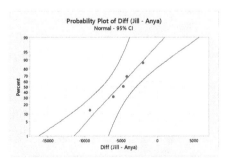

Mechanics (test statistic / p-value approach)

$$\bar{x} = \frac{-6050 - 2000 - 9200 - 4200 - 4700}{5} = -5,230$$

$$s^2 = \frac{\left(-6050 - (-5230)\right)^2 + \left(-2000 - (-5230)\right)^2 + \left(-9200 - (-5230)\right)^2 + \left(-4200 - (-5230)\right)^2 + \left(-4700 - (-5230)\right)^2}{5-1}$$

$$= 7,052,000$$

$$s = \sqrt{s^2} \approx 2,655.56$$

$$t = \frac{\bar{x} - \mu_0}{s/\sqrt{n}} = \frac{-5230 - 0}{2655.56/\sqrt{5}} = -4.404 \qquad\qquad df = n - 1 = 5 - 1 = 4$$

The data provide evidence for H_a in the form of *low or high* values of $\bar{x}$ and t. Since the data suggest lower values of μ, the p-value is the probability of observing a value of t as low or lower than -4.404 or as high or higher than +4.404 if H_0 were true.

Using the t-table and 4 degrees of freedom, we can only approximate the p-value because our t-value is not shown exactly; it's between -4.60 and -3.75:

$$p - value = 2P(t < -4.404)$$
$$= 2(something\ between\ 0.005\ and\ 0.01)$$

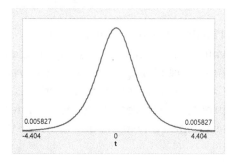

So the p-value is between 0.01 and 0.02. Using some kind of technology (e.g. Minitab or Excel) you can get the $p - value$ more precisely:

$$p - value = 2(0.005827) = 0.0117$$

If there is no mean difference in Jill's and Anya's appraised values for all properties currently for sale, there is a 1.17% chance of finding a mean difference at least as much as $5,230 in a random sample of 5 properties.

Mechanics (confidence interval approach)

You can also calculate a 95% confidence interval for μ as a way to assess the evidence for H_a:

For 95% confidence, $\alpha = 1 - 0.95 = 0.05$ so t^* is the value having an area of 0.025 to its right (0.975 to its left) in the t-distribution with $df = 4$. Using Table II we get $t^* = 2.78$.

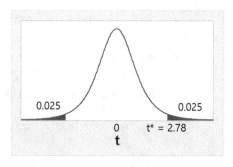

Now just substitute into the one-sample t-interval formula to get the endpoints:

$$\bar{x} \pm t^* \frac{s}{\sqrt{n}}$$

$$-5230 \pm 2.78\frac{2655.56}{\sqrt{5}}$$

$$-5230 \pm 3301.535$$

$$(-8531.54, -1928.46)$$

We are 95% confident that the mean difference in appraised values (Jill – Anya) for all properties currently for sale is between -$8,531.54 and -$1,928.46. That is, we are 95% confident that Anya would appraise properties higher than Jill by between $1,928.46 and $8,531.54, on average.

Conclusion

The p-value is quite small (using the typical 5% significance level), so we reject H_0 at that level. The 95% confidence interval for μ does not include 0. Since H_0: $\mu = 0$ is not plausible based on the data, we reject H_0 using the confidence interval approach, which tells the same story as the test statistic / p-value approach. We have enough evidence at the 5% significance level of a mean difference in appraised values (Jill – Anya) for all properties currently for sale. In fact, the data convincingly suggest that Anya's valuations are, on average, above Jill's.

Potential Efficiency of Paired Samples

I want to revisit the note I made earlier about the considerable variation in values across properties. Here are the data again along with summary statistics:

Property	Value (Jill)	Value (Anya)	Diff (Jill – Anya)
1	$175,600	$181,650	-$6,050
2	$94,800	$96,800	-$2,000
3	$227,300	$236,500	-$9,200
4	$419,650	$423,850	-$4,200
5	$275,200	$279,900	-$4,700
mean	$238,510	$243,740	-$5,230
sd	$121,340	$121,703	$2,656

Notice how the variation across properties (in the hundreds of thousands) is much larger than the relatively subtle difference between Jill and Anya (only in the thousands). What if we had not accounted for this by calculating the differences first and then using the paired t-method? Here's a summary of the analysis which treats the two samples of appraised values as independent, ignoring the pairing by property:

$$H_0\text{: } \mu_1 = \mu_2$$
$$H_a\text{: } \mu_1 \neq \mu_2$$

μ_1 = mean appraised value (by Jill) for all properties currently for sale
μ_2 = mean appraised value (by Anya) for all properties currently for sale

Welch's t-method: $t = -0.068$ $p-value = 0.9476$
95% confidence interval for $\mu_1 - \mu_2$: (-$186,968 , $176,508)

Look at that! Welch's method uncovers hardly any evidence of a difference! Why not? Let's compare how the test statistic is calculated in both Welch's t-method and the paired t-method using the differences:

Welch's t-method: $t = \dfrac{\bar{x}_1 - \bar{x}_2}{\sqrt{\dfrac{s_1^2}{n_1} + \dfrac{s_2^2}{n_2}}} = \dfrac{238510 - 243740}{\sqrt{\dfrac{121340^2}{5} + \dfrac{121703^2}{5}}} = \dfrac{-5230}{76856.96} = -0.068$

paired t-method: $t = \dfrac{\bar{x} - \mu_0}{s/\sqrt{n}} = \dfrac{-5230 - 0}{2655.56/\sqrt{5}} = \dfrac{-5230}{1187.60} = -4.404$

The numerators in both statistics are the same. Calculating the means of both samples first and then taking the difference ($\bar{x}_1 - \bar{x}_2$) gives the same number as calculating the paired differences first and then taking their mean ($\bar{x}$). The numerator captures the subtle difference in how the two appraisers valued the sample of properties.

But check out the huge difference in the denominators. Why is the denominator of Welch's t so high compared to the paired t? The large variability across properties gets captured in the denominator of Welch's t-method because that method treats the data set as 10 different properties (two independent SRSs of properties). Since the paired t-method uses the paired differences as the data, the variability across properties is essentially filtered out. For example, look at properties 4 and 5. The difference in appraised values is between $4,000 and $5,000 for both; looking at only the differences, you would hardly guess that property 4 is valued much higher than property 5. The paired differences effectively eliminate much of the variability across properties from the analysis. It's still there in the data, but the paired t-method disregards it for the purpose of comparing Jill and Anya. How about that!

At the end of section 8.1 I alluded to the importance of paired vs. independent samples. The previous example illustrates how a paired design can yield data which can be analyzed much more efficiently. If you can anticipate a variable that will have a big effect on response variability, it will generally be to your advantage to pair responses by that variable.

On the other hand, a disadvantage of using a paired design is that of data reduction: the paired t-method effectively cuts the sample size in half and so you generally have fewer degrees of freedom. Here's how that works for the appraised property values:

Welch's t-method: $df = \dfrac{\left(\dfrac{s_1^2}{n_1} + \dfrac{s_2^2}{n_2}\right)^2}{\dfrac{\left(\dfrac{s_1^2}{n_1}\right)^2}{n_1 - 1} + \dfrac{\left(\dfrac{s_2^2}{n_2}\right)^2}{n_2 - 1}} = \dfrac{\left(\dfrac{121340^2}{5} + \dfrac{121703^2}{5}\right)^2}{\dfrac{\left(\dfrac{121340^2}{5}\right)^2}{5-1} + \dfrac{\left(\dfrac{121703^2}{5}\right)^2}{5-1}} = 7.99993$ so use $df = 7$

paired t-method: $df = n - 1 = 5 - 1 = 4$

There are almost twice as many degrees of freedom for conducting Welch's t-method as the paired t-method. All else equal, more degrees of freedom means smaller p-values (more evidence for Ha). But in this example, all else is not equal! The t-statistic value is so much less extreme when not accounting for variability across properties, so more degrees of freedom doesn't help much.

A related advantage of Welch's method is that it will be more sensitive to differences if the population of objects is very similar with respect to the response. Let's look at what would happen in this same problem if there was very little variability across properties. Suppose we had these data instead:

Property	Value (Jill)	Value (Anya)	Diff (Jill – Anya)
1	$375,600	$381,650	-$6,050
2	$376,900	$378,900	-$2,000
3	$373,100	$382,300	-$9,200
4	$375,500	$379,700	-$4,200
5	$375,750	$380,450	-$4,700
mean	$375,370	$380,600	-$5,230
sd	$1,388	$1,389	$2,656

You'd get the same paired differences and so the result of the paired t-method would be exactly the same. But now notice how similar the properties are. For these properties, the variability across properties is smaller than the variability of the differences. You'd actually get a more significant result (smaller p-value and narrower confidence interval) using Welch's t-method:

Welch's t-method:
$$t = \frac{\bar{x}_1 - \bar{x}_2}{\sqrt{\frac{s_1^2}{n_1} + \frac{s_2^2}{n_2}}} = \frac{375370 - 380600}{\sqrt{\frac{1388^2}{5} + \frac{1389^2}{5}}} = \frac{-5230}{878.15} = -5.956$$

$$df = \frac{\left(\frac{s_1^2}{n_1} + \frac{s_2^2}{n_2}\right)^2}{\frac{\left(\frac{s_1^2}{n_1}\right)^2}{n_1 - 1} + \frac{\left(\frac{s_2^2}{n_2}\right)^2}{n_2 - 1}} = \frac{\left(\frac{1388}{5} + \frac{1389^2}{5}\right)^2}{\frac{\left(\frac{1388^2}{5}\right)^2}{5 - 1} + \frac{\left(\frac{1389^2}{5}\right)^2}{5 - 1}} = 7.999998$$

p-value = 0.0006

95% CI for $\mu_1 - \mu_2$: $(-\$7,302.47\,, -\$3,157.53)$

paired t-method:
$$t = \frac{\bar{x} - \mu_0}{s/\sqrt{n}} = \frac{-5230 - 0}{2655.56/\sqrt{5}} = \frac{-5230}{1187.60} = -4.404$$

$$df = n - 1 = 5 - 1 = 4$$

p-value = 0.0117

95% CI for μ: $(-\$8,531.54, -\$1,928.46)$

The bottom line: If you can anticipate a variable that greatly affects variation in the response, it's a good idea to collect paired samples using that variable as the pairing mechanism, if possible. Then use the paired t-method to analyze the data and filter that variation out when making an inference about the two groups. Otherwise, independent sampling may be the better design option, so that you can use Welch's t-method to analyze the data and avoid the data reduction that happens in the paired t-method.

Notice how closely data collection issues (independent vs. paired samples) and data analysis issues (Welch's t-method vs. paired t-method) are related here. Good statistical practice gives careful consideration to both the collection and analysis of data.

Categorical Data (Independent Samples)

Comparing two populations using categorical data requires analyzing counts and proportions instead of means and standard deviations.

Example: In a study of snoring in Taiwan[6], a simple random sample of adults were contacted by telephone. Among the questions asked was, "Do you or your family members hear you snore when you sleep?" Each person's sex was also obtained. You'd like to know if there is a difference between the fraction of all adult Taiwanese males who snore and the fraction of all adult Taiwanese females who snore.

> *Populations*: all adult Taiwanese males and all adult Taiwanese females
> *Samples*: the sets of adult Taiwanese males and females that were randomly selected
> *Response*: snoring status (yes or no)
> *Parameters*: p_1 = proportion of all adult Taiwanese males who snore
> p_2 = proportion of all adult Taiwanese females who snore

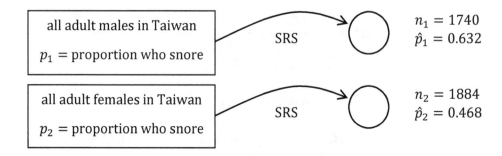

Shown below are data summaries based on the study.

	snore	don't snore	total
males	1100 (63.2%)	640	1740
females	882 (46.8%)	1002	1884
total	1982	1642	3624

In this sample, 63.2% of the males snored while only 46.8% of the females snored. What would you conclude? Is it possible that you just happened to randomly select more of the male snorers and fewer of the female snorers? Of course! A better question is, *even if the proportions of all adult*

Taiwanese male and females snorers are the same, how likely is it to get two random samples where the proportions of snorers are at least 0.164 (0.632 – 0.468) apart?

You can answer this question by making an inference about $p_1 - p_2$, the difference between the population proportions of snorers. To do that, consider the sampling distribution of $\hat{P}_1 - \hat{P}_2$, the difference between the sample proportions, under the assumption that the null hypothesis is true.

Shown below are the features of the sampling distribution of $\hat{P}_1 - \hat{P}_2$. These features are based on two <u>independent</u> SRS's from populations with success proportions p_1 and p_2.

	$\hat{P}_1 - \hat{P}_2$
Center	$E(\hat{P}_1 - \hat{P}_2) = p_1 - p_2$
Variance	$Var(\hat{P}_1 - \hat{P}_2) = \dfrac{p_1(1 - p_1)}{n_1} + \dfrac{p_2(1 - p_2)}{n_2}$
Standard Deviation	$SD(\hat{P}_1 - \hat{P}_2) = \sqrt{\dfrac{p_1(1 - p_1)}{n_1} + \dfrac{p_2(1 - p_2)}{n_2}}$
Shape	Nearly-normal if $n_1 p_1,\ n_1(1 - p_1),\ n_2 p_2,\ n_2(1 - p_2)\ are\ all \geq 15$

The number we use for $p_1 - p_2$ is 0, the one usually specified by the null (no difference) hypothesis. If the independent SRS's condition and the nearly-normal conditions are satisfied, then the statistic

$$Z = \frac{\hat{P}_1 - \hat{P}_2}{\sqrt{\frac{p_1(1 - p_1)}{n_1} + \frac{p_2(1 - p_2)}{n_2}}}$$

has a standard normal distribution under the null hypothesis. Now if p_1 and p_2 are the same, say some common proportion p, then the denominator simplifies a bit to

$$Z = \frac{\hat{P}_1 - \hat{P}_2}{\sqrt{\frac{p(1 - p)}{n_1} + \frac{p(1 - p)}{n_2}}} = \frac{\hat{P}_1 - \hat{P}_2}{\sqrt{p(1 - p)\left(\frac{1}{n_1} + \frac{1}{n_2}\right)}}.$$

The problem is that the common proportion p is unknown, so we'll have to estimate it based on the data. A good estimate of p is the overall proportion of successes if both samples are combined into one:

$$\hat{p} = \frac{total\ number\ of\ successes}{n_1 + n_2}.$$

We'll call the following method based on this analysis the **two-proportion z-method**.

Two-proportion z-method

H_0: $p_1 = p_2$
H_a: $p_1 < p_2$ or $p_1 > p_2$ or $p_1 \neq p_2$

Test Statistic: $z = \dfrac{\hat{p}_1 - \hat{p}_2}{\sqrt{\hat{p}(1-\hat{p})\left(\frac{1}{n_1}+\frac{1}{n_2}\right)}}$ $\hat{p} = \dfrac{total\ number\ of\ successes}{n_1 + n_2}$

$100(1-\alpha)\%$ CI for $p_1 - p_2$: $\hat{p}_1 - \hat{p}_2 \pm z^* \sqrt{\dfrac{\hat{p}_1(1-\hat{p}_1)}{n_1} + \dfrac{\hat{p}_2(1-\hat{p}_2)}{n_2}}$

where z^* is the value having an area of $\alpha/2$ to its right under the standard normal distribution

Conditions
  • independent SRS's
  • $n_1\hat{p}_1$, $n_1(1-\hat{p}_1)$, $n_2\hat{p}_2$, $n_2(1-\hat{p}_2)$ are all ≥ 15

Let's apply the two-proportion z-method to the snoring data.

<u>Hypotheses</u>

H_0: $p_1 = p_2$
H_a: $p_1 \neq p_2$

p_1 = proportion of all adult Taiwanese males who snore
p_2 = proportion of all adult Taiwanese females who snore

<u>Method/Conditions</u>: We'll use the two-proportion z-method.

  • independent SRS's: The data are from a simple random sample so we consider each group an independent SRS.
  • $n_1\hat{p}_1 = 1740\frac{1100}{1740} = 1100$ $n_1(1-\hat{p}_1) = 1740\left(1 - \frac{1100}{1740}\right) = 640$
    $n_2\hat{p}_2 = 1884\frac{882}{1884} = 882$ $n_2(1-\hat{p}_2) = 1884\left(1 - \frac{882}{1884}\right) = 1002$

    are all ≥ 15

<u>Mechanics (test statistic / p-value approach)</u>

$\hat{p} = \dfrac{1100+882}{1740+1884} = \dfrac{1982}{3624} \approx 0.547$

$z = \dfrac{\hat{p}_1 - \hat{p}_2}{\sqrt{\hat{p}(1-\hat{p})\left(\frac{1}{n_1}+\frac{1}{n_2}\right)}} = \dfrac{\frac{1100}{1740} - \frac{882}{1884}}{\sqrt{\frac{1982}{3624}\left(1-\frac{1982}{3624}\right)\left(\frac{1}{1740}+\frac{1}{1884}\right)}} = \dfrac{0.164}{0.01655} = 9.91$

The data provide evidence for H_a in the form of *low or high* values of z. Since the data gave a high value of z, the p-value is the probability of observing a value of z as high or higher than 9.91 or as low or lower than – 9.91 if H_0 were true.

Using Table I, we can only approximate the p-value because our z-value is not shown; it's higher than 3.99:

$$p - value = 2P(z \geq 9.91)$$
$$= 2(1 - something\ greater\ than\ 0.999967)$$
$$= 2(something\ less\ than\ 0.000033)$$

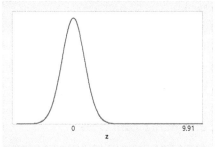

So the p-value is less than 0.000066. Using some kind of technology (e.g. Minitab or Excel) you can get the $p - value$ more precisely:

$$p - value = 2(1.873\ x\ 10^{-23}) = 3.75\ x\ 10^{-23}$$

If there is no difference in the proportions of all male and female Taiwanese adults who snore, it is nearly impossible to observe a difference of at least 0.164 in the sample proportions of snorers from random samples of 1740 males and 1884 females.

<u>Mechanics (confidence interval approach)</u>

You can also calculate a 95% confidence interval for $p_1 - p_2$ as a way to assess the evidence for H_a:

For 95% confidence, $\alpha = 1 - 0.95 = 0.05$ so z^* is the value having an area of 0.025 to its right (0.975 to its left) in the standard normal distribution. Using Table I we get $z^* = 1.96$.

Now just substitute into the formula to get the endpoints:

$$\hat{p}_1 - \hat{p}_2 \pm z^* \sqrt{\frac{\hat{p}_1(1 - \hat{p}_1)}{n_1} + \frac{\hat{p}_2(1 - \hat{p}_2)}{n_2}}$$

$$\frac{1100}{1740} - \frac{882}{1884} \pm 1.96 \sqrt{\frac{\frac{1100}{1740}\left(1 - \frac{1100}{1740}\right)}{1740} + \frac{\frac{882}{1884}\left(1 - \frac{882}{1884}\right)}{1884}}$$

$$0.164 \pm 0.032$$
$$(0.132, 0.196)$$

We are 95% confident that the difference in the proportions of all male and female Taiwanese adults who snore is between 0.132 and 0.196. That is, we are 95% confident that between 13.2 and 19.6 percentage points *more* male than female adults snore in Taiwan.

Conclusion

The p-value is microscopically small, so we reject H_0 at any significance level. The 95% confidence interval for $p_1 - p_2$ does not include 0. Since H_0: $p_1 = p_2$ ($p_1 - p_2 = 0$) is not plausible based on the data, we reject H_0 using the confidence interval approach, which tells the same story as the test statistic / p-value approach. We have enough evidence at the 5% significance level (or any reasonable significance level for that matter) of a difference in the proportions of all male and female Taiwanese adults who snore. In fact, the data convincingly suggest that at least 13.2 percentage points more males snore.

Section 8.2 Exercises

1. Suppose a simple random sample of size 20 is selected from each of two populations of numeric data that are nearly-normal with $\mu_1 = 50.4$ and $\mu_2 = 48.7$. Is Welch's t-method appropriate to test H_0: $\mu_1 = \mu_2$ vs. H_a: $\mu_1 \neq \mu_2$? Explain.

2. Suppose a simple random sample of size 20 is selected from each of two populations of numeric data that are nearly-normal resulting in $\bar{x}_1 = 50.4$ and $\bar{x}_2 = 48.7$. Is Welch's t-test appropriate to test H_0: $\mu_1 = \mu_2$ vs. H_a: $\mu_1 \neq \mu_2$? Explain.

3. Suppose a convenience sample of size 2000 is selected from each of two populations of numeric data that are right skewed resulting in $\bar{x}_1 = 50.4$ and $\bar{x}_2 = 48.7$. Is Welch's t-test appropriate to test H_0: $\mu_1 = \mu_2$ vs. H_a: $\mu_1 \neq \mu_2$? Explain.

4. Suppose a simple random sample of size 375 is selected from each of two populations of categorical data with $p_1 = 0.264$ and $p_2 = 0.352$. Is the two-proportion z-test appropriate to test H_0: $p_1 = p_2$ vs. H_a: $p_1 \neq p_2$? Explain.

5. Suppose a simple random sample of size 375 is selected from each of two populations of categorical data resulting in $\hat{p}_1 = 0.264$ and $\hat{p}_2 = 0.352$. Is the two-proportion z-test appropriate to test H_0: $p_1 = p_2$ vs. H_a: $p_1 \neq p_2$? Explain.

6. Suppose a simple random sample of size 375 is selected from each of two populations of categorical data resulting in $\hat{p}_1 = 0.016$ and $\hat{p}_2 = 0.352$. Is the two-proportion z-test appropriate to test H_0: $p_1 = p_2$ vs. H_a: $p_1 \neq p_2$? Explain.

7. Suppose a simple random sample of 12 pairs of objects is selected from a population of pairs of numeric data having nearly-normal differences with $\mu = 6.7$. Is the paired t-test appropriate to test H_0: $\mu = 0$ vs. H_a: $\mu > 0$?

8. Suppose a simple random sample of 12 pairs of objects is selected from a population of pairs of numeric data having nearly-normal differences resulting in $\bar{x} = 6.7$. Is the paired t-test appropriate to test H_0: $\mu = 0$ vs. H_a: $\mu > 0$?

9. Suppose a simple random sample of 1200 pairs of objects is selected from a population of pairs of numeric data having right skewed differences resulting in $\bar{x} = 6.7$. Is the paired t-test appropriate to test H_0: $\mu = 0$ vs. H_a: $\mu > 0$?

10. Consider these summaries of numeric data from two simple random samples:

$n_1 = 10$	$\bar{x}_1 = 5.5$	$s_1 = 1.6$	unimodal, symmetric shape
$n_2 = 13$	$\bar{x}_2 = 6.2$	$s_2 = 1.1$	unimodal, symmetric shape

    a. Find the p-value for the test of H_0: $\mu_1 = \mu_2$ vs. H_a: $\mu_1 < \mu_2$.
    b. Construct a 95% confidence interval for $\mu_1 - \mu_2$.

11. Consider these summaries of categorical data from two simple random samples:

$$n_1 = 1,500 \qquad \hat{p}_1 = 0.63$$
$$n_2 = 1,300 \qquad \hat{p}_2 = 0.58$$

    a.   Find the p-value for the test of H_0: $p_1 = p_2$ vs. H_a: $p_1 > p_2$.
    b.   Construct a 95% confidence interval for $p_1 - p_2$.

12. Consider these summaries of differences of numeric data from a simple random sample of pairs:

$$n = 50 \qquad\qquad \bar{x} = 22.6 \qquad\qquad s = 46.5 \qquad\qquad \text{right skewed shape}$$

    a.   Find the p-value for the test of H_0: $\mu = 0$ vs. H_a: $\mu \neq 0$.
    b.   Construct a 95% confidence interval for μ.

13. In a survey of U.S. parents of children under 18 years old, the Pew Research Center asked a sample of parents, "Do you think you spend too much time with your child/children, too little time, or about the right amount of time?" Suppose the responses summarized below are from a simple random sample of 1,807 U.S. parents of children under 18.[7]

	too much	right amount	too little
Mothers	70	579	221
Fathers	19	472	446

We'd like to use these data to see if there's a difference in the proportions of all U.S. mothers and all U.S. fathers who think they spend too little time with their child(ren).

    a.   Write down the hypotheses and define the parameters in context.
    b.   Check the conditions. Do you need to make any assumptions to proceed with the test? Explain.
    c.   Find and interpret the p-value.
    d.   Construct a 95% confidence interval for $p_1 - p_2$.
    e.   Make a conclusion in context.

14. Consider the data on median student loan debt (thousands of dollars) for the samples of institutions from exercise 14 in section 2.3. The data are grouped by the predominant degree granted.

Associate's: 9.5   5.8   6.3   9.6   4.0   9.9   6.8   9.0   4.0   15.7
                 3.5

Bachelor's: 15.6   17.0   12.6   12.0   17.3   20.1   15.0   16.0   19.8

Is there a difference between Associate's and Bachelor's institutions in the U.S. with respect to the average median student loan debt?

    a.   Write down the hypotheses and define the parameters in context.
    b.   Check the conditions. Do you need to make any assumptions to proceed with the test? Explain.
    c.   Find and interpret the p-value.
    d.   Construct a 95% confidence interval for $\mu_1 - \mu_2$.
    e.   Make a conclusion in context.

15. Jack is a waiter at a restaurant A and Jill waits tables at restaurant B. Jill always introduces herself by name but Jack does not. They decide to collect data to see if introductions increase tip percentages. Knowing that randomness is important for making inferences, they select a SRS of 7 days and each reports their tips as a percentage of the total receipts on those days. The data are shown here.

Day	Jack's Tip %	Jill's Tip %
1	12.4	15.5
2	8.8	8.1
3	18.2	21.5
4	22.3	24.9
5	10.4	11.7
6	14.8	19.3
7	21.1	23.6

   a.   Write down the hypotheses and define the parameter in context.
   b.   Check the conditions. Do you need to make any assumptions to proceed with the test? Explain.
   c.   Find and interpret the p-value.
   d.   Construct a 95% confidence interval for μ.
   e.   Make a conclusion in context. Be careful how you word your conclusion!
   f.   Can we conclude that Jill's introductions resulted in higher average tip percentages? Explain.
   g.   Now conduct the analysis using Welch's t-method (the wrong method). How would your conclusion change?

16. You may notice that "plain math" questions tend to be easier than word problems. In a simple random sample of 90 U.S. adults, 74 answered this question correctly: "What is 3.5% of 2000?" In another simple random sample of 180 U.S. adults, 112 answered this question correctly: "A nurse needs to add some salt to a bag of water in the amount of 3.5% of the weight of a bag. If the bag weighs 2000 grams, how much salt must the nurse add?"[8] Do these data provide convincing statistical evidence that U.S. adults would get the simpler question correct more frequently than the word problem?

17. Two different methods for measuring the amount of milk ingested by a newborn baby are being compared. A simple random sample of newborns at a certain hospital was studied and the amount of milk as measured by both methods was recorded for each baby. The data are shown here.[9]

Baby	Method 1 (ml)	Method 2 (ml)
1	1481	1506
2	1416	1264
3	1546	1336
4	1563	1579
5	2169	1983
6	1761	1315
7	1102	1404
8	1187	1123

   a.   Do these data provide convincing statistical evidence of a difference between the two methods?
   b.   To what population can these results be generalized? Be very specific.

18. In two simple random samples of outdoor public real estate auctions, the winning bid amount was recorded. In one sample, the auctions were held in July and in the other sample, the auctions were held in November of the same year. The winning bids (in thousands of dollars) are shown below.

July:	163	160	121	324	231	130	197	145	200	242
	170	154	162	79	218	283	268	284	276	189
	196	312	231	343	153					

Nov:	273	338	148	176	180	237	300	248	313	294
	206	243	293	324	358	253	239	304	251	177
	144	387	223	427	271					

   a.   Do these data provide convincing statistical evidence that the average winning bid amount for all auctions in November of that year was higher than that for July?
   b.   Does your result show that lower temperatures result in higher bid amounts? Explain.

19. Consider exercise 14 where you compared Associate's and Bachelor's institutions with respect to average median student loan debt.

    a. Change the sample sizes to 3 each but keep all other statistics the same. Recalculate the p-value and the endpoints of the 95% confidence interval. Then make a general statement about how the sample sizes affect the results of inferences.
    b. True or false: Any (non-zero) difference in sample means, no matter how small, can be statistically significant if the sample sizes are large enough.
    c. Change sample mean for the Bachelor's sample to 10 but keep all other statistics the same. Recalculate the p-value and the endpoints of the 95% confidence interval. Then make a general statement about how the difference between the sample means affects the results of inferences.
    d. Change sample standard deviations for both samples to 8 but keep all other statistics the same. Recalculate the p-value and the endpoints of the 95% confidence interval. Then make a general statement about how the sample standard deviations affect the results of inferences.

20. Consider exercise 16 where you compared proportions who would get a simple math question correct to that for a word problem.

    a. Change the number who got the word problem correct to 144 but keep all other statistics the same. Recalculate the p-value and the endpoints of the 95% confidence interval. Then make a general statement about how the difference between the sample proportions affects the results of inferences.
    b. Using the two $\hat{p}$'s from part a, find the two equal sample sizes n_1 and n_2 that would result in the same p-value as the original data. Then make a general statement about how the sample sizes affect the results of inferences.
    c. True or false: Any (non-zero) difference in sample proportions, no matter how small, can be statistically significant if the sample sizes are large enough.

8.3 Causal Inference About Two Treatments

In this section, we'll explore how to use data to draw cause/effect conclusions. One of the very important keys to doing this is isolating the variable thought to be doing the causing from all other variables in the universe. The gold standard for doing this well is random assignment. You would have a tough time justifying causal inferences based on data where the treatments have not been randomly assigned!

In this section we'll consider only two treatment groups based on random assignment. The section is divided into three parts based on the kinds of response data used to compare the two treatments:

- Numeric Response (Independent Samples)
- Numeric Response (Paired Samples)
- Categorical Response (Independent Samples)

It turns out, rather surprisingly, that we can use the same analysis methods that we used for generalization inferences in the previous section. However, there are two very important differences in using the methods for causal inference. I will emphasize these differences as we go through this section:

- Random assignment generally permits a conclusion about the *cause* of a difference between two groups. Random assignment should result in groups that are balanced with respect to the effects of confounding variables. Therefore, any differences in responses can be potentially attributed to the difference between the two treatments.
- The data are generally NOT randomly selected from a larger population. We are not guaranteed that the sample represents any such larger population. Therefore the results can only be generalized to, at most, a group of objects *similar to those in the study*.

Numeric Response (Independent Samples)

Let's consider an experiment designed to compare two treatments. We randomly assign the objects available to us to the two treatments using SRA (simple random assignment) and then observe a numeric response on each object. The goal is to see if the difference in the treatments *caused changes in* or *affected* the objects' responses.

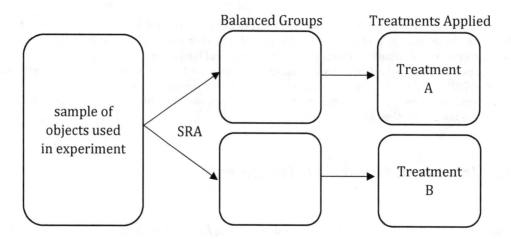

Since the data structure is the same as that from two SRS's of numeric data, we can use Welch's t-method to analyze the data with two important differences.

- The random mechanism is random assignment, not random sampling. Therefore, the data permit cause/effect conclusions.
- The parameters, μ_1 and μ_2, represent the average responses on treatments 1 and 2 for all objects *similar to those in the study*. If there is no random sampling from a larger population, you must restrict the scope of the inference accordingly.

Welch's t-method (causal inference)

H_0: $\mu_1 = \mu_2$
H_a: $\mu_1 < \mu_2$ or $\mu_1 > \mu_2$ or $\mu_1 \neq \mu_2$

Test Statistic: $t = \dfrac{\bar{x}_1 - \bar{x}_2}{\sqrt{\frac{s_1^2}{n_1} + \frac{s_2^2}{n_2}}}$ $\qquad df = \dfrac{\left(\frac{s_1^2}{n_1} + \frac{s_2^2}{n_2}\right)^2}{\frac{\left(\frac{s_1^2}{n_1}\right)^2}{n_1 - 1} + \frac{\left(\frac{s_2^2}{n_2}\right)^2}{n_2 - 1}}$ (round down)

$100(1 - \alpha)\%$ CI for $\mu_1 - \mu_2$: $\bar{x}_1 - \bar{x}_2 \pm t^* \sqrt{\dfrac{s_1^2}{n_1} + \dfrac{s_2^2}{n_2}}$

where t^* is the value having an area of $\alpha/2$ to its right under the t-distribution with df shown above.

Conditions
- independent SRA to treatment
- two nearly-normal data distributions (can be relaxed for large sample sizes or roughly symmetric distributions with no outliers)

Example: In the fall of 2014 I conducted an experiment in my statistics classes to see if homework sets containing review problems, called "distributed practice" (DP) would be better than homework sets containing all new problems, called "massed practice" (MP).[10] Students who chose to participate were randomly assigned to the two treatments. All other aspects of the course were the same for these students except for the homework sets. The DP group got 5 new problems and 5 review problems in each homework set. The MP group got 10 new problems in each homework set. The primary response variable was the final exam score.

Factor: homework type
Levels: DP, MP
Treatments: DP, MP
Response: final exam score

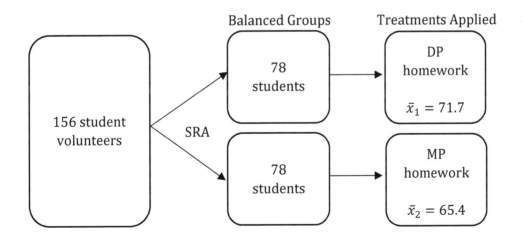

Here are some summaries.

Final Exam Score (DP)	Final Exam Score (MP)
$n_1 = 78$	$n_2 = 78$
$\bar{x}_1 = 71.7$	$\bar{x}_2 = 65.4$
$s_1 = 16.1$	$s_2 = 18.8$

In this study, students with the DP homework scored 6.3 percentage points higher (71.7 – 65.4), on average, than the students with the MP homework. What would you conclude? Is it possible that the DP group did better on average just by chance? Of course! A better question is, *even if the average final exam score for students like these doing DP homework is the same as that for students like these doing MP homework, how likely is it to observe two groups of students assigned at random where the DP group's mean is at least 6.3 percentage points higher?*

Let's apply Welch's method to answer that question.

<u>Hypotheses</u>

$H_0: \mu_1 = \mu_2$
$H_a: \mu_1 > \mu_2$

μ_1 = average final exam score for students like those in the study doing DP homework
μ_2 = average final exam score for students like those in the study doing MP homework

<u>Method/Conditions</u>: We'll use Welch's t-method.

Conditions:
- independent SRA: Students were assigned at random to the two treatments. There is no natural, one-to-one way to match the final exam scores across the two treatments, so they are independent samples.
- two nearly-normal data distributions: 78 is considered large enough for both samples so that we don't need to worry about this condition too much.

<u>Mechanics</u>

$$t = \frac{\bar{x}_1 - \bar{x}_2}{\sqrt{\frac{s_1^2}{n_1} + \frac{s_2^2}{n_2}}} = \frac{71.7 - 65.4}{\sqrt{\frac{16.1^2}{78} + \frac{18.8^2}{78}}} = 2.248 \qquad df = \frac{\left(\frac{s_1^2}{n_1} + \frac{s_2^2}{n_2}\right)^2}{\frac{\left(\frac{s_1^2}{n_1}\right)^2}{n_1 - 1} + \frac{\left(\frac{s_2^2}{n_2}\right)^2}{n_2 - 1}} = \frac{\left(\frac{16.1^2}{78} + \frac{18.8^2}{78}\right)^2}{\frac{\left(\frac{16.1^2}{78}\right)^2}{78 - 1} + \frac{\left(\frac{18.8^2}{78}\right)^2}{78 - 1}} = 150.4$$

We'll round down and use $df = 150$.

The data provide evidence for H_a in the form of *high* values of t so the p-value is the probability of observing a value of t as high or higher than 2.248 if H_0 were true.

Using the t-table and 150 degrees of freedom, we can only approximate the p-value because our t-value is not shown exactly; it's between 1.98 and 2.35:

$$p - value = P(t \geq 2.248)$$
$$= 1 - P(t \leq 2.248)$$
$$= 1 - something\ between\ 0.975\ and\ 0.99$$

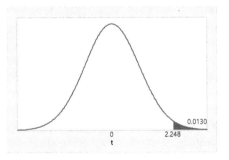

So the p-value is between 0.01 and 0.025. Using some kind of technology (e.g. Minitab or Excel) you can get the $p - value$ more precisely:

$$p - value = 0.0130$$

If the average final exam score for students like those in the study doing DP homework is the same as that for students like those in the study doing MP homework, there is a 1.3% chance of finding a difference at least as high as 6.3 percentage points in the mean final exam score for the two treatments where 78 students are randomly assigned to each treatment. That is, it would be fairly unlikely to have observed data like ours by chance alone. SRA is not a good explanation for the data.

Conclusion

The p-value is small (using the typical 5% significance level), so we reject H_0 at that level. We have enough evidence at the 5% significance level that average final exam score for students *like those in the study* doing DP homework would be higher than that for students *like those in the study* doing MP homework. Because students were randomly assigned to the two treatments, we can therefore potentially conclude that the homework difference is the *cause* of higher final exam scores.

I italicized two important parts of the conclusion. The first is that we can conclude that the difference in the types of homework sets was potentially the *cause* of higher final exam scores. Since the study was an experiment with random assignment to homework type, the two groups of students should have been balanced with respect to the effects of confounding variables. Therefore any differences in the final exam scores can be reasonably attributed to the homework differences.

The second important part of the conclusion is that this effect of homework type can be generalized only to students *like the ones who participated in my study*. Since the students were not randomly selected from any larger population of students, we are not guaranteed that the sample represents any such larger population. I'd have to replicate the study in many different settings (different schools, different subjects, etc.) in order to get a better idea of where this strategy might work well.

In addition to the limitation on the scope of the inference, can you think of any other limitations in this study? I can! Here are two more:

1) While the random assignment permits a causal inference, it's not a slam dunk. Since students were not blinded to the treatment (how could they be?), there is the potential for a Hawthorne effect here. Perhaps the students who got the DP homework sets were

somehow motivated to do better simply because they were being treated differently than usual.

2) Long-term retention was not observed. Would the DP and MP students perform more similarly if we had tested them six months later? A year later? Five years later? The impact of DP vs. MP is somewhat limited if it lasts only a short time.

Numeric Response (Paired Samples)

As you saw in the previous section, the analysis of paired samples involves using the paired differences as the data in a one-sample t-method (called the paired t-method). We can also use the paired t-method to make causal inference in the context of an experiment where the treatments have been randomly assigned in a way that results in paired samples. However, keep in mind that there are important differences in how the method is used:

- The random mechanism is random assignment, not random sampling. Therefore, the data permit cause/effect conclusions.
- The parameter, μ, represents the average difference in responses for treatments 1 and 2 for all pairs *similar to those in the study*. If there is no random sampling from a larger population, you must restrict the scope of the inference accordingly.

Paired t-method (causal inference)

H_0: $\mu = 0$
H_a: $\mu > 0$ or $\mu < 0$ or $\mu \neq 0$

Calculate $x = $ difference in responses for all n pairs.

Test Statistic: $t = \dfrac{\bar{x}}{s/\sqrt{n}}$ $\qquad df = n - 1$

$100(1 - \alpha)\%$ CI for μ: $\qquad \bar{x} \pm t^* \dfrac{s}{\sqrt{n}}$

where t^* is the value having an area of $\alpha/2$ to its right under the t-distribution with df shown above.

Conditions
- SRA to treatment within each pair
- nearly-normal distribution of differences (can be relaxed for a large sample size or a roughly symmetric distribution with no outliers)

Example: In order to determine which variety of corn (A or B) is better for yield, a farmer divides each of her 10 fields into 2 plots. She tosses a coin for each field to assign a plot to either variety A or variety B. At harvest time she observes the yield (in bushels) for each plot. The data are shown here.

Remember this one? These samples of yields are paired by field.

Field	Yield (variety A)	Yield (variety B)	Difference (A – B)
1	144	150	-6
2	96	99	-3
3	125	126	-1
4	117	122	-5
5	87	85	2
6	133	137	-4
7	109	117	-8
8	139	140	-1
9	140	146	-6
10	125	129	-4

Factor: variety
Levels: A, B
Treatments: A, B
Response: yield (bushels)

Since the data structure is the same as paired data from a single SRS, we can use the paired t-test to analyze the data.

<u>Hypotheses</u>

H_0: $\mu = 0$
H_a: $\mu \neq 0$

μ = mean difference in yield (A – B) for fields similar to those in the study

<u>Method/Conditions</u>: We'll use the paired t-test on the differences.

Conditions:
- SRA to treatment within each pair: varieties were assigned to plots within each field by tossing a coin
- nearly-normal distribution of differences: A normal probability plot of the differences shows a fairly linear pattern, so yes, this condition is satisfied.

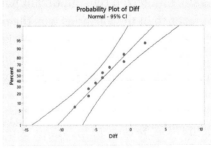

<u>Mechanics</u>

$$\bar{x} = \frac{-6-3-1-5+2-4-8-1-6-4}{10} = -3.6$$

$$s^2 = \frac{\left(-6-(-3.6)\right)^2+\left(-3-(-3.6)\right)^2+\cdots+\left(-4-(-3.6)\right)^2}{10-1} = 8.7\overline{1}$$

$$s = \sqrt{s^2} \approx 2.951$$

$$t = \frac{\bar{x}}{s/\sqrt{n}} = \frac{-3.6}{2.951/\sqrt{10}} = -3.857 \qquad df = n - 1 = 10 - 1 = 9$$

The data provide evidence for H_a in the form of either *low* or *high* values of t so the p-value is the probability of observing a value of t at least as low as –3.857 or at least as high as 3.857 if H_0 were true.

Using the t-table and 9 degrees of freedom, we can only approximate the p-value because our t-value is not shown exactly; it's between –4.30 and –3.25:

$$p - value = 2P(t \leq -3.857)$$
$$= 2(something \; between \; 0.001 \; and \; 0.005)$$

So the p-value is between 0.002 and 0.01. Using some kind of technology (e.g. Minitab or Excel) you can get the $p - value$ more precisely:

$$p - value = 0.0039$$

If there is no mean difference (A – B) in yield for fields like those in the study, there is a 0.39% chance of finding an average difference at least as high as 3.6 bushels. It would have been very unlikely to have observed data like ours by chance alone. Random assignment is not a good explanation for the data.

Conclusion

The p-value is very small, so we reject H_0. We have enough evidence that there is a mean difference in yield (A – B) for all fields *like those in this study*. Because plots were randomly assigned within field, we can therefore conclude that the variety *would affect* the yield for fields *like those in this study*. The data suggest that variety A results in higher average yield than variety B.

Again, I italicized two important parts of the conclusion. The first is that we can conclude that the difference in the varieties was the *cause* of higher yields. Since the study was an experiment with random assignment to variety within field, the two groups of plots should have been balanced with respect to the effects of confounding variables. Therefore any differences in the yields can be reasonably attributed to the variety differences.

The second important part of the conclusion is that this effect of variety can be generalized only to fields *like the ones in the study*. Since the fields were not randomly selected from any larger population of fields, we are not guaranteed that the sample represents any such larger population. The study should be repeated in different locations and under different soil and weather conditions to determine how the two varieties would compare more generally.

Categorical Data (Independent Samples)

Are you seeing a pattern yet? In the previous section, we used the two-proportion z-method to compare two populations using categorical responses. We'll use the same method in this section to compare two treatments in a randomized experiment using categorical responses. We'll use the same mechanics but since the application will be to a causal inference problem, there are two very important differences:

- The random mechanism is random assignment, not random sampling. Therefore, the data permit cause/effect conclusions.
- The parameters, p_1 and p_2, represent the proportion of successes on treatments 1 and 2 for all objects *similar to those in the study*. If there is no random sampling from a larger population, you must restrict the scope of the inference accordingly.

Two-proportion z-method (causal inference)

$H_0: p_1 = p_2$

$H_a: p_1 < p_2$ or $p_1 > p_2$ or $p_1 \neq p_2$

Test Statistic: $z = \dfrac{\hat{p}_1 - \hat{p}_2}{\sqrt{\hat{p}(1-\hat{p})\left(\frac{1}{n_1}+\frac{1}{n_2}\right)}}$
    $\qquad \hat{p} = \dfrac{total\ number\ of\ successes}{n_1 + n_2}$

$100(1-\alpha)\%$ CI for $p_1 - p_2$:
    $\qquad \hat{p}_1 - \hat{p}_2 \pm z^* \sqrt{\dfrac{\hat{p}_1(1-\hat{p}_1)}{n_1} + \dfrac{\hat{p}_2(1-\hat{p}_2)}{n_2}}$

where z^* is the value having an area of $\alpha/2$ to its right under the standard normal distribution

Conditions
- independent SRA to treatment
- $n_1\hat{p}_1$, $n_1(1-\hat{p}_1)$, $n_2\hat{p}_2$, $n_2(1-\hat{p}_2)$ *are all* ≥ 15

Example: You are an analyst for a company who wants to know whether putting the "Buy Now" button at the top or bottom of the web page for a product works better. Does the location of the button affect sales? You design an experiment where you randomly assign all visits to the product page to one of two versions of the page: one where the "Buy Now" button is at the top and one where the "Buy Now" button is at the bottom. For each visit, you determine whether the customer purchased the product or not. Summaries of the data are shown below.

	purchased	didn't purchase	total
top	75 (14.7%)	435	510
bottom	104 (19.9%)	418	522
total	179	853	1032

Factor: "Buy Now" button location
Levels: top, bottom
Treatments: top, bottom
Response: purchase status

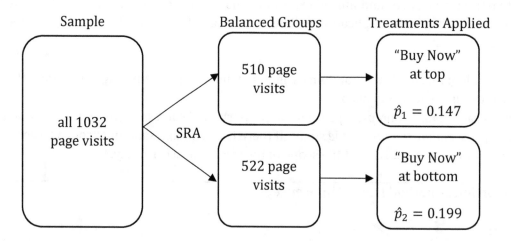

What would you conclude if 14.7% of customers who saw the "top" version purchased the product vs. 19.9% in the "bottom" group? The company should put the button at the bottom of the page, right? Hold your horses. Is it possible that the "bottom" group did better just by chance? Of course! A better question is, *even if the proportion of purchases for customers like these who see the "bottom" version of the page is the same as that for customers like these who see the "top" version, how likely is it to randomly assign customers to the two button locations and get purchase percentages at least 5.2 percentage points apart?*

<u>Hypotheses</u>

H_0: $p_1 = p_2$
H_a: $p_1 \neq p_2$

p_1 = proportion of purchases among all customers like those in the study seeing the "top" version of the page
p_2 = proportion of purchases among all customers like those in the study seeing the "bottom" version of the page

<u>Method/Conditions</u>: We'll use the two-proportion z-method.

- independent SRA to treatment: The version of the webpage viewed by each customer was determined at random.

- $n_1 \hat{p}_1 = 510 \frac{75}{510} = 75$ $n_1(1 - \hat{p}_1) = 510 \left(1 - \frac{75}{510}\right) = 435$
  $n_2 \hat{p}_2 = 522 \frac{104}{522} = 104$ $n_2(1 - \hat{p}_2) = 522 \left(1 - \frac{104}{522}\right) = 418$

  are all ≥ 15

Mechanics (test statistic / p-value approach)

$$\hat{p} = \frac{75+104}{510+522} = \frac{179}{1032} \approx 0.173$$

$$z = \frac{\hat{p}_1-\hat{p}_2}{\sqrt{\hat{p}(1-\hat{p})\left(\frac{1}{n_1}+\frac{1}{n_2}\right)}} = \frac{\frac{75}{510}-\frac{104}{522}}{\sqrt{\frac{179}{1032}\left(1-\frac{179}{1032}\right)\left(\frac{1}{510}+\frac{1}{522}\right)}} = \frac{-0.052}{0.0236} = -2.21$$

The data provide evidence for H_a in the form of *low or high* values of z. The p-value is the probability of observing a value of z as low or lower than -2.21 or as high or higher than 2.21 if H_0 were true.

Using Table I, we get:

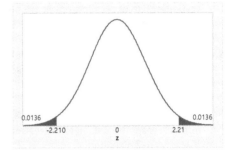

$$p-value = 2P(z \le -2.21)$$
$$= 2(0.0136)$$
$$= 0.0272$$

If the proportions of purchases are the same ("bottom button" vs. "top button") for customers like the ones in the study, there is only a 2.72% chance of getting groups of 510 and 522 customers assigned at random where the purchase percentages are at least 5.2 percentage points apart. It would have been quite unlikely to have observed data like these by chance alone. Random assignment is not a good explanation for the data.

Conclusion

The p-value is quite small (using the typical 5% significance level), so we reject H_0 at that level. We have enough evidence at the 5% significance level that the proportion of purchases among all customers like those in the study who see the "Buy Now" button at the top of the page would differ from that for customers like those in the study who see the "Buy Now" button at the bottom of the page. Because customers were randomly assigned to the two version of the page, we can therefore conclude that the "Buy Now" button location difference is the *cause* of different purchase proportions for customers *like those in the study*. The data suggest that placing the button at the bottom of the page increases sales.

I hope you are beginning to catch on to the difference in how the methods are used for causal inferences. Here we can conclude that the difference in the "Buy Now" button location was the *cause* of different purchase proportions. Since the study was an experiment with random assignment to button location, the two groups of customers should have been balanced with respect to the effects of confounding variables. Therefore any differences in the purchase statuses can be reasonably attributed to the button location differences.

The second important part of the conclusion is that this effect of button location can be generalized only to customers *like the ones who participated in the study*. Since the customers were not randomly selected from any larger population of customers, we are not guaranteed that the sample represents any such larger population. The study should be repeated with customers in different demographics and with different products/services to determine how the two button locations would compare more generally.

Section 8.3 Exercises

1. Suppose a group of volunteers is split into two treatment groups of 20 by simple random assignment. The numeric response data in each treatment are nearly-normal. Is Welch's t-test appropriate to test H_0: $\mu_1 = \mu_2$ vs. H_a: $\mu_1 \neq \mu_2$? Explain.

2. Suppose a group of volunteers is split into two treatment groups of 20 by simple random assignment. The numeric response data in each treatment are severely skewed. Is Welch's t-test appropriate to test H_0: $\mu_1 = \mu_2$ vs. H_a: $\mu_1 \neq \mu_2$? Explain.

3. Suppose a group of kindergarteners is split into two treatment groups of 200 by cluster random assignment. The numeric response data in each treatment are nearly-normal. Is Welch's t-test appropriate to test H_0: $\mu_1 = \mu_2$ vs. H_a: $\mu_1 \neq \mu_2$? Explain.

4. Suppose all 25 objects in a population each get two different treatments. The order in which each object gets each treatment is determined at random. The distribution of differences in the responses are symmetric. Is the paired t-test appropriate to test H_0: $\mu = 0$ vs. H_a: $\mu > 0$?

5. Suppose all 150 objects in a population each get two different treatments. The order in which each object gets each treatment is determined at random. The distribution of differences in the responses is mildly skewed. Is the paired t-test appropriate to test H_0: $\mu = 0$ vs. H_a: $\mu > 0$?

6. Suppose viewers of a webpage are assigned (by simple random assignment) to one of two versions of the webpage. Of the 1,500 people who saw the first version, 550 clicked the call to action button. Of the 1,500 people who saw the second version, 300 clicked the call to action button. Is the two-proportion z-test appropriate to test H_0: $p_1 = p_2$ vs. H_a: $p_1 \neq p_2$? Explain.

7. Suppose a group of volunteers are assigned (by simple random assignment) to either an active flu vaccine or placebo. Of the 50 people who got the active vaccine, 10% got the flu. Of the 50 people who got placebo, 12% got the flu. Is the two-proportion z-test appropriate to test H_0: $p_1 = p_2$ vs. H_a: $p_1 \neq p_2$? Explain.

8. Consider these summaries of numeric response data from an experiment where 60 subjects were assigned by simple random assignment to one of two treatments:

$n_1 = 30$	$\bar{x}_1 = 22.1$	$s_1 = 2.9$	mildly skewed
$n_2 = 30$	$\bar{x}_2 = 24.5$	$s_2 = 3.2$	mildly skewed

    a. Find the p-value for the test of H_0: $\mu_1 = \mu_2$ vs. H_a: $\mu_1 \neq \mu_2$.
    b. Construct a 95% confidence interval for $\mu_1 - \mu_2$.

9. Consider these summaries of categorical response data from an experiment where 2,000 objects were assigned by simple random assignment to one of two treatments:

$n_1 = 1000$	645 successes
$n_2 = 1000$	667 successes

    a. Find the p-value for the test of H_0: $p_1 = p_2$ vs. H_a: $p_1 \neq p_2$.
    b. Construct a 95% confidence interval for $p_1 - p_2$.

10. Consider these summaries of the differences (treatment 1 – treatment 2) in numeric response data from an experiment where 100 objects were paired together. A coin was tossed to decide which object within a pair got which treatment.

    $\bar{x} = -42.8$ $s = 10.7$ unimodal, symmetric shape

    a. Find the p-value for the test of H_0: $\mu = 0$ vs. H_a: $\mu < 0$.
    b. Construct a 95% confidence interval for μ.

11. "An orange a day slashes risk of failing eyesight, scientists say" was the headline in a report of a study that followed over 2,000 adults over 50 for 15 years. Those who ate at least one serving of oranges a day had a 60% lower risk of developing macular degeneration.[11]

    a. Identify the response variable and the treatments.
    b. Is this study an experiment or an observational study?
    c. Is the headline justified? Explain.

12. A study of gun violence was conducted in a certain city. Each call to police concerning domestic violence involving a gun was observed. For each case it was noted whether or not the responding officer removed the gun from the premises. The average number of subsequent domestic violence calls was higher in cases where the gun was removed compared to those cases where the officer elected not to remove the gun.

    a. Identify the response variable and the treatments.
    b. Is this study an experiment or an observational study?
    c. Do the data suggest that removing the gun increases the number of subsequent calls to police? Explain.

13. A group of volunteer obese adults with diabetes were randomly assigned to either gastric bypass surgery or major dietary and lifestyle changes. Thirty percent of those who had surgery went into remission; none in the other group did.

    a. Identify the response variable and the treatments.
    b. Is this study an experiment or an observational study?
    c. Do the data suggest that gastric bypass surgery increases the changes of remission from diabetes for adults like these? Explain.

14. In one high school, students were classified as "music students" (instrumental or vocal) or "non-music students." The music students were found to have an average GPA of 3.42 compared to an average GPA of 3.17 for the non-music students. Based on these data, the school developed a campaign to encourage more students to take music lessons or join a musical group.

    a. Identify the response variable and the treatments.
    b. Is this study an experiment or an observational study?
    c. Is the campaign justified by the data? Explain.

15. An energy company claims that cars will get better gas mileage on its brand of regular, 87 octane, unleaded gasoline. A test is conducted where the 40 cars in one rental car company's fleet are used. Twenty are randomly chosen and driven through a test course using a tank of the company's gas and the other 20 use a tank of a competitor's gas. The gas mileage data (in mpg) are shown below.

Company's Gas:	26.4	29.2	27.9	29.7	27.3	29.2	27.3	27.3	28.2	29.8
	30.3	30.1	26.9	25.7	25.3	26.4	23.3	28.2	28.3	28.1
Competitor's Gas:	22.8	24.7	26.8	26.2	23.5	25.0	27.6	27.7	24.5	29.1
	24.6	25.7	26.0	26.1	24.5	23.5	30.4	26.4	29.1	24.9

Do these data provide convincing statistical evidence that the company's gas results in higher average gas mileage compared to the competitor's gas for cars like these?

   a. Write down the hypotheses and define the parameters in context.
   b. Check the conditions and say which method you'll use.
   c. Find and interpret the p-value.
   d. Make a conclusion in context.

16. An experiment was conducted to see whether a nasal spray flu vaccine would be more effective in preventing the flu compared to the injection version. A group of 3,000 adults was divided at random (using simple random assignment) into two groups. One group was administered the flu vaccine via nasal spray and the other group via an injection. Sixty-two of the adults who got the nasal spray contracted the flu compared to 130 who got the injection. Do these data provide convincing statistical evidence that the nasal spray vaccine would be more effective in preventing the flu compared to the injection for adults like these?

   a. Write down the hypotheses and define the parameters in context.
   b. Check the conditions and say which method you'll use.
   c. Find and interpret the p-value.
   d. Make a conclusion in context.

17. A gardener wants to compare an organic fertilizer to a conventional fertilizer. She divides each of her 6 garden plots in half and plants the same number of potato plants in each half. Using a coin toss for each plot, she assigns one half to get the organic fertilizer and the other half to get the conventional fertilizer. Other than the fertilizer, she treats each half-plot in the same way (e.g. watering, weeding). The potato yields (in pounds) are shown below.

Plot	Yield (Organic)	Yield (Conventional)
1	23	32
2	45	40
3	33	30
4	46	39
5	58	62
6	27	26

Do these data provide convincing statistical evidence of an average difference in the effects of the two fertilizer types on the yield for potato plants like these?

   a. Write down the hypotheses and define the parameter in context.
   b. Check the conditions and say which method you'll use.
   c. Find and interpret the p-value.
   d. Make a conclusion in context.

18. Consider exercise 17. Suppose the gardener used 600 plots and got the same results (same sample mean difference and standard deviation).

   a. Recalculate the p-value.
   b. Would your conclusion be different?
   c. Calculate the two 95% confidence intervals for μ, one based on the original sample size and one based on 600 plots.
   d. Make at least two observations about how sample size affects the information extracted from the data.

19. An experiment used simple random assignment to divide a group of 180 student volunteers into two groups and then each student was given an online shopping task. In one group, students got a $20 coupon to use toward their purchase. The percent change (before vs. after shopping) in the levels of a stress hormone was measured for each student. The data are summarized here.[12]

Group	n	mean	stdev
Coupon	90	– 8.1	36.9
No coupon	90	– 1.2	32.2

Do these data provide convincing statistical evidence that a $20 coupon would affect the average percent change in stress hormone levels for students like these?

20. A new automotive wax is advertised to "bead up" longer than a leading brand. To investigate, an experiment is conducted using four vehicle models. Two identical vehicles of each model are used; one is randomly assigned to be treated with the new wax and the other is treated with the leading brand. Weather conditions and washings are simulated on each vehicle and the time (months) until the wax no longer beads up is recorded. The data are shown below.

Vehicle	Months (New wax)	Months (Leading brand)
1	15.5	13.2
2	18.1	19.7
3	12.6	12.8
4	14.0	14.3

Do these data provide convincing statistical evidence that the new wax increases bead-up time over the leading brand for vehicles like these?

21. A survey designed to assess the effect of question wording on responses asked a set of U.S. adults about their level of agreement with this statement: "There are plenty of [good] jobs available in my area." Respondents were divided into two groups using simple random assignment. Half of the adults saw the question with the word "good" in it; the other half saw the question without the word "good." The response data are summarized below.

Group	Definitely Agree	Somewhat Agree	Neutral	Somewhat Disagree	Definitely Disagree
"good jobs"	55	245	42	122	36
"jobs"	67	173	37	202	21

Do these data provide convincing statistical evidence that including the word "good" in the question would change the proportion of adults like these who definitely or somewhat agree with the statement?

Notes

[1] Brody, J. (December 4, 2017), "Cataract Surgery May Prolong Your Life," *New York Times*.

[2] Based on Pew Research Center (2015). "Raising kids and running a household: how working parents share the load."

[3] Stickle, et al, 2008. "An experimental evaluation of teen courts," Journal of Experimental Criminology, 4:137 – 163.

[4] Welch, B. L. (1947), "The Generalization of "Student's" problem when several different population variances are involved," *Biometrika*, 34, pgs 28 – 35. Note the reference to "Student's" t-distribution.

[5] Satterthwaite, F. E. (1946), "An Approximate Distribution of Estimates of Variance Components," *Biometrics Bulletin*, 2, pgs. 110 – 114.

[6] Chuang, L. et al. (2017), "The gender difference of snore distribution and increased tendency to snore in women with menopausal syndrome: a general population study," *Sleep Breath*, 21, 543 – 547.

[7] Based on Pew Research Center (2015). "Raising kids and running a household: how working parents share the load."

[8] Based on Schmeiser, et al. (2013). "Using the Right Yardstick: Assessing Financial Literacy Measures by Way of Financial Well-Being," Journal of Consumer Affairs, 243 – 262.

[9] Based on Butte, et al. (1983). "Evaluation of the Deuterium Dilution Technique Against the Test Weighing Procedure for the Determination of Breast Milk Intake." American Journal of Clinical Nutrition, 996 – 1003.

[10] Crissinger, B. (2015), "The Effect of Distributed Practice in Undergraduate Statistics Homework Sets: A Randomized Trial," *Journal of Statistics Education*, 23 (3).

[11] July 18, 2018 edition of *The Telegraph*.

[12] Based on Alexander, et al., (2015). "Preliminary Evidence for the Neurophysiologic Effects of Online Coupons: Changes in Oxytocin, Stress, and Mood," *Psychology & Marketing*, 32, 977 – 986.

Appendix Tables

Table I: Areas Under the Standard Normal Curve

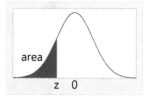

z	area	z	area	z	area	z	area	z	area	z	area	z	area	z	area
-3.99	0.000033	-3.49	0.00024	-2.99	0.0014	-2.49	0.0064	-1.99	0.0233	-1.49	0.0681	-0.99	0.1611	-0.49	0.3121
-3.98	0.000034	-3.48	0.00025	-2.98	0.0014	-2.48	0.0066	-1.98	0.0239	-1.48	0.0694	-0.98	0.1635	-0.48	0.3156
-3.97	0.000036	-3.47	0.00026	-2.97	0.0015	-2.47	0.0068	-1.97	0.0244	-1.47	0.0708	-0.97	0.1660	-0.47	0.3192
-3.96	0.000037	-3.46	0.00027	-2.96	0.0015	-2.46	0.0069	-1.96	0.0250	-1.46	0.0721	-0.96	0.1685	-0.46	0.3228
-3.95	0.000039	-3.45	0.00028	-2.95	0.0016	-2.45	0.0071	-1.95	0.0256	-1.45	0.0735	-0.95	0.1711	-0.45	0.3264
-3.94	0.000041	-3.44	0.00029	-2.94	0.0016	-2.44	0.0073	-1.94	0.0262	-1.44	0.0749	-0.94	0.1736	-0.44	0.3300
-3.93	0.000042	-3.43	0.00030	-2.93	0.0017	-2.43	0.0075	-1.93	0.0268	-1.43	0.0764	-0.93	0.1762	-0.43	0.3336
-3.92	0.000044	-3.42	0.00031	-2.92	0.0018	-2.42	0.0078	-1.92	0.0274	-1.42	0.0778	-0.92	0.1788	-0.42	0.3372
-3.91	0.000046	-3.41	0.00032	-2.91	0.0018	-2.41	0.0080	-1.91	0.0281	-1.41	0.0793	-0.91	0.1814	-0.41	0.3409
-3.90	0.000048	-3.40	0.00034	-2.90	0.0019	-2.40	0.0082	-1.90	0.0287	-1.40	0.0808	-0.90	0.1841	-0.40	0.3446
-3.89	0.000050	-3.39	0.00035	-2.89	0.0019	-2.39	0.0084	-1.89	0.0294	-1.39	0.0823	-0.89	0.1867	-0.39	0.3483
-3.88	0.000052	-3.38	0.00036	-2.88	0.0020	-2.38	0.0087	-1.88	0.0301	-1.38	0.0838	-0.88	0.1894	-0.38	0.3520
-3.87	0.000054	-3.37	0.00038	-2.87	0.0021	-2.37	0.0089	-1.87	0.0307	-1.37	0.0853	-0.87	0.1922	-0.37	0.3557
-3.86	0.000057	-3.36	0.00039	-2.86	0.0021	-2.36	0.0091	-1.86	0.0314	-1.36	0.0869	-0.86	0.1949	-0.36	0.3594
-3.85	0.000059	-3.35	0.00040	-2.85	0.0022	-2.35	0.0094	-1.85	0.0322	-1.35	0.0885	-0.85	0.1977	-0.35	0.3632
-3.84	0.000062	-3.34	0.00042	-2.84	0.0023	-2.34	0.0096	-1.84	0.0329	-1.34	0.0901	-0.84	0.2005	-0.34	0.3669
-3.83	0.000064	-3.33	0.00043	-2.83	0.0023	-2.33	0.0099	-1.83	0.0336	-1.33	0.0918	-0.83	0.2033	-0.33	0.3707
-3.82	0.000067	-3.32	0.00045	-2.82	0.0024	-2.32	0.0102	-1.82	0.0344	-1.32	0.0934	-0.82	0.2061	-0.32	0.3745
-3.81	0.000069	-3.31	0.00047	-2.81	0.0025	-2.31	0.0104	-1.81	0.0351	-1.31	0.0951	-0.81	0.2090	-0.31	0.3783
-3.80	0.000072	-3.30	0.00048	-2.80	0.0026	-2.30	0.0107	-1.80	0.0359	-1.30	0.0968	-0.80	0.2119	-0.30	0.3821
-3.79	0.000075	-3.29	0.00050	-2.79	0.0026	-2.29	0.0110	-1.79	0.0367	-1.29	0.0985	-0.79	0.2148	-0.29	0.3859
-3.78	0.000078	-3.28	0.00052	-2.78	0.0027	-2.28	0.0113	-1.78	0.0375	-1.28	0.1003	-0.78	0.2177	-0.28	0.3897
-3.77	0.000082	-3.27	0.00054	-2.77	0.0028	-2.27	0.0116	-1.77	0.0384	-1.27	0.1020	-0.77	0.2206	-0.27	0.3936
-3.76	0.000085	-3.26	0.00056	-2.76	0.0029	-2.26	0.0119	-1.76	0.0392	-1.26	0.1038	-0.76	0.2236	-0.26	0.3974
-3.75	0.000088	-3.25	0.00058	-2.75	0.0030	-2.25	0.0122	-1.75	0.0401	-1.25	0.1056	-0.75	0.2266	-0.25	0.4013
-3.74	0.000092	-3.24	0.00060	-2.74	0.0031	-2.24	0.0125	-1.74	0.0409	-1.24	0.1075	-0.74	0.2296	-0.24	0.4052
-3.73	0.000096	-3.23	0.00062	-2.73	0.0032	-2.23	0.0129	-1.73	0.0418	-1.23	0.1093	-0.73	0.2327	-0.23	0.4090
-3.72	0.000100	-3.22	0.00064	-2.72	0.0033	-2.22	0.0132	-1.72	0.0427	-1.22	0.1112	-0.72	0.2358	-0.22	0.4129
-3.71	0.000104	-3.21	0.00066	-2.71	0.0034	-2.21	0.0136	-1.71	0.0436	-1.21	0.1131	-0.71	0.2389	-0.21	0.4168
-3.70	0.000108	-3.20	0.00069	-2.70	0.0035	-2.20	0.0139	-1.70	0.0446	-1.20	0.1151	-0.70	0.2420	-0.20	0.4207
-3.69	0.000112	-3.19	0.00071	-2.69	0.0036	-2.19	0.0143	-1.69	0.0455	-1.19	0.1170	-0.69	0.2451	-0.19	0.4247
-3.68	0.000117	-3.18	0.00074	-2.68	0.0037	-2.18	0.0146	-1.68	0.0465	-1.18	0.1190	-0.68	0.2483	-0.18	0.4286
-3.67	0.000121	-3.17	0.00076	-2.67	0.0038	-2.17	0.0150	-1.67	0.0475	-1.17	0.1210	-0.67	0.2514	-0.17	0.4325
-3.66	0.000126	-3.16	0.00079	-2.66	0.0039	-2.16	0.0154	-1.66	0.0485	-1.16	0.1230	-0.66	0.2546	-0.16	0.4364
-3.65	0.000131	-3.15	0.00082	-2.65	0.0040	-2.15	0.0158	-1.65	0.0495	-1.15	0.1251	-0.65	0.2578	-0.15	0.4404
-3.64	0.000136	-3.14	0.00084	-2.64	0.0041	-2.14	0.0162	-1.64	0.0505	-1.14	0.1271	-0.64	0.2611	-0.14	0.4443
-3.63	0.000142	-3.13	0.00087	-2.63	0.0043	-2.13	0.0166	-1.63	0.0516	-1.13	0.1292	-0.63	0.2643	-0.13	0.4483
-3.62	0.000147	-3.12	0.00090	-2.62	0.0044	-2.12	0.0170	-1.62	0.0526	-1.12	0.1314	-0.62	0.2676	-0.12	0.4522
-3.61	0.000153	-3.11	0.00094	-2.61	0.0045	-2.11	0.0174	-1.61	0.0537	-1.11	0.1335	-0.61	0.2709	-0.11	0.4562
-3.60	0.000159	-3.10	0.00097	-2.60	0.0047	-2.10	0.0179	-1.60	0.0548	-1.10	0.1357	-0.60	0.2743	-0.10	0.4602
-3.59	0.000165	-3.09	0.00100	-2.59	0.0048	-2.09	0.0183	-1.59	0.0559	-1.09	0.1379	-0.59	0.2776	-0.09	0.4641
-3.58	0.000172	-3.08	0.00104	-2.58	0.0049	-2.08	0.0188	-1.58	0.0571	-1.08	0.1401	-0.58	0.2810	-0.08	0.4681
-3.57	0.000178	-3.07	0.00107	-2.57	0.0051	-2.07	0.0192	-1.57	0.0582	-1.07	0.1423	-0.57	0.2843	-0.07	0.4721
-3.56	0.000185	-3.06	0.00111	-2.56	0.0052	-2.06	0.0197	-1.56	0.0594	-1.06	0.1446	-0.56	0.2877	-0.06	0.4761
-3.55	0.000193	-3.05	0.00114	-2.55	0.0054	-2.05	0.0202	-1.55	0.0606	-1.05	0.1469	-0.55	0.2912	-0.05	0.4801
-3.54	0.000200	-3.04	0.00118	-2.54	0.0055	-2.04	0.0207	-1.54	0.0618	-1.04	0.1492	-0.54	0.2946	-0.04	0.4840
-3.53	0.000208	-3.03	0.00122	-2.53	0.0057	-2.03	0.0212	-1.53	0.0630	-1.03	0.1515	-0.53	0.2981	-0.03	0.4880
-3.52	0.000216	-3.02	0.00126	-2.52	0.0059	-2.02	0.0217	-1.52	0.0643	-1.02	0.1539	-0.52	0.3015	-0.02	0.4920
-3.51	0.000224	-3.01	0.00131	-2.51	0.0060	-2.01	0.0222	-1.51	0.0655	-1.01	0.1562	-0.51	0.3050	-0.01	0.4960
-3.50	0.000233	-3.00	0.00135	-2.50	0.0062	-2.00	0.0228	-1.50	0.0668	-1.00	0.1587	-0.50	0.3085	0.00	0.5000

Table I: Areas Under the Standard Normal Curve

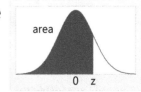

area

0 z

z	area	z	area	z	area	z	area	z	area	z	area	z	area	z	area
0.00	0.5000	0.50	0.6915	1.00	0.8413	1.50	0.9332	2.00	0.9772	2.50	0.9938	3.00	0.99865	3.50	0.999767
0.01	0.5040	0.51	0.6950	1.01	0.8438	1.51	0.9345	2.01	0.9778	2.51	0.9940	3.01	0.99869	3.51	0.999776
0.02	0.5080	0.52	0.6985	1.02	0.8461	1.52	0.9357	2.02	0.9783	2.52	0.9941	3.02	0.99874	3.52	0.999784
0.03	0.5120	0.53	0.7019	1.03	0.8485	1.53	0.9370	2.03	0.9788	2.53	0.9943	3.03	0.99878	3.53	0.999792
0.04	0.5160	0.54	0.7054	1.04	0.8508	1.54	0.9382	2.04	0.9793	2.54	0.9945	3.04	0.99882	3.54	0.999800
0.05	0.5199	0.55	0.7088	1.05	0.8531	1.55	0.9394	2.05	0.9798	2.55	0.9946	3.05	0.99886	3.55	0.999807
0.06	0.5239	0.56	0.7123	1.06	0.8554	1.56	0.9406	2.06	0.9803	2.56	0.9948	3.06	0.99889	3.56	0.999815
0.07	0.5279	0.57	0.7157	1.07	0.8577	1.57	0.9418	2.07	0.9808	2.57	0.9949	3.07	0.99893	3.57	0.999822
0.08	0.5319	0.58	0.7190	1.08	0.8599	1.58	0.9429	2.08	0.9812	2.58	0.9951	3.08	0.99896	3.58	0.999828
0.09	0.5359	0.59	0.7224	1.09	0.8621	1.59	0.9441	2.09	0.9817	2.59	0.9952	3.09	0.99900	3.59	0.999835
0.10	0.5398	0.60	0.7257	1.10	0.8643	1.60	0.9452	2.10	0.9821	2.60	0.9953	3.10	0.99903	3.60	0.999841
0.11	0.5438	0.61	0.7291	1.11	0.8665	1.61	0.9463	2.11	0.9826	2.61	0.9955	3.11	0.99906	3.61	0.999847
0.12	0.5478	0.62	0.7324	1.12	0.8686	1.62	0.9474	2.12	0.9830	2.62	0.9956	3.12	0.99910	3.62	0.999853
0.13	0.5517	0.63	0.7357	1.13	0.8708	1.63	0.9484	2.13	0.9834	2.63	0.9957	3.13	0.99913	3.63	0.999858
0.14	0.5557	0.64	0.7389	1.14	0.8729	1.64	0.9495	2.14	0.9838	2.64	0.9959	3.14	0.99916	3.64	0.999864
0.15	0.5596	0.65	0.7422	1.15	0.8749	1.65	0.9505	2.15	0.9842	2.65	0.9960	3.15	0.99918	3.65	0.999869
0.16	0.5636	0.66	0.7454	1.16	0.8770	1.66	0.9515	2.16	0.9846	2.66	0.9961	3.16	0.99921	3.66	0.999874
0.17	0.5675	0.67	0.7486	1.17	0.8790	1.67	0.9525	2.17	0.9850	2.67	0.9962	3.17	0.99924	3.67	0.999879
0.18	0.5714	0.68	0.7517	1.18	0.8810	1.68	0.9535	2.18	0.9854	2.68	0.9963	3.18	0.99926	3.68	0.999883
0.19	0.5753	0.69	0.7549	1.19	0.8830	1.69	0.9545	2.19	0.9857	2.69	0.9964	3.19	0.99929	3.69	0.999888
0.20	0.5793	0.70	0.7580	1.20	0.8849	1.70	0.9554	2.20	0.9861	2.70	0.9965	3.20	0.99931	3.70	0.999892
0.21	0.5832	0.71	0.7611	1.21	0.8869	1.71	0.9564	2.21	0.9864	2.71	0.9966	3.21	0.99934	3.71	0.999896
0.22	0.5871	0.72	0.7642	1.22	0.8888	1.72	0.9573	2.22	0.9868	2.72	0.9967	3.22	0.99936	3.72	0.999900
0.23	0.5910	0.73	0.7673	1.23	0.8907	1.73	0.9582	2.23	0.9871	2.73	0.9968	3.23	0.99938	3.73	0.999904
0.24	0.5948	0.74	0.7704	1.24	0.8925	1.74	0.9591	2.24	0.9875	2.74	0.9969	3.24	0.99940	3.74	0.999908
0.25	0.5987	0.75	0.7734	1.25	0.8944	1.75	0.9599	2.25	0.9878	2.75	0.9970	3.25	0.99942	3.75	0.999912
0.26	0.6026	0.76	0.7764	1.26	0.8962	1.76	0.9608	2.26	0.9881	2.76	0.9971	3.26	0.99944	3.76	0.999915
0.27	0.6064	0.77	0.7794	1.27	0.8980	1.77	0.9616	2.27	0.9884	2.77	0.9972	3.27	0.99946	3.77	0.999918
0.28	0.6103	0.78	0.7823	1.28	0.8997	1.78	0.9625	2.28	0.9887	2.78	0.9973	3.28	0.99948	3.78	0.999922
0.29	0.6141	0.79	0.7852	1.29	0.9015	1.79	0.9633	2.29	0.9890	2.79	0.9974	3.29	0.99950	3.79	0.999925
0.30	0.6179	0.80	0.7881	1.30	0.9032	1.80	0.9641	2.30	0.9893	2.80	0.9974	3.30	0.99952	3.80	0.999928
0.31	0.6217	0.81	0.7910	1.31	0.9049	1.81	0.9649	2.31	0.9896	2.81	0.9975	3.31	0.99953	3.81	0.999931
0.32	0.6255	0.82	0.7939	1.32	0.9066	1.82	0.9656	2.32	0.9898	2.82	0.9976	3.32	0.99955	3.82	0.999933
0.33	0.6293	0.83	0.7967	1.33	0.9082	1.83	0.9664	2.33	0.9901	2.83	0.9977	3.33	0.99957	3.83	0.999936
0.34	0.6331	0.84	0.7995	1.34	0.9099	1.84	0.9671	2.34	0.9904	2.84	0.9977	3.34	0.99958	3.84	0.999938
0.35	0.6368	0.85	0.8023	1.35	0.9115	1.85	0.9678	2.35	0.9906	2.85	0.9978	3.35	0.99960	3.85	0.999941
0.36	0.6406	0.86	0.8051	1.36	0.9131	1.86	0.9686	2.36	0.9909	2.86	0.9979	3.36	0.99961	3.86	0.999943
0.37	0.6443	0.87	0.8078	1.37	0.9147	1.87	0.9693	2.37	0.9911	2.87	0.9979	3.37	0.99962	3.87	0.999946
0.38	0.6480	0.88	0.8106	1.38	0.9162	1.88	0.9699	2.38	0.9913	2.88	0.9980	3.38	0.99964	3.88	0.999948
0.39	0.6517	0.89	0.8133	1.39	0.9177	1.89	0.9706	2.39	0.9916	2.89	0.9981	3.39	0.99965	3.89	0.999950
0.40	0.6554	0.90	0.8159	1.40	0.9192	1.90	0.9713	2.40	0.9918	2.90	0.9981	3.40	0.99966	3.90	0.999952
0.41	0.6591	0.91	0.8186	1.41	0.9207	1.91	0.9719	2.41	0.9920	2.91	0.9982	3.41	0.99968	3.91	0.999954
0.42	0.6628	0.92	0.8212	1.42	0.9222	1.92	0.9726	2.42	0.9922	2.92	0.9982	3.42	0.99969	3.92	0.999956
0.43	0.6664	0.93	0.8238	1.43	0.9236	1.93	0.9732	2.43	0.9925	2.93	0.9983	3.43	0.99970	3.93	0.999958
0.44	0.6700	0.94	0.8264	1.44	0.9251	1.94	0.9738	2.44	0.9927	2.94	0.9984	3.44	0.99971	3.94	0.999959
0.45	0.6736	0.95	0.8289	1.45	0.9265	1.95	0.9744	2.45	0.9929	2.95	0.9984	3.45	0.99972	3.95	0.999961
0.46	0.6772	0.96	0.8315	1.46	0.9279	1.96	0.9750	2.46	0.9931	2.96	0.9985	3.46	0.99973	3.96	0.999963
0.47	0.6808	0.97	0.8340	1.47	0.9292	1.97	0.9756	2.47	0.9932	2.97	0.9985	3.47	0.99974	3.97	0.999964
0.48	0.6844	0.98	0.8365	1.48	0.9306	1.98	0.9761	2.48	0.9934	2.98	0.9986	3.48	0.99975	3.98	0.999966
0.49	0.6879	0.99	0.8389	1.49	0.9319	1.99	0.9767	2.49	0.9936	2.99	0.9986	3.49	0.99976	3.99	0.999967

Table II: Areas Under the t-Curve

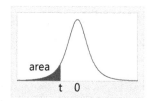

Values of t are inside the table.

df	0.0005	0.001	0.005	0.01	0.025	0.05	0.1	0.5	0.9	0.95	0.975	0.99	0.995	0.999	0.9995
1	-636.62	-318.31	-63.66	-31.82	-12.71	-6.31	-3.08	0	3.08	6.31	12.71	31.82	63.66	318.31	636.62
2	-31.60	-22.33	-9.92	-6.96	-4.30	-2.92	-1.89	0	1.89	2.92	4.30	6.96	9.92	22.33	31.60
3	-12.92	-10.21	-5.84	-4.54	-3.18	-2.35	-1.64	0	1.64	2.35	3.18	4.54	5.84	10.21	12.92
4	-8.61	-7.17	-4.60	-3.75	-2.78	-2.13	-1.53	0	1.53	2.13	2.78	3.75	4.60	7.17	8.61
5	-6.87	-5.89	-4.03	-3.36	-2.57	-2.02	-1.48	0	1.48	2.02	2.57	3.36	4.03	5.89	6.87
6	-5.96	-5.21	-3.71	-3.14	-2.45	-1.94	-1.44	0	1.44	1.94	2.45	3.14	3.71	5.21	5.96
7	-5.41	-4.79	-3.50	-3.00	-2.36	-1.89	-1.41	0	1.41	1.89	2.36	3.00	3.50	4.79	5.41
8	-5.04	-4.50	-3.36	-2.90	-2.31	-1.86	-1.40	0	1.40	1.86	2.31	2.90	3.36	4.50	5.04
9	-4.78	-4.30	-3.25	-2.82	-2.26	-1.83	-1.38	0	1.38	1.83	2.26	2.82	3.25	4.30	4.78
10	-4.59	-4.14	-3.17	-2.76	-2.23	-1.81	-1.37	0	1.37	1.81	2.23	2.76	3.17	4.14	4.59
11	-4.44	-4.02	-3.11	-2.72	-2.20	-1.80	-1.36	0	1.36	1.80	2.20	2.72	3.11	4.02	4.44
12	-4.32	-3.93	-3.05	-2.68	-2.18	-1.78	-1.36	0	1.36	1.78	2.18	2.68	3.05	3.93	4.32
13	-4.22	-3.85	-3.01	-2.65	-2.16	-1.77	-1.35	0	1.35	1.77	2.16	2.65	3.01	3.85	4.22
14	-4.14	-3.79	-2.98	-2.62	-2.14	-1.76	-1.35	0	1.35	1.76	2.14	2.62	2.98	3.79	4.14
15	-4.07	-3.73	-2.95	-2.60	-2.13	-1.75	-1.34	0	1.34	1.75	2.13	2.60	2.95	3.73	4.07
16	-4.01	-3.69	-2.92	-2.58	-2.12	-1.75	-1.34	0	1.34	1.75	2.12	2.58	2.92	3.69	4.01
17	-3.97	-3.65	-2.90	-2.57	-2.11	-1.74	-1.33	0	1.33	1.74	2.11	2.57	2.90	3.65	3.97
18	-3.92	-3.61	-2.88	-2.55	-2.10	-1.73	-1.33	0	1.33	1.73	2.10	2.55	2.88	3.61	3.92
19	-3.88	-3.58	-2.86	-2.54	-2.09	-1.73	-1.33	0	1.33	1.73	2.09	2.54	2.86	3.58	3.88
20	-3.85	-3.55	-2.85	-2.53	-2.09	-1.72	-1.33	0	1.33	1.72	2.09	2.53	2.85	3.55	3.85
21	-3.82	-3.53	-2.83	-2.52	-2.08	-1.72	-1.32	0	1.32	1.72	2.08	2.52	2.83	3.53	3.82
22	-3.79	-3.50	-2.82	-2.51	-2.07	-1.72	-1.32	0	1.32	1.72	2.07	2.51	2.82	3.50	3.79
23	-3.77	-3.48	-2.81	-2.50	-2.07	-1.71	-1.32	0	1.32	1.71	2.07	2.50	2.81	3.48	3.77
24	-3.75	-3.47	-2.80	-2.49	-2.06	-1.71	-1.32	0	1.32	1.71	2.06	2.49	2.80	3.47	3.75
25	-3.73	-3.45	-2.79	-2.49	-2.06	-1.71	-1.32	0	1.32	1.71	2.06	2.49	2.79	3.45	3.73
26	-3.71	-3.43	-2.78	-2.48	-2.06	-1.71	-1.31	0	1.31	1.71	2.06	2.48	2.78	3.43	3.71
27	-3.69	-3.42	-2.77	-2.47	-2.05	-1.70	-1.31	0	1.31	1.70	2.05	2.47	2.77	3.42	3.69
28	-3.67	-3.41	-2.76	-2.47	-2.05	-1.70	-1.31	0	1.31	1.70	2.05	2.47	2.76	3.41	3.67
29	-3.66	-3.40	-2.76	-2.46	-2.05	-1.70	-1.31	0	1.31	1.70	2.05	2.46	2.76	3.40	3.66
30	-3.65	-3.39	-2.75	-2.46	-2.04	-1.70	-1.31	0	1.31	1.70	2.04	2.46	2.75	3.39	3.65
31	-3.63	-3.37	-2.74	-2.45	-2.04	-1.70	-1.31	0	1.31	1.70	2.04	2.45	2.74	3.37	3.63
32	-3.62	-3.37	-2.74	-2.45	-2.04	-1.69	-1.31	0	1.31	1.69	2.04	2.45	2.74	3.37	3.62
33	-3.61	-3.36	-2.73	-2.44	-2.03	-1.69	-1.31	0	1.31	1.69	2.03	2.44	2.73	3.36	3.61
34	-3.60	-3.35	-2.73	-2.44	-2.03	-1.69	-1.31	0	1.31	1.69	2.03	2.44	2.73	3.35	3.60
35	-3.59	-3.34	-2.72	-2.44	-2.03	-1.69	-1.31	0	1.31	1.69	2.03	2.44	2.72	3.34	3.59
36	-3.58	-3.33	-2.72	-2.43	-2.03	-1.69	-1.31	0	1.31	1.69	2.03	2.43	2.72	3.33	3.58
37	-3.57	-3.33	-2.72	-2.43	-2.03	-1.69	-1.30	0	1.30	1.69	2.03	2.43	2.72	3.33	3.57
38	-3.57	-3.32	-2.71	-2.43	-2.02	-1.69	-1.30	0	1.30	1.69	2.02	2.43	2.71	3.32	3.57
39	-3.56	-3.31	-2.71	-2.43	-2.02	-1.68	-1.30	0	1.30	1.68	2.02	2.43	2.71	3.31	3.56
40	-3.55	-3.31	-2.70	-2.42	-2.02	-1.68	-1.30	0	1.30	1.68	2.02	2.42	2.70	3.31	3.55

Table II: Areas Under the t-Curve

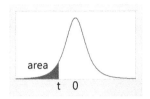

area
t   0

Values of t are inside the table.

							Area								
df	0.0005	0.001	0.005	0.01	0.025	0.05	0.1	0.5	0.9	0.95	0.975	0.99	0.995	0.999	0.9995
41	-3.54	-3.30	-2.70	-2.42	-2.02	-1.68	-1.30	0	1.30	1.68	2.02	2.42	2.70	3.30	3.54
42	-3.54	-3.30	-2.70	-2.42	-2.02	-1.68	-1.30	0	1.30	1.68	2.02	2.42	2.70	3.30	3.54
43	-3.53	-3.29	-2.70	-2.42	-2.02	-1.68	-1.30	0	1.30	1.68	2.02	2.42	2.70	3.29	3.53
44	-3.53	-3.29	-2.69	-2.41	-2.02	-1.68	-1.30	0	1.30	1.68	2.02	2.41	2.69	3.29	3.53
45	-3.52	-3.28	-2.69	-2.41	-2.01	-1.68	-1.30	0	1.30	1.68	2.01	2.41	2.69	3.28	3.52
46	-3.51	-3.28	-2.69	-2.41	-2.01	-1.68	-1.30	0	1.30	1.68	2.01	2.41	2.69	3.28	3.51
47	-3.51	-3.27	-2.68	-2.41	-2.01	-1.68	-1.30	0	1.30	1.68	2.01	2.41	2.68	3.27	3.51
48	-3.51	-3.27	-2.68	-2.41	-2.01	-1.68	-1.30	0	1.30	1.68	2.01	2.41	2.68	3.27	3.51
49	-3.50	-3.27	-2.68	-2.40	-2.01	-1.68	-1.30	0	1.30	1.68	2.01	2.40	2.68	3.27	3.50
50	-3.50	-3.26	-2.68	-2.40	-2.01	-1.68	-1.30	0	1.30	1.68	2.01	2.40	2.68	3.26	3.50
51	-3.49	-3.26	-2.68	-2.40	-2.01	-1.68	-1.30	0	1.30	1.68	2.01	2.40	2.68	3.26	3.49
52	-3.49	-3.25	-2.67	-2.40	-2.01	-1.67	-1.30	0	1.30	1.67	2.01	2.40	2.67	3.25	3.49
53	-3.48	-3.25	-2.67	-2.40	-2.01	-1.67	-1.30	0	1.30	1.67	2.01	2.40	2.67	3.25	3.48
54	-3.48	-3.25	-2.67	-2.40	-2.00	-1.67	-1.30	0	1.30	1.67	2.00	2.40	2.67	3.25	3.48
55	-3.48	-3.25	-2.67	-2.40	-2.00	-1.67	-1.30	0	1.30	1.67	2.00	2.40	2.67	3.25	3.48
56	-3.47	-3.24	-2.67	-2.39	-2.00	-1.67	-1.30	0	1.30	1.67	2.00	2.39	2.67	3.24	3.47
57	-3.47	-3.24	-2.66	-2.39	-2.00	-1.67	-1.30	0	1.30	1.67	2.00	2.39	2.66	3.24	3.47
58	-3.47	-3.24	-2.66	-2.39	-2.00	-1.67	-1.30	0	1.30	1.67	2.00	2.39	2.66	3.24	3.47
59	-3.46	-3.23	-2.66	-2.39	-2.00	-1.67	-1.30	0	1.30	1.67	2.00	2.39	2.66	3.23	3.46
60	-3.46	-3.23	-2.66	-2.39	-2.00	-1.67	-1.30	0	1.30	1.67	2.00	2.39	2.66	3.23	3.46
65	-3.45	-3.22	-2.65	-2.39	-2.00	-1.67	-1.29	0	1.29	1.67	2.00	2.39	2.65	3.22	3.45
70	-3.44	-3.21	-2.65	-2.38	-1.99	-1.67	-1.29	0	1.29	1.67	1.99	2.38	2.65	3.21	3.44
75	-3.43	-3.20	-2.64	-2.38	-1.99	-1.67	-1.29	0	1.29	1.67	1.99	2.38	2.64	3.20	3.43
80	-3.42	-3.20	-2.64	-2.37	-1.99	-1.66	-1.29	0	1.29	1.66	1.99	2.37	2.64	3.20	3.42
85	-3.41	-3.19	-2.63	-2.37	-1.99	-1.66	-1.29	0	1.29	1.66	1.99	2.37	2.63	3.19	3.41
90	-3.40	-3.18	-2.63	-2.37	-1.99	-1.66	-1.29	0	1.29	1.66	1.99	2.37	2.63	3.18	3.40
95	-3.40	-3.18	-2.63	-2.37	-1.99	-1.66	-1.29	0	1.29	1.66	1.99	2.37	2.63	3.18	3.40
100	-3.39	-3.17	-2.63	-2.36	-1.98	-1.66	-1.29	0	1.29	1.66	1.98	2.36	2.63	3.17	3.39
110	-3.38	-3.17	-2.62	-2.36	-1.98	-1.66	-1.29	0	1.29	1.66	1.98	2.36	2.62	3.17	3.38
120	-3.37	-3.16	-2.62	-2.36	-1.98	-1.66	-1.29	0	1.29	1.66	1.98	2.36	2.62	3.16	3.37
130	-3.37	-3.15	-2.61	-2.36	-1.98	-1.66	-1.29	0	1.29	1.66	1.98	2.36	2.61	3.15	3.37
140	-3.36	-3.15	-2.61	-2.35	-1.98	-1.66	-1.29	0	1.29	1.66	1.98	2.35	2.61	3.15	3.36
150	-3.36	-3.15	-2.61	-2.35	-1.98	-1.66	-1.29	0	1.29	1.66	1.98	2.35	2.61	3.15	3.36
160	-3.35	-3.14	-2.61	-2.35	-1.97	-1.65	-1.29	0	1.29	1.65	1.97	2.35	2.61	3.14	3.35
170	-3.35	-3.14	-2.61	-2.35	-1.97	-1.65	-1.29	0	1.29	1.65	1.97	2.35	2.61	3.14	3.35
180	-3.35	-3.14	-2.60	-2.35	-1.97	-1.65	-1.29	0	1.29	1.65	1.97	2.35	2.60	3.14	3.35
190	-3.34	-3.13	-2.60	-2.35	-1.97	-1.65	-1.29	0	1.29	1.65	1.97	2.35	2.60	3.13	3.34
200	-3.34	-3.13	-2.60	-2.35	-1.97	-1.65	-1.29	0	1.29	1.65	1.97	2.35	2.60	3.13	3.34
300	-3.32	-3.12	-2.59	-2.34	-1.97	-1.65	-1.28	0	1.28	1.65	1.97	2.34	2.59	3.12	3.32
400	-3.32	-3.11	-2.59	-2.34	-1.97	-1.65	-1.28	0	1.28	1.65	1.97	2.34	2.59	3.11	3.32
500	-3.31	-3.11	-2.59	-2.33	-1.96	-1.65	-1.28	0	1.28	1.65	1.96	2.33	2.59	3.11	3.31
1000	-3.30	-3.10	-2.58	-2.33	-1.96	-1.65	-1.28	0	1.28	1.65	1.96	2.33	2.58	3.10	3.30
∞	-3.49	-3.26	-2.68	-2.40	-2.01	-1.68	-1.30	0	1.30	1.68	2.01	2.40	2.68	3.26	3.49

Answers to Odd-Numbered Exercises

Chapter 1

1. a. numeric b. numeric c. categorical d. categorical e. numeric **3.** objects are refrigerators; variables are make (categorical), model (categorical), price (numeric), size (numeric), energy cost (numeric), freezer type (categorical) **5.** objects are properties currently for sale; variables are asking price (numeric), value estimate (numeric), house size (numeric), number of BRs (numeric), number of BAs (numeric), heating type (categorical), central air (categorical), lot size (numeric) **7.** subjects are babies born; variables are T18 status (categorical), Down's status (categorical) **9.** objects are galaxies; variables are distance from us (numeric), speed (numeric) **11.** subjects are asthma patients; variables are treatment group (categorical), peak flow rate (numeric) **13.** a. false; subjects are the U.S. adults polled b. false; the variable is the characteristic of a single adult polled (their opinion about the President's job) c. true

Section 2.1

1. a.

Marital Status	Count	%
Divorced	6	15.0
Married	11	27.5
Separated	5	12.5
Single	17	42.5
Widowed	1	2.5
Total	40	100.0

b. 60%

c.

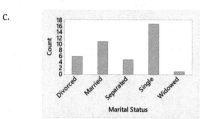

d.

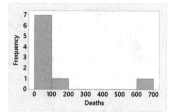

3. a. subjects are people who died in 2016; variable is cause of death b. 2,744,248 rows, one for each person

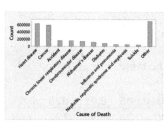

who died; one column for the cause of death variable c. i. 23.1% ii. 44.9% iii. 25.9% d. 0.485 e. a bar graph is shown here

5. a.

Ethnicity	Count	%
White	115	70.1
Asian or Pacific Islander	9	5.5
Black	24	14.6
Hispanic	13	7.9
Other	3	1.8
Total	164	100.0

b. subjects are asthma patients c. ethnicity **7.** a. objects are aircrafts/flights in Feb 2019 b. percentages are rounded c. 74.43, 5.91, 0.82, 7.57, 0.04, 8.07, 2.86, 0.30, sum is 100.00 d.

Reason for Delay	Count	%
Air carrier delay	31,505	23.1
Weather delay	4,356	3.2
National Aviation System delay	40,374	29.6
Security delay	220	0.2
Aircraft arriving late	43,022	31.6
Cancelled	15,255	11.2
Diverted	1,606	1.2
Total	136,338	100.0

Section 2.2

1. a.

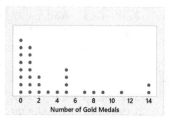

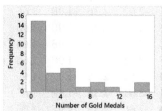

b. histogram is shown here c. 3.433, 1.5 d. since these countries are all those earning medals, use μ and η for the mean and median

3. a. symmetric, bell shaped b. uniform c. right skewed d. bimodal
5. a. objects are earthquakes; variable is number of deaths b. 9 c. too few intervals d. histogram is shown

here e. mean would be higher than median, $\bar{x} = 119.3$, $m = 58.0$ **7.** 66.25 **9.** a. right skewed

b. histogram is shown here c. mean would decrease d. median would likely decrease (because the location of the middle value will change) but not

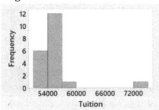

by as much e. with ACAST: $\bar{x} = 55490, m = 54314$; without ACAST: $\bar{x} = 54489, m = 55310$ **11.** a. 11 b. 500 c. 4% d. median e. mean **13.** a. I largest, II smallest b. I: $s^2 = 25, s = 5, range = 10$ II: $s^2 = 0$, $s = 0, range = 0$ III: $s^2 = 1, s = 1, range = 2$ c. $\bar{x} = 8, s^2 = 1, range = 2$

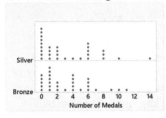

15. dotplots are shown here; The distribution of the number of silver medals is slightly more variable than that of the number of bronze medals, but with a similar center. Both distributions are right skewed. **21.** a. 6 b. 3/4 are women **23.** a. 0 and 80, 30 and 50, -1 and -2 b. 40 and 40, 38 and 42 **25.** a. about 91% of the math scores are below 680 b. $z = -0.30$; The student's score of 480 is 0.30 standard deviations below the mean math score. c. scores below 283 and above 747 d. no **27.** a. 3, 20, 56 b. 3, 5, 22 c. 53, 19 d. $z = 1.57$; The player's number of fouls committed is 1.57 standard deviations above the mean number of fouls committed. e. $\bar{x} = 31.09, s = 32.93$, $m = 20, range = 100, IQR = 53$ **29.** a. -1.8, 7.304 b. $z = 0.84$; The stock screener's return of 5.75% is 0.84 standard deviations above the mean return. e. $z = -1.65$; The stock screener's return of -3% is 1.65 standard deviations below the mean return. **31.** a. close to 5,000,000 b. higher because the distribution is right skewed **33.** a. -3.79, 10.05 b. -7.47, -6.52, 10.65

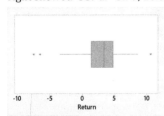

c. boxplot is shown here d. -2.92, -2.65, 2.23; yes **35.** a. years b. number of Atlantic storms c. 17 d. 19 e. 14, 19 f. 75 g. none; upper fence is 26.5

Section 2.3

1. a. Both distributions of full-time faculty rates are left skewed. The rates for the Bachelor's schools have a higher center (mean) and are less variable (lower SD) than for schools granting graduate degrees. **3.** a. 11 b. 12 c. 6 **5.** a. In order, the medians are $36,000, $4,300, $62,000, $16,500, $60,900. b. In order, the age

groups are 40-49, 20-29, 60-69, 30-39, 50-59. **7.** a. parental role (categorical), amount of time spent with children (categorical) b. 51.9% c. 24.7% d. 47.6% e. A higher fraction of the fathers thought they spent too little time with their children (47.6%) compared to the mothers (25.4%). Smaller proportions of the fathers thought they spent the right amount (50.4%) and too much time (2.0%) compared to the mothers: 66.6% and 8.0%, respectively. f. if there would be the same number of fathers and mothers in the data set **9.** a. 77.2% b. 9.1% c. 22.6% d. The lowest fraction of preterm twin births (40.0%) was in the group of mothers who received less than adequate prenatal care. 53.8% of twin mothers who received intensive prenatal care had preterm births. 57.7% of twin mothers who received adequate prenatal care had preterm births. e. segmented bar chart is shown here f. 301,623 in Table 1 vs. 278,511 in Table 2

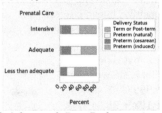

11. 1.86% **13.** a. graph A b. graph B c. B; the y-axis starts at 0 d. The minumums show the worst result on the posttest compared to the pretest. In the no assistance group, the worst result was 3 points lower on the postest; in the check figures group, the worst result was the same on both tests; in the complete solution group, the worst result was 5 points lower on the posttest. e. The check figures group had the highest mean score difference, followed by the no assistance group and the complete solution group, in order. Using the medians shows the no assistance group with the highest median score difference of 3 compared to 2 for the other two groups. The complete solutions group had the most variability in the score differences, then the no assistance group, and the check figures group had the least variability (ranges were 13, 10, and 7, respectively). f. Shape is tough to determine without a graph. Using the relationship between the mean and median in each group, we might suspect the no assistance differences to be left skewed, the check figures differences to be right skewed, and the complete solution differences to be fairly symmetric.

15. a.

	$\bar{x}$	m	s	IQR
Jill	238,510	227,300	121,340	212,225
Anya	243,740	236,500	121,703	212,650

b. boxplots are shown here; No, these boxplots don't show much difference in the valuations by Jill and Anya.

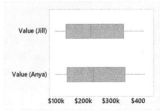

d. $\bar{x} = -\$5230, m = -\$4700, s = \$2656, IQR = \$4525$

e. boxplots is shown here; Yes! The distribution of differences shows that Anya appraised every property higher than Jill.

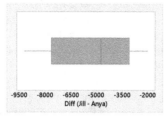

f. Anya's subtly higher valuations are obscured by the relatively high variability of the house values, which dominates the graph of both distributions separately. Taking the differences allows a more precise comparison of Jill's and Anya's valuations on the same properties.

Section 2.4

1. a. very strong (perfect), positive, linear b. strong, positive, linear c. weak-to-no relationship d. positive, non-linear **3.** A: strong, positive, linear B: weak-to-no relationship C: moderate, negative, linear D: moderate, positive, linear **5.** a. plot is shown here b. 0.99977 **7.** a. $\hat{y} = 0.19x + 38.64$ b. A player with one more stolen base is predicted to have been caught stealing

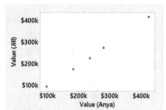

0.19 more times. c. 115 d. Henderson's number of stolen bases is well above the range of the data used to obtain the least-squares line; the linear pattern may not hold in that range. **9.** a. x is number of ads per capita, y is percentage of registered voters who voted b. This interpretation implies that more ads *caused* higher voter turnout. Since the number of ads per capita was not randomly assigned to the counties, a causal inference like this is not justified by the data.

11. a. fairly strong, negative, linear relationship b. $\hat{y} = -4.95x + 86.7$ c. A country having a births per woman rate of 1 higher is predicted to have a

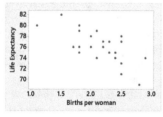

life expectancy of 4.95 years lower. d. The headline implies that a higher births per woman rate *causes* higher life expectancy. Since birth rates cannot be randomly assigned to the countries, a causal inference like this is not justified by the data. e. 65.4 years; The prediction is an extrapolation of both kinds: 1) This birth rate is well above those observed in the data used to obtain the least-squares line so the linear pattern may not hold in that range. 2) Unlike all the countries in the data set, Iraq is not a country in the Western Hemisphere. Perhaps there is a different relationship between life expectancy and birth rates for Middle Eastern countries. **13.** a. 89.37 minutes b. The

relationship is very strong (almost perfect), positive and linear, like we'd expect for the same variable expressed in two different units. c. DIM was calculated as DIS/60 and then rounded off a bit. **15.** a. A player with one more foul is predicted to have 2.839 more points. b. No, such a conclusion implies that more fouls committed *causes* more points to be scored. Since the number of fouls cannot be randomly assigned to players, a causal inference like this is not justified by the data.

19. a. The revenue axis does not start at 0 which distorts the quarterly changes. b. plot is shown here c. $3.244 billion

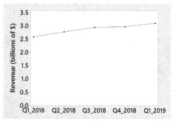

Section 3.1

1. pose a question, make a hypothesis, collect data, analyze the data, make a conclusion **3.** Predictive inference: an estimate/prediction of a characteristic of a single object. Generalization inference: a statement about characteristics of a large groups of objects. Causal inference: a claim about the efficacy of an intervention. **5.** a. predictive b. generalization c. generalization d. causal e. generalization f. causal g. predictive

7. a. How does the skill level of Maryland drivers compare with that of drivers in other states? (generalization inference) b. Does not wearing a coat in cold weather increase the risk of getting a cold? (causal inference) c. Will XYZ earnings be higher next quarter? (predictive inference)

Section 3.2

1. a. The population is all people in the U.S. who use alternative medical treatments. The sample is your 10 friends. The variable is status of condition (improved or not). b. The population is all bank transactions in a day. The sample is the 100 transactions at the end of Friday. The variable is status of the transaction (processed correctly or not). c. The population is all parents in the district. The sample is the 678 parents who responded to the online poll. The variable is opinion of the policy (favor, oppose). d. The population is all bags of chips produced in an hour. The sample is the 30 bags selected. The variable is the weight of a bag.

3. Number the farmers from 1 to 82. Use a random number generator to select 10 integers without replacement from 1 to 82. The farmers with those numbers constitute the SRS. **5.** a. While it's possible that a SRS would not have any students in a certain class (e.g. no seniors), a STRS would guarantee that the sample would include students from each class, allowing a more precise way to estimate the proportion who own their own car. b. Number the students from 1 to 34,000. Start with the freshmen and end with the seniors. Use a random number generator to select 100 integers without replacement from 1 to 10,000. The freshmen with those numbers are included in the sample. Use a random number generator to select 90 integers without replacement from 10,001 to 19,000. The sophomores with those numbers are included in the sample. Use a random number generator to select 80 integers without replacement from 19,001 to 27,000. The juniors with those numbers are included in the sample. Use a random number generator to select 70 integers without replacement from 27,001 to 34,000. The seniors with those numbers are included in the sample. **7.** a. Compiling a list of all people in the country in order to select a SRS would be very difficult. It would be easier to compile a list of addresses or households. b. Let's say there are 200 million households in the U.S. Number them from 1 to 200 million. Use a random number generator to select 1000 integers without replacement from 1 to 200 million. The people living in the households with those numbers are included in the sample. **9.** a. CRS b. SRS c. STRS **11.** An advantage is that a STRS guarantees that objects from each strata are included in the sample which allows for a more precise way to estimate a population feature. Disadvantages are that both the sampling process and data analysis are more complex.

13.

	mean	SD
SRS	45,250	11,074
STRS	45,438	3,175

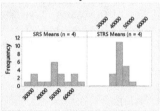

c. Both distributions of sample means are centered at about the same place (a bit over $45k). The sample means from the STRS are much less variable than those from the SRS. d. The STRS gave estimates that are much more consistent than the SRS. Using a STRS to estimate the population mean salary would be more efficient than using a SRS. **15.** Number the patients from 1 to 2000 starting with the ED, then the NICU, and end with the MICU. Use a random number generator to select 40 integers without replacement from 1 to 1000. The ED patients with those numbers are included in the sample. Use a random number generator to select 40 integers without replacement from 1001 to 1300. The NICU patients with those numbers are included in the sample. Use a random number generator to select 40 integers without replacement from 1301 to 2000. The MICU patients with those numbers are included in the sample. b. $0.4\overline{6}$ c. 0.36 d. The estimate based on the STRS sample is over 10 percentage points higher than the population proportion who were satisfied. The problem is that the STRS disproportionately sampled patients, over-representing the more satisfied NICU patients and under-representing the least-satisfied ED patients. We can correct the estimate in part b. by

weighting the sample proportions in each unit by their relative strata sizes. **17.** Selection bias is likely present since your friends' experiences may not represent those of all people in the U.S. who use alternative medical treatments. Maybe your friends had better experiences than the general population. A better approach would have been to select the sample of people at random. **19.** Selection bias is likely present since the transactions processed during that time on Friday may not represent all daily transactions. Maybe there are more errors in processing transactions late on Fridays (when everyone wants to go home!). A better approach would have been to select the sample of transactions at random. **21.** a. Non-response bias is likely present since the feelings of those who did not respond may be different than the feelings of those who did respond. Maybe those who didn't respond are more disgruntled and wouldn't be as quick to recommend. A better approach would have been to select the sample of seniors at random. b. 20.4% c. The variable (how likely to recommend) is categorical with 5 choices and they are numbered in a non-ordinal way; "not sure" should not be numbered lower than "not likely." Therefore the mean is not an appropriate summary. Means are appropriate only for numeric data. **23.** Response bias is likely present since the wording of the question may tend to elicit more yes's than no's. A question that may tend to elicit more no's would be, "Do you think we should divert resources from education to curb development in the rainforests?"

Section 3.3
1. a. factor: activity rating (levels are "a lot less active," "a little less active," "about as active," "a little more active," "a lot more active"), response: time until death b. The study is observational because activity ratings were reported by the subjects, not intentionally assigned by the researcher. c. Only headlines ii. and iii. are appropriate based on the data as they do not imply that activity rating causes changes in lifespan.
3. a. factor: AA group (levels are "AA before transport," "AA after transport," "saline transport," "saline no transport"), response: weight after transport b. The study is an experiment because the AA group was intentionally assigned to the goats by the researchers. c. Since there is only one factor, the treatments are the levels: "AA before transport," "AA after transport," "saline transport," "saline no transport." d. Yes, since the goats were randomly assigned to the treatments, the groups of goats should have been similar before applying the treatments and so any differences in average weights can be potentially attributed to the treatment effects. **5.** Since there is no replication in this experiment, there is no way to measure and account for subject-to-subject variability and causal inferences (i.e. that it was the candy that caused Jill to run faster) are not justified by the data. **7.** This study was

observational since the farmer did not intentionally assign the rooster's status to each morning; it was simply observed. Therefore causal inferences (i.e. the rooster crowing makes the sun rise) are not justified by the data. **9.** Sex would be a confounding variable if, for example, guys tend to cut smaller pieces than girls and also be less likely to recycle. By randomly assigning the cutting treatments to the subjects, the study controls the effect of sex on recycle status. **11.** Season is related to the flavor name and would be a confounding variable if, for example, ice cream sales are higher in early summer than in late spring. The study does not control the effect of season on sales because the flavor names are not randomly assigned to the days. **13.** In a spreadsheet, list the names of the 40 students in a column. In the next column, enter "new problems only" in the first 20 rows and "new+review" in the next 20 rows. In the next column, use a random number function to fill in 40 random numbers. Sort only the treatment and random number column by the random numbers. Assign to each student the treatment label that now is next to their name. **15.** Let's use the hat method. Number the goats from 1 to 24. Number equal-sized slips of paper from 1 to 24. Put the slips in a hat and mix well. Selecting slips without replacement, assign the first 6 goats to the "AA before" treatment, the next 6 to "AA after," the next 6 to "saline transport," and the last 6 to "saline no transport." **17.** A CRA may be easier to implement than an SRA since it may be easier to obtain a list of clusters than a list of individual objects. One disadvantage is that the data analysis for an experiment using a CRA is more complex than for a SRA. **19.** a. factor: sound type (levels are "Mozart piano sonata," "relaxation instructions," "silence"), response: spatial reasoning test score b. student c. Write the names of the three treatments on equal-sized slips of paper. Put the slips in a hat and mix well. For the first student, draw out the three slips and record the order. Administer the three treatments in that order to the first student. Put the slips back in the hat and mix well. Repeat this process for all 36 students. d. The RBA would likely be more efficient than the SRA because it would provide more consistent estimates of the differences in mean test score. This would allow for more precise comparisons of the treatments. **21.** a. The differences in means produced by the two methods have similar centers (around 0) and shapes (symmetric, bell shaped) but the differences from method 2 are more variable. b. Both distributions of differences are centered at around 0. c. The distribution of differences labeled "Method 1" was from the RBA since it has smaller variability. The larger variability of differences in "Method 2" is from the SRA which allows a sex imbalance in the treatments. A larger proportion of men in treatment A would tend to result in large positive differences. A larger proportion

of men in treatment B would tend to result in small negative differences. These extremes would not be possible in the RBA. d. 0.32, 1.68, RBA **23.** a. Number the communities from 1 to 96. Use a random number generator to generate 48 integers between 1 and 96, without replacement. Assign the women in the communities having these numbers to the bed net treatment and the rest to the no bed net treatment. b. Number the women from 1 to 1961. Use a random number generator to generate 980 integers between 1 and 1961, without replacement. Assign the women having these numbers to the bed net treatment and the rest to the no bed net treatment.
25. a.

	$\bar{x}_A$	$\bar{x}_B$	$\bar{x}_A - \bar{x}_B$
Original CRA Assignment	4.5	0.5	4.0
Other CRA Assignment	0.5	4.5	-4.0
Original SRA Assignment	2.5	2.5	0.0
Other SRA Assignment 1	4.5	0.5	4.0
Other SRA Assignment 2	2.0	3.0	-1.0
Other SRA Assignment 3	0.5	4.5	-4.0
Other SRA Assignment 4	2.5	2.5	0.0
Other SRA Assignment 5	3.0	2.0	1.0

b. plot shown here
c. The distributions of possible differences are centered at 0 for both SRA and CRA. CRA results in differences that are more variable compared to SRA.
d.

	Mean	SD
SRA	0	2.61
CRA	0	5.66

e. Both means are 0 since this analysis assumes that the programs do not differ in their effect on AP score. Since the two schools have very different scores, and students are assigned by school under CRA, the differences in means are either low negative or high positive. There is no possible outcome having a difference near 0. Since students are assigned individually under SRA, it's possible for the treatment groups to have a mix of low and high scores and so many differences in means are at or near 0. **27.** a. The women could not be blinded to the treatment. It's possible that a placebo or Hawthorne effect produced the lower malaria incidence among women who got bed nets. b. The study would not have external validity for people in Burma since the participants were pregnant women in Ghana. We cannot be certain that the effect would be the same for people elsewhere (exposed to perhaps different species of mosquitos) or even for non-pregnant women and men in Ghana.

Section 4.1

1. a. $S = \{HHH, THH, HTH, HHT, TTH, THT, HTT, TTT\}$ b. 0.125 each c. 0.375 d. 0.5 e. $\{0, 1, 2, 3\}$ f. 0.875
3. a. H=hit, N=no hit, $S = \{NNNN, HNNN, NHNN, NNHN, NNNH, HHNN, HNHN, HNNH, NHHN, NHNH, NNHH, HHHN, HHNH, HNHH, NHHH, HHHH\}$ b. listed in the same order as S outcomes: 81/256, 27/256, 27/256, 27/256, 27/256, 9/256, 9/256, 9/256, 9/256, 9/256, 3/256, 3/256, 3/256, 3/256, 1/256 c. 0.2109 d. 0.7383 e. $\{0, 1, 2, 3, 4\}$ f. 0.0508 **5.** a. S=stop, N=not $S = \{SS, SN, NS, NN\}$ b. listed in the same order as S outcomes: 0.06, 0.24, 0.14, 0.56 c. 0.44 d. 0.94 e. 0.38 f. $\{20, 21, 22, 23\}$

g.

x	20	21	22	23
$p(x)$	0.56	0.24	0.14	0.06

7. b. and d. are legitimate. The $p(x)$'s in a. add to more than 1 and there is a negative $p(x)$ in c. **9.** a. 0.0625 b. 0.9375 c. 0.375 **11.**

x	$3000	$1500	-$17,000	-$297,000
$p(x)$	0.80	0.15	0.045	0.005

13. a. 0.25
b.

x	4	6	8	10	12
$p(x)$	0.25	0.25	0.3125	0.125	0.0625

c. .75 **15.** a. N=pen, C=pencil, $S = \{NNCC, NCNC, NCCN, CNNC, CNCN, CCNN\}$ b. 0.25 **17.** $0.\overline{3}$ **19.** a. discrete b. continuous c. continuous d. discrete

Section 4.2

1. a. All 4 distributions are centered at 1. Distribution II has the least variability, followed by III, then I, and IV with the most variability.

b.
	$E(X)$	$Var(X)$	$SD(X)$
I	1	$0.\overline{6}$	0.816
II	1	0.2	0.447
III	1	0.4	0.632
IV	1	1.0	1.0

3. $a = 0.05, b = 0.75$ **5.** a. $E(X) = 1$; Over many games, Darryl's average number of hits per game will be close to 1. b. $SD(X) = 0.866$; Over many games, Darryl's hits per game will have a standard deviation close to 0.866. **7.** a. $E(X) = 20.7$; Over many commutes, the average of Maria's commute times will be close to 20.7 minutes. b. $SD(X) = 0.922$; Over many commutes, Maria's commute times will have a standard deviation close to 0.922 minutes. **9.** a. 2 b. Over many auditions of 4 people, the average number who get a spot on the show will be close to 2. **11.** Only d. is correct **13.** $5.904 **15.** a. 1 b. 0.577 c. Using the coin method, you get the same expected value but a smaller standard deviation (0.5). Over many applications, both methods result in gender-balanced groups, on average

(1 boy with a pen) but the number of boys with a pen varies less around 1 for the coin method (i.e. smaller chance of a gender imbalance). It turns out though, that this result for the coin method is highly dependent on the order of assignment! **17.** a. 1 b. 0 c. Using a STRS, you get the same expected value but no variability. Over many applications, SRS's will represent both departments accurately (1 patient from labor and delivery) but the number from labor and delivery can vary from 1. Using the STRS in this case will always give samples with patients from both departments; the number from labor and delivery will never vary from 1. **19.** a. 0.3286 b. 0.0071, 0.084 **21.** a. 26.77 b. 3.875, 1.969 **23.** a. −$0.316 US b. −$0.421 Canadian at an exchange rate of $1 US = $4/3 Canadian. **25.** a. $13.41\overline{6}$°C b. $13.41\overline{6}$°C

Section 4.3

1. Since a continuous random variable has infinitely-many possible values in any interval, we cannot list them in table form. **3.** Probabilities are calculated by finding areas under curves. **5.** a. $0.1\overline{6}$ b. 0.25 c. 0.25 d. 0.5 e. 0.25 **7.** a. false; $E(Z) = 0$ b. false; $Var(Z) = 1$ c. true d. true e. true f. false; $P(Z < 1.4) = P(Z > -1.4)$ **9.** a. 0.0495 b. 0.9949 c. 0.2005 d. 0.8997 e. 0.6836 f. 0.3549 g. 0.4412 h. 0.9973 **11.** a. 1.28 b. -1.75 c. 1.28 d. -1.65 e. 2.40 f. 1.28 g. 1.16 h. -0.26 **13.** a. 0 b. 0.5 c. 0.5 d. 0.6064 e. 0.2514 f. 0.7019 g. 0.716 h. 0.365 **15.** a. 0.0116 b. 0.9768 c. 0.99926 d. 0.8186 e. 0.50126 f. 0.41381 g. 0.42184 h. 0.46848 **17.** 1460 **19.** 4.34 **21.** a. 0.2743 b. 0.0808 c. 0.6% d. 10.35% **23.** a. 0.1867 b. 0.0409 c. 127.32 **25.** a. 0.9525 b. 0.999904 **27.** a. 0.4207 b. 0.176 c. 44.7 mpg d. 4.56% e. 28.8 **29.** a. 0.1056 b. 12.164 oz. **31.** $\mu = 504.\overline{54}, \sigma = 113.\overline{63}$

Section 4.4

1. 5.88, 33.33, 82.35 **3.** 25, 50, 75 **5.** a. 442.48 b. 493.72 c. 560.92 d. 640.72 e. 745.72 **7.** a. 49.316 b. 57.884 c. 49.622 d. 56.864 **9.** a. $\bar{x} = 28.\overline{81}, s = 28.223$ b. 33.33 c. 16.7 d. (5, 16.7) **11.** a. $\bar{x} = 102.5$, $s = 28.54$ b. plot shown here c. The distribution of speeds in this data set is nearly-normal since the normal probability plot is fairly linear.

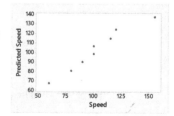

13. a. The distribution of transaction amounts is not normal since the normal probability plot is clearly non-linear. b. -$80.86 c. ($3500, $1855.86) Minitab uses a different y-axis scale

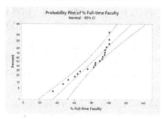

than our "predicted amount" scale. **15.** Since the plot shows a non-linear pattern, the distribution of full-time faculty rates is not normal.

Section 4.5
1. No. Knowing that the temperature is high (during the summer) would increase the chances that there would be a home game and many more people using the subway. **3.** Yes. Knowing whether the Chinese plant produced more than one ear (perhaps reflecting good growing conditions there) would likely not change the chances that the plant in Kansas does or does not produce more than one ear. **5.** No. Knowing that you did well in algebra may indicate that you also may do well in calculus. Both are likely related to innate math ability. **7.** a. 50 b. $16.\overline{6}, 4.082$ **9.** a. $12, 3.464$ b. $8.\overline{3}$, 2.887 c. $5.08\overline{3}, 2.255$ **11.** a. 5, 7.81 b. 0.7389 c. 0.2006 **13.** a. 0.36, 0.6 b. 0.16, 0.4 c. 0.81, 0.9 d. 0.49, 0.7 e. 0.11375, 0.337 **15.** a. 990,000; 56,250,000; 7500 b. 0.0918 c. 21 **17.** a. \$980,000 b. \$307,310.9175 c. No, not all that likely. The probability of a loss is 0.3974, but this is too high for most insurance companies to tolerate. c. \$1,594,621.84

Section 4.6
1. i. $\{0, 1, 2, 3\}$ ii. binomial iii. $n = 3, p = 0.5$ **3.** i. $\{0, 1, 2\}$ ii. not binomial; p does not remain constant **5.** i. $\{0, 1, 2, ...\}$ ii. not binomial; the number of trials n is not fixed **7.** i. $\{0, 1, 2, 3, 4\}$ ii. binomial iii. $n = 4$, $p = 0.5$ **9.** i. $\{-\$297,000, -\$17,000, \$1500, \$3000\}$ ii. not binomial; X is not the number of successes out of n trials **11.** i. $\{1, 2, 3, 4\}$ ii. not binomial; X is not the number of successes out of n trials **13.** a. 184,756 b. 31,824 c. 13 d. 3,268,760 **15.** 1, 5, 10, 10, 5, 1 **17.** any odd n, e.g. $n = 7$ **19.** a. 0.3087 b. 0.02835 c. 0.52822 d. 0.16308 e. 0.66885 f. 0.441 g. 1.5 h. 1.05, 1.0247 **21.** a. 0.1101 b. 0.0100 c. 0.000038 d. 0.6302 e. 0.3950 f. 0.6132 g. 4 h. 3, 1.7321 **23.** $p(0) = 0.125, p(1) = 0.375, p(2) = 0.375, p(3) = 0.125$ **25.** $p(0) = 0.69\overline{4}, p(1) = 0.2\overline{7}, p(2) = 0.02\overline{7}$ **27.** a. 0.375 b. 0.875 c. $E(X) = 1.5$; Over many sets of 3 tosses, the average number of tails will get close to 1.5. d. 0.75, 0.866 **29.** b. For many such companies having 25 members in class B with the same probability of death, the average number of deaths would be close to 0.75. d. 0.3146, 0.8529, 0.6164 **31.** a. 0.3646 b. 0.4797 c. $E(X) = 1.\overline{459}$; Over many gamblers playing this game (spin 3 times, bet on red each spin), the average number of wins will be close to $1.\overline{459}$. d. 0.749, 0.8657 **33.** a. -2.15 b. 0.0192 c. Since 3 hits in 30 at-bats is considered unusually low for a 0.275 hitter, we would conclude that Javier is not as good as a 0.275 hitter. **35.** a. 3.5 b. 0.0015 c. Since 12 correct out of 25 is unusually high for a guesser, we would conclude that Tom is not just guessing. **37.** a. 0.76 b. 0.2858 c. Since winning the toss 16 times out of 28 is not unusual for a fair coin, these data do not convincingly support accusing Tanya of using a biased

coin. **39.** a. $p(0) = 0.0156, p(1) = 0.1012, p(2) = 0.2628, p(3) = 0.3413, p(4) = 0.2216, p(5) = 0.0576$ **41.** a. 0.1762 b. 0.0005

Section 4.7
1. a. no b. no c. no d. yes e. no **3.** a. 0.5339 b. No since $np = 1.5 < 15$ c. 1.5, 1.06066 d. 0.3192 e. no **5.** a. 0.2480 b. Yes since $np = 175 \geq 15$ and $n(1 - p) = 75 \geq 15$ c. 175, 7.4257 d. 0.2259 e. yes **7.** > 0.999967 **9.** 0.0057 **11.** a. \$37,500 b. \$3,038.811445 c. 0.7939 d. Yes, very likely. The probability that total replacement costs will exceed \$50,000 is less than 0.000033. **13.** a. exact: 0.0455 approx: 0.0430 b. exact: 0.0324 approx: 0.0336

Section 5.1
1. population: all people in the U.S. who use alternative medical treatments; sample: your 10 friends; parameter: p = fraction of all people in the U.S. who use alternative medical treatments that improve in their condition; statistic: $\hat{P} = 0.7$
3. population: all parents in the district; sample: the 678 parents who responded; parameter: p = proportion of all parents in the district who are in favor of the new policy; statistic: $\hat{P} = 68/678 \approx 0.10$
5. population: most recent batches of milk from all local farmers; sample: 10 local farmers; parameter: μ = average antibiotic level in the most recent batches of milk for all local farmers; statistic: $\bar{X} = 22 \mu g/kg$
7. population: all students at this university; sample: the 340 students selected; parameter: p = proportion of all students at this university who own their own car; statistic: $\hat{P} = 168/340 \approx 0.494$ **9.** population: all people in the country; sample: the 1693 people selected; parameter: p = proportion of all people in the country who have their own smartphone; statistic: $\hat{P} = 1405/1693 \approx 0.83$ **11.** $\bar{X}, S$ **13.** a. correct b. incorrect; A sampling distribution shows the distribution of a statistic's values over many samples. c. correct d. incorrect; Sampling distributions describe how frequently the possible values of data summaries (statistics) occur over many samples. **15.** Non-random samples are prone to selection bias so that the sampling distribution of the statistic of interest is not centered at the population parameter. Moreover, we cannot even describe the sampling distribution of a statistic based on a non-random sample. Random sampling generates structured uncertainty. The uncertainty associated with non-random samples is unstructured. **17.** a. The sampling distributions of percentages based on SRS's and CRS's have similar shapes (symmetric, bell-shaped), centers (around 10%), and variabilities. b. The sampling distribution of percentages based on SRS's has less variation than that based on CRS's. The centers of the two distributions are similar (around 10%) as are the shapes (symmetric, bell-shaped). c. Scenario I goes with part a. Whether you take a SRS of 480 bottles or a CRS of 40 cases

shouldn't make a difference in terms of the variability of the sample % of bottles not sealed correctly. Scenario II goes with part b. Whether someone attends weekly religious services or not is likely to be similar for people in the same household (e.g. either all do are all do not) and there would be much variation across different households. A CRS of households (e.g. CRS of 160 households averaging 3 people per household) would subject the sample % of people who attend to more variability than a SRS of 480 people. **19.** Step 1: Take a SRS of 1000 U.S. adults, ask each one the question, and record the results. Step 2: Calculate the value of $\hat{P}$, the sample proportion opposed as $\frac{\#\ opposed}{1000}$. Repeat steps 1 and 2 500 times to generate 500 values of $\hat{P}$. A histogram of the 500 $\hat{p}$'s is the sampling distribution (approximately) of the sample proportion. **21.** a. $\eta = 3$ b. 10,626 c. The sampling distribution of sample medians of number texted is centered close to the population median but has less variability. The shapes of both distributions are right skewed. d. 7.12×10^{16} e. The sampling distribution of sample medians of number texted based on SRS's of 14 is also centered close to the population median as that for SRS's of 3, but has less variability than both the population distribution of number texted and the sampling distribution of medians for SRS's of 3. All three distributions are right skewed.

23. a.

m	5	10
$p(m)$	0.5	0.5

b. 7.5; Over many SRS's of 3 bills from the box, the average median amount would be close to $7.50. This matches the population median but is lower than the population mean of $9.

25. a.

$\hat{p}$	1/3	2/3	1
$p(\hat{p})$	0.3	0.6	0.1

b. 0.6; Over many SRS's of 3 people, the average proportion who need stamps will be close to 0.6.
c. 0.583.

27. a.

$\hat{p}$	0	1/3	2/3	1
$p(\hat{p})$	0.045536	0.245995	0.442830	0.265639

b. 9/14, 0.27662
c.

$\hat{p}$	0	1/3	2/3	1
$p(\hat{p})$	0.045554	0.245991	0.442784	0.265671

9/14, 0.27664 d. No, because the sample size is very small compared to the population size.

29. a.

$\bar{x}$	50	60	70
$p(\bar{x})$	0.25	0.5	0.25

b. 60, 7.071 c. Both the SRS and STRS have the same expected value of 60, the same as the population mean, which is good. The STRS has a much smaller standard deviation, meaning sample means vary less around 60 for the STRS than for the SRS. A STRS will produce sample means that are generally closer to the population mean than will a SRS.

31. a.

$\hat{p}$	0	0.5	1
$p(\hat{p})$	1/6	2/3	1/6

b. 0.5, 0.289

$\hat{p}$	0.5
$p(\hat{p})$	1

c. 0.5, 0

d. In a scenario where all the clusters look identical, the CRS will perform better than the SRS because there will be no uncertainty in using the sample proportion to estimate the population proportion.

Section 5.2

1. b and d **3.** a. 5, 2.5, 1.581, no b. 15, 7.5, 2.739, yes c. 5, 4.5, 2.121, no d. 135, 13.5, 3.674, yes e. 1980, 19.8, 4.450, yes **5.** a. 0.3, 0.00525, 0.0725, no b. 0.5, 0.0025, 0.05, yes c. 0.7, 0.00014, 0.0118, yes d. 0.05, 0.00095, 0.0308, no e. 0.8, 0.0000032, 0.00179, yes **7.** a. 9, 16, 21, 24, 25, 24, 21, 16, 9 c. 0.5 **9.** a. $E(X) = 120, SD(X) = 4.899$, approximately normal b. $E(\hat{P}) = 0.8, SD(\hat{P}) = 0.0327$, approximately normal **11.** a. $E(X) = 2, SD(X) = 1.3416$, binomial b. $E(\hat{P}) = 0.1, SD(\hat{P}) = 0.0671$, binomial **13.** Since the sample was not obtained by a SRS, we do not know what the sampling distribution looks like! The uncertainty associated with your sampling method is not structured. **15.** Since the sample was not obtained by a SRS, we do not know what the sampling distribution looks like! The uncertainty associated with volunteer samples like this is not structured. **17.** a. 0.3557 b. 0.0078 c. 0.9875 d. 0.1742 **19.** a. 0.0749 b. 0.2706 c. 0.8749 d. 0.0749 **21.** a. approximately normal, centered at 251.6 with standard deviation 8.088 b. approximately normal, centered at 0.74 with standard deviation 0.0238 c. 0.9236 d. 0.000078 **23.** a. 0.0475 b. 0.9544 **25.** a. 0.0096 b. Since that result would be unusually low ($z = -3.24$) for a SRS of 300 from a population with a default rate of 10.8%, we'd conclude that the national default rate has changed for the most recent cohort. In fact, these data would be convincing statistical evidence for a drop in the default rate. **27.** a. 0.0102 b. Since that result would be unusually low ($z = -3.17$) for a SRS from a population with a contribution rate of 85%, we'd conclude that your company's contribution rate is not 85%. In fact, these data would be convincing statistical evidence that your company's contribution rate is lower than 85%. **29.** a. 0.3570 b. Since that result would be unusually high ($z = 3.98$) for a SRS of 30 from a population with a default rate of 10.8%, we'd conclude that the national default rate has changed for the most recent cohort. In fact, these data would be convincing statistical evidence for an increase in the default rate.

Section 5.3

1. a. $32.80, $38.04 d. $32.80, $23.30 e. $\sigma/\sqrt{n}$ is a good approximation as long as the sample size is small relative to the population size, i.e. when $\frac{n}{N} < 0.05$. Here $\frac{n}{N} = \frac{2}{5} = 0.40$. g. The center of the distribution of all of the possible sample means is the population mean, but sample means vary less than the population cash amounts. Also notice that while the population cash amounts are markedly right skewed, the distribution of sample means is less so and more symmetric. **3.** a. I, II, II, V, and certain cases of IV b. II **5.** a. S b. μ c. $\frac{\sigma}{\sqrt{n}}$ d. σ e. $\bar{X}$ **7.** c

9. a. false; Sample *means* vary less for larger sample sizes. b. false; The standard deviation of sample means is equal to the population standard deviation divided by the square root of the sample size. c. true d. true e. false; When $n = 1$, $Var(\bar{X}) = \sigma^2$.

11. a. 183, 321.03, 17.92, no b. 515, 269.12, 16.40, yes c. 10, 0.643, 0.802, yes d. –16, 4.84, 2.2, yes **13.** The ages of U.S. senators is not a random sample so the sample mean age does not have a sampling distribution. (What would you get for $\bar{X}$ if you repeatedly observed the sample of U.S. Senators?) **15.** The distribution of sample means would be centered at 0.344 but we have not studied other properties of STRSs. Sample means would likely vary less than they would for SRSs.

17. a. 0.0174 b. 0.2188 c. 0.9808 d. 72.95, 77.45 **19.** a. 0.7019 b. 0.9959 c. 0.0041 d. Since that result would be highly unlikely ($z = -2.73$) if the U.K. population of drivers looks like the U.S. population (i.e. has the same mean and standard deviation), we'd conclude that the U.K. population looks different. Perhaps drivers in the U.K. haven't driven as fast as 104.1 mph, on average. **21.** a. 0.1814 b. Yes, it would unless the product weights are severely skewed. A sample size of 30 is usually considered sufficiently large for sample means to be normally distributed. **23.** a. No. I'd expect a left-skewed distribution of weights: many adult weights but a few children's and even infants' weights. b. 0.9656 c. 0.0104 **25.** a. 6.25, 2.5 b. 9, 3; 6.25, 2.5; 20.25, 4.5; 12.25; 3.5 c. 2.984, 1.728 **27.** a. 0.36 b. 0.4$\bar{6}$ c. The expected value of the sample proportion is NOT the population proportion. The sample proportion would generally overestimate the population proportion satisfied by more than 10%! This is because the STRS used the same sample sizes for each department and $\hat{P}$ weights each department's data equally as well but departments have very different numbers of patients where the largest department (ED) has the lowest satisfaction rate and the smallest department (NICU) has the highest satisfaction rate. Therefore the NICU will be overrepresented and the ED underrepresented in the sample without any adjustments for that problem. d. 0.36 Great! The "fixed up" estimator $\tilde{P}$ gets it right, on average.

29. Exercise 27 showed that if the STRS uses the same sample size from each stratum, the estimate must be weighted in order for it to have the correct expected value. Exercise 28 showed that if the STRS uses sample sizes proportionate to the strata sizes, the simpler unweighted estimate will have the correct expected value.

Section 6.1

1. the alternative hypothesis **3.** null hypothesis **5.** H_0: The average range of all electric cars is currently 150 miles. H_a: The average range of all electric cars is currently more than 150 miles. **7.** H_0: There is no difference in music preference between rule-focused people and people who are not. H_a: There is a difference in music preference between rule-focused people and people who are not. **9.** H_0: $p = 0.60$ H_a: $p > 0.60$ where p = proportion of all seniors at the university who have attended a career fair. The objects are seniors at this university and the variable is whether or not a senior has attended a career fair (categorical). **11.** H_0: $p = 0.30$ vs. H_a: $p > 0.30$ where p = proportion of all students at this college who have all required course materials within the first two weeks of the semester. The objects are students at this college and the variable is whether or not a student has all required course materials within the first two weeks of the semester (categorical). **13.** H_0: $p = 0.5$ vs. H_a: $p > 0.5$ where p = proportion of all likely voters who favor candidate X. The objects are likely voters and the variable is whether or not a likely voter favors candidate X (categorical). **15.** H_0: $\mu = 4.5$ vs. H_a: $\mu > 4.5$ where μ = average FEV1 for all male athletes. The objects are male athletes and the variable is FEV1 (numeric). **17.** a. 0.01 and 0.00005 b. 0.999 and 0.50 c. 0.01 and 0.00005 d. 0.999 and 0.50 **19.** a. 4, 5, and 6 for sure...perhaps 3 since they are unusual values under the null hypothesis b. 2 and perhaps 3 since both are more than 20% of 6 but not unusually so c. 0, 1 since neither are more than 20% of 6 **21.** a. −1.10 b. 0.2241; If there are 75% successes in the population, there is a 22.41% chance of getting 6 or fewer successes in a SRS of 10. c. No, since that result is not very unlikely for a population of 75% successes. **23.** a. 2.24 b. 0.0207; If Tanya's coin is fair, there is only a 2.07% chance of getting at least 15 heads in 20 tosses. c. Perhaps, since this result is quite unlikely if the coin is fair. However, your significance level might be very small since you don't want to jeopardize your relationship with Tanya by falsely accusing her. In that case, this result may not be convincing enough to outweigh possibly damaging your relationship. **25.** a. 1 b. 0 c. I'd only need to see one non-black crow to convincingly discredit H_0! d. No. It means only that the data are in perfect agreement with the null hypothesis. There may be non-black crows out there that we did not observe in the sample. **27.** a. 0.0024; If there is no difference in average runtimes for all sci-fi and all romance movies, there's only a 0.24% chance of getting a sci-fi sample mean runtime of at least 18.2 minutes higher in two SRS's of 30 movies each. b. Yes, since this result is highly unlikely if there is no difference in average runtimes between all sci-fi and romance movies. **29**. a. There is no context here and the layperson likely doesn't understand p-values. A better conclusion would be, "The data do not provide convincing statistical evidence that a majority of people in my market segment prefer my product over my main competitor's product." b. Avoid concluding that H_0 is true. A better conclusion would be, "The data are inconclusive about the proportion of people in my market segment prefer my product over my main competitor's product." c. Never use the word "prove" when making conclusions where there is uncertainty/randomness. A better conclusion would be, "The data provide convincing statistical evidence that a majority of people in my market segment prefer my product over my main competitor's product." d. There is no context here and he layperson likely doesn't understand what "H_a" means. A better conclusion would be, "There is not enough statistical evidence that a majority of people in my market segment prefer my product over my main competitor's product." e. It's very rare that the data do not provide *any* evidence for H_a; there is almost always *some* evidence. A better conclusion would be, "The data are inconclusive about the proportion of people in my market segment prefer my product over my main competitor's product." **31**. a. 3.13 b. 0.00087; If half of all customers prefer the company's product over the competition's, then there's only a 0.087% chance of getting at least 1,070 who do in a SRS of 2,000 customers. c. Yes, since this result is highly unlikely if only half of all customers do.

Section 6.2

1. a. yes b. no; Hypotheses are never about a sample statistic like $\hat{p}$. c. no; Hypotheses are never about a single data value like x. d. no; Hypotheses are never about a z-score. e. no; Hypotheses are never about a p-value. **3.** a. one-proportion z-test b. neither c. neither d. neither e. one-proportion binomial test **5.** a. neither b. neither c. neither **7.** a. 0.1091 b. 0.0015 c. 0.1737 d. 0.6846 **9.** a. 0.0102 b. 0.5686 c. 0.0078 d. 0.7490 **11.** a. subjects: students at this university; variable: course materials status (categorical) b. i. H_0: $p = 0.30$ H_a: $p > 0.30$ where $p =$ proportion of all students at this university who have all required course materials within the first two weeks of the semester ii. SRS? Yes, given. $np_0 = 75 \geq$ 15 and $n(1 - p_0) = 175 \geq 15$ so use the one-proportion z-test iii. $z = 3.86$ $p - value = 0.000057$ iv. If 30% of all students at this university have all required course materials within the first two weeks of the semester,

then there's only a 0.0057% chance of obtaining data like these (at least 103 who had in a SRS of 250). v. Since the p-value is very small, there is very convincing statistical evidence that more than 30% of all students at this university have all required course materials within the first two weeks of the semester. **13.** a. subjects: likely voters; variable: opinion about candidate X (categorical) b. i. H_0: $p = 0.50$ H_a: $p >$ 0.50 where $p =$ proportion of all likely voters who are in favor of candidate X ii. SRS? Yes, given. $np_0 =$ $750 \geq 15$ and $n(1 - p_0) = 750 \geq 15$ so use the one-proportion z-test iii. $z = 3.61$ $p - value = 0.000153$ If 50% of all likely voters are in favor of candidate X, then there's only a 0.0153% chance of obtaining data like these (at least 820 who do in a SRS of 1500). iv. Since the p-value is very small, there is very convincing statistical evidence that a majority (more than 50%) of all likely voters are in favor of candidate X. **15.** The p-value would be 0.0526, which is fairly small, but not nearly as small as 0.000153. Therefore there would be only marginal statistical evidence that a majority (more than 50%) of all likely voters are in favor of candidate X. **17.** a. subjects: U.S. adults; variable: religious views of parents, (categorical) b. i. H_0: $p = 0.20$ H_a: $p > 0.20$ where $p =$ proportion of all U.S. adults who were raised in interfaith homes ii. SRS? We'd have to assume the sample of 5,000 was a SRS. $np_0 = 1000 \geq 15$ and $n(1 - p_0) = 4000 \geq 15$ so use the one-proportion z-test iii. $z = 1.77$ $p - value = 0.0384$ If 20% of all U.S. adults were raised in interfaith homes, then there's a 3.84% chance of obtaining data like these (at least 21% who were in a SRS of 5000). iv. Since the p-value is marginally small, there is only marginal statistical evidence that more than 20% of all U.S. adults were raised in interfaith homes. **19.** a. subjects: adults in my county; variable: ethnicity (categorical) b. i. H_0: $p = 0.57$ H_a: $p \neq 0.57$ where $p =$ proportion of all adults in your county who are white ii. SRS? Yes, given. $np_0 = 34.2 \geq 15$ and $n(1 - p_0) = 25.8 \geq 15$ so use the one-proportion z-test iii. $z = -1.10$ $p -$ $value = 0.2714$ If 57% of all adults in your county are white, then there's a 27.14% chance of obtaining data like these (a sample proportion who are white at least 7% points away from 57% in a SRS of 60). iv. Since the p-value is not considered small, the data are inconclusive. There is not convincing statistical evidence that your county is different than the urban county in my state in terms of the proportion of white adults. **21.** If we consider a p-value of less than 0.05 to be convincing statistical evidence, we'd need to see at least 28 with lost bags. **23.** a. iv b. Since the p-value is not what we'd consider small, the data are inconclusive. No, we do not have convincing statistical evidence that the U.S. national high school dropout rate has decreased since 2017. **25.** i. H_0: $p = 0.10$ H_a: $p > 0.10$ where $p =$ proportion of all 5-year-old cars with back-up

camera malfunctions ii. SRS? Yes, given. $np_0 = 5 < 15$ and $n(1 - p_0) = 45 \geq 15$ so use the one-proportion binomial test iii. $p - value = 0.122$ If 10% of all 5-year-old cars have back-up camera malfunctions, then there's a 12.2% chance of obtaining data like these (at least 8 with problems in a SRS of 50). iv. Since the p-value is only marginally small, there is only marginal statistical evidence that more than 10% of all 5-year-old cars have back-up camera malfunctions.
27. i. H_0: $p = 0.057$ H_a: $p < 0.057$ where $p =$ current proportion of all late-teen to early-20-year-olds who have dropped out of high school ii. SRS? Yes, given. $np_0 = 57 \geq 15$ and $n(1 - p_0) = 943 \geq 15$ so use the one-proportion z-test iii. $z = -0.68$ $p - value = 0.248$ If the current dropout rate is 5.7%, then there's a 24.8% chance of obtaining data like these (5.2% or fewer who've dropped out in a SRS of 1000). iv. Since the p-value is not considered small, the data are inconclusive. There is not convincing statistical evidence that the current U.S. national high school dropout rate for all late-teen to early-20-year olds is lower than 5.7%.

Section 6.3
1. a. no; Hypotheses are never about a single data value b. no; Hypotheses are never about a sample statistic like $\bar{x}$. c. yes d. no; Hypotheses are never about a t-score. e. no; Hypotheses are never about a p-value. **3.** a. no b. yes c. no d. no e. yes **5.** a. no b. no c. no **7.** a. p-value < 0.0005; 0.00003 b. 0.1 < p-value < 0.5; 0.161 c. 0.1 < p-value < 0.2; 0.138 d. 0.1 < p-value < 0.5; 0.259 **9.** Likely not because the data distribution is severely skewed. **11.** Yes. The sample size is sufficiently large and the data distribution is not severely skewed. **13.** a. objects: German families; variable: number of children under 18 (numeric) b. i. H_0: $\mu = 1.8$ H_a: $\mu < 1.8$ where $\mu =$ average number of children under 18 for all German families ii. SRS? Yes, given. $n = 40$ and a histogram shows the data are not severely skewed, so use the one-sample t-test iii. $t = -0.12$ $p - value = 0.453$ iv. If the average number of children under 18 for all German families is 1.8, then there's a 45.3% chance of obtaining data like these (a mean of 1.775 or less in a SRS of 40). v. Since the p-value is not small, the data are inconclusive. There is not convincing statistical evidence that the average number of children under 18 for all German families is less than 1.8. **15.** a. objects: checked bags of current customers; variable: weight in pounds (numeric) b. i. H_0: $\mu = 34$ H_a: $\mu < 34$ where $\mu =$ average weight of all checked bags of current customers ii. SRS? Yes, given. $n = 101$ and the fairly linear normal probability plot shows the data are nearly normal, so use the one-sample t-test iii. $t = -3.05$ $p - value = 0.0014$ iv. If the mean weight of all checked bags of current customers is 34 pounds, then there's only a 0.14% chance of obtaining data like these (a

mean of 32.287 or less in a SRS of 101). v. Since the p-value is extremely small, there is very convincing statistical evidence that the mean weight of all checked bags of current customers is less than 34 pounds.
17. a. The new p-value = 0.103 so there would be only marginal statistical evidence that the mean weight of all checked bags of current customers is less than 34 pounds. b. The new p-value = 0.194 so the data would be fairly inconclusive. There wouldn't be convincing statistical evidence that the mean weight of all checked bags of current customers is less than 34 pounds.
19. a. subjects: residents of this town; variable: annual income (numeric) b. i. H_0: $\mu = 60$ H_a: $\mu > 60$ where $\mu =$ average income (in thousands) for all residents of this town ii. SRS? Yes, given. The data distribution is not severely skewed, so use the one-sample t-test iii. $t = 1.04$ $p - value = 0.173$ If the average income for all residents of this town is $60,000 per year, then there's a 17.3% chance of obtaining data like these (a mean of $67,167 or more in a SRS of 6). iv. No. Since the p-value is not very small, the data are inconclusive. There is not convincing statistical evidence that the average income for all residents in this town is above $60,000 per year. **21.** a. i. b. Yes. Since the p-value = 0.019 is somewhat small, there is marginal statistical evidence that the average expense ratio for all small cap equity funds differs from 1.25%. **23.** The symbol in the hypotheses should be μ, not $\bar{x}$ and H_a should show $\neq$. Use the one-sample t-test, not z-test. The test statistic should be a t, the terms in the numerator are switched, and the denominator should be $0.6/\sqrt{91}$. The $p - value = 2P(t \geq 2.38) = 0.019$. Incorrect p-value interpretation. Incorrect conclusion that H_0 is true for a large p-value. **25.** The parameter is defined incorrectly as the sample mean. H_0 should be $\mu = 4$ and H_a should be <. The t-distribution is valid for smaller sample sizes as long as the data distribution is fairly symmetric with no outliers, which is given. The t-statistic is calculated incorrectly. T.DIST using "false" gives the height of the curve, not the area to the left. Incorrect p-value interpretation. Incorrect conclusion.

Section 6.4
1. Type II **3.** a. neither b. Type I c. Type I **5.** α
7. a. Type I b. Type II **9.** a. Type II b. Type I
11. a. Type I b. Type II **13.** a. A Type I error would be to conclude that the drug is more effective than placebo when, in fact, it is not. b. A Type II error would be to conclude that the drug is not more effective than placebo when, in fact, it is. c. A consequence of a Type I error would be to prescribe the drug when it is really not effective, thereby subjecting patients to the bad side without treating their insomnia. A consequence of a Type II error would be to not prescribing the drug when it really is effective, thereby preventing patients from

both the bad side effects and effective treatment for insomnia. Which consequence seems worse to you? d. If you think the consequences of a Type I error are more serious, it would be more important to make sure α (the probability of a Type I error) is small. If you think the consequences of a Type II error are more serious, it would be more important to make sure β (the probability of a Type II error) is small. **15.** a. A Type I error would be to conclude that more than 10% of all cars are currently experiencing the problem when, in fact, only 10% are. b. A Type II error would be to conclude that only 10% of all cars are currently experiencing the problem when, in fact, more than 10% are. c. A consequence of a Type I error would be to conduct the recall when only 10% of all cars are currently experiencing the problem, thereby incurring perhaps unnecessary expense. A consequence of a Type II error would be to not conduct the recall when it really should be done, thereby leaving too many problem cameras in service. Which consequence seems worse to you? d. If you think the consequences of a Type I error are more serious, it would be more important to make sure α (the probability of a Type I error) is small. If you think the consequences of a Type II error are more serious, it would be more important to make sure β (the probability of a Type II error) is small. **17.** a. H_0: $\mu = 30$ vs. H_a: $\mu < 30$ where μ = mean nitrate level (in ppm) for all soil specimens b. A Type I error would be to conclude that the mean nitrate level for all soil specimens is less than 30 ppm when, in fact, it is 30 ppm. c. A Type II error would be to conclude that the mean nitrate level for all soil specimens is 30 ppm when, in fact, it is less than 30 ppm. d. A consequence of a Type I error would be to apply the fertilizer when it really isn't necessary, thereby over-fertilizing and incurring unnecessary expense. A consequence of a Type II error would be to not apply the fertilizer when it really is necessary, thereby saving fertilizer expense but perhaps giving up crop yield. Which consequence seems worse to you? **19.** a. H_0: $p = 0.40$ vs. H_a: $p > 0.40$ where p = proportion of all eligible voters in the state that prefer her over her closest competitor b. A Type I error would be to conclude that more than 40% of all eligible voters in the state prefer her over her closest competitor when, in fact, only 40% do. c. A Type II error would be to conclude that only 40% of all eligible voters in the state prefer her over her closest competitor when, in fact, more than 40% do. d. A consequence of a Type I error would be to campaign in the state when there is little chance of winning, thereby wasting time and money. A consequence of a Type II error would be to not campaign in the state when there would be a chance of winning, thereby perhaps forfeiting the state. Which consequence seems worse to you?

Section 6.5

1. $power = 1 - P(Type\ II\ error) = 1 - \beta$
3. a. $z > 1.645$ b. $z < -1.645$ c. $|z| > 1.96$
5. a. $z > 1.28$ b. $z < -1.28$ c. $|z| > 1.645$
7. a. $z > 1.28$ b. 0.3587 **9.** a. $|z| > 1.645$ b. 0.3823, 0.4977 **11.** $z < -1.75$ b. 6.8626 **13.** a. 0.5714, 0.4286 b. 0.2005, 0.7995 **15.** a. 0.5596, 0.4404 b. 0.1335, 0.8665 **17.** a. The farther p_a is above 0.3, the lower the probability of a Type II error. The closer p_a is to 0.3, the higher the probability of a Type II error. b. The farther p_a is above 0.3, the higher the power of the test. The closer p_a is to 0.3, the lower the power of the test. **19.** a. The farther μ_a is below 1.8, the lower the probability of a Type II error. The closer μ_a is to 1.8, the higher the probability of a Type II error. b. The farther μ_a is below 1.8, the higher the power of the test. The closer μ_a is to 1.8, the lower the power of the test. **21.** a. 0.3446, 0.6554 b. 0.2206, 0.7794 c. All else equal, the larger the sample size, the higher the power of the test. **23.** a. 1,708 b. 1,298 c. 238 d. 61,635 **25.** a. 118 b. 214 c. 11,719 **27.** a. H_0: $p = 0.3$, H_a: $p > 0.3$ b. $z > 1.645$ c. 0.6026, 0.3974 d. 830 **29.** a. H_0: $p = 0.05$, H_a: $p > 0.05$ b. $z > 2.33$ c. 0.4840, 0.5160 d. 694 **31.** a. H_0: $\mu = 60,000, H_a$: $\mu > 60,000$ b. $z > 1.645$ c. 0.7019, 0.2981 d. 126 **33.** a. H_0: $\mu = 3, H_a$: $\mu < 3$ b. $z < -1.645$ c. 0.3300, 0.6700 d. 37

Section 6.6

1. a. – c. No, we cannot generalize these results to any larger population because the sample was not obtained randomly. The sample consisted only of those who volunteered their response to the online poll. 3. e
5. Only when we've determined that the power of the test is high to detect the smallest meaningful difference from the null hypothesis. **7.** a. 0.32, 0.24 b. The performance of the test is compromised by peeking at the data to determine the hypotheses. In the scientific methods, that's why setting up the hypotheses comes before collecting data. **9.** a. any significance level greater than 0.049 b. No. A commonly-used significance level is 0.05 and this p-value is very close to that threshhold, indicating only marginal evidence for H_a. **11.** a. No, not necessarily. The p-value could be very close to 0.05, indicating only marginal evidence for H_a. b. No, not necessarily. The p-value could be very close to 0.05, indicating marginal evidence for H_a. **13.** a. with $n = 40$, $t = -4.117$ and $p - value = 0.0001$ b. with $\bar{x} = 26.4$, $t = -4.632$ and $p - value = 0.0006$ **15.** a. 5 b. No, Dave must consider the multiple testing problem. Since Dave conducted 100 tests each at the 5% significance level, that means the expected number of tests for which the fair coin hypothesis will be falsely rejected is 5, which matches his results exactly. It is quite likely that those 5 tests are false rejections. **17.** a. Yes, highly significant! b. No. Although these results show highly convincing

statistical evidence that the average one-month growth is more than 0.46" for the population of one-year-old boys, the data suggest that it's not that much more than 0.46" and creating separate sizes by month for such a small difference would not be necessary.

Section 7.1

1. a. μ = population mean; s = sample standard deviation, $\hat{p}$ = sample proportion of successes; σ^2 = population variance; s^2 = sample variance; p = population proportion of successes; $\bar{x}$ = sample mean; σ = population standard deviation b. $(\mu, \bar{x})$, (σ, s), (σ^2, s^2), $(p, \hat{p})$ **3.** a. 90% b. 40% c. 97% d. 99% **5.** a. α = 0.6 b. α = 0.1 c. α = 0.01 d. α = 0.2 **7.** a. μ = average number of children under 18 per family for all German families b. $\bar{x}$ = 1.775 c. 0.001, 0.009, 0.199, 0.906, 0.291, 0.001 d. 1.5, 1.8, 2 **9.** a. p = proportion of all current subscribers who would be interested in an online version b. $\hat{p} = \frac{36}{200} =$ 0.18 c. 0.000164, 0.2340, 0.4778, 0.0220 d. 0.15, 0.2 **11.** There are other plausible values of μ other than 45 that would also be supported by the data. **13.** There are other plausible values of p other than 0.8 that would also be supported by the data. **15.** We are 95% confident that the average number of children under 18 per family for all German families is between 1.35 and 2.20. **17.** We are 95% confident that between 13.3% and 23.9% of all current subscribers would be interested in an online version. **19.** a. no context b. about an individual fund, not about a population parameter c. no language of confidence d. about an individual fund, not about a population parameter, and no language of confidence e. about individual funds, not a population parameter, and no language of confidence f. about the sample, not the population g. best **21.** a. best b. no context c. about the sample, not about the population d. no language of confidence e. the parameter is a proportion, not a mean **23.** If we'd collect many, many SRS's of 91 small cap equity funds, and construct a 95% confidence interval for each, about 95% of those intervals would contain the mean expense ratio for all small cap equity funds and about 5% would not. **25.** If we'd collect many, many SRS's of 60 adults from that county, and construct a 95% confidence interval for each, about 95% of those intervals would contain the proportion of all adults in the county who are white and about 5% would not.

Section 7.2

1. $\bar{x}$ **3.** 1) a SRS and 2) a nearly-normal distribution of data, which can be relaxed somewhat if n is sufficiently large **5.** a. 4.01 b. 1.83 c. 3.75 d. 2.11 **7.** t^* gets closer to 0 **9.** 0% **11.** Yes. We have a SRS and the data distribution is nearly normal by the fairly linear normal probability plot. **13.** They are the same. **15.** a. $\bar{x}$ = 22.8 b. 2.71575 c. (20.08, 25.52) d. We are not given the data values or a histogram, dotplot, or normal probability plot, so we are unable to check the nearly-

normal data distribution condition. For a skewed data distribution, n = 16 would not be considered sufficiently large. e. It would be narrower. **17.** a. $\bar{x}$ = 96.8 b. 4.5774 c. (92.22, 101.38) d. If the data are severely skewed, n = 38 may not be considered sufficiently large. e. It would be narrower. **19.** a. objects: electric cars; variable: 0 – 60mph acceleration time (numeric) b. i. μ = mean 0 – 60mph acceleration time for all electric cars ii. 1) SRS? Yes, given. 2) The times are right skewed but not severely so; n = 29 is considered sufficiently large. iii. (6.75, 9.16) iv. We are 95% confident that the mean 0 – 60mph acceleration time for all electric cars is between 6.75 seconds and 9.16 seconds. **21.** a. objects: ginger beer bottle; variable: amount of product (numeric) b. i. μ = mean amount of ginger beer in all bottles currently being filled ii. 1) SRS? Yes, given. 2) A histogram shows a nearly-normal distribution of amounts. iii. (11.80, 12.10) iv. We are 95% confident that the mean amount of ginger beer in all bottles currently being filled is between 11.80 ounces and 12.10 ounces. c. No, the data are inconclusive about a problem in the filling process since the target amount (12 oz.) is inside the confidence interval. **23.** a. objects: U.S. small cap mutual funds; variable: YTD return (numeric) b. We are 95% confident that the mean YTD return for all U.S. small cap mutual funds is between 20.57% and 24.57%. **25.** a. subjects: mild asthmatics; variable: FEV1 (numeric) b. We are 95% confident that the mean FEV1 for all mild asthmatics is between 3.10 and 3.90. **27.** These data are the population data so we can calculate the value of μ directly from the data set. There is no need for the one-sample t-interval. **29.** Since we have the data on all the players and can calculate the value of μ (10.09), there is no need for the one-sample t-interval. **31.** a. We are 95% confident that the mean appraised value (by Jill) for all houses currently for sale is between $87,653 and $389,367. We are 95% confident that the mean appraised value (by Anya) for all houses currently for sale is between $92,433 and $395,047. b. No, since the two confidence intervals overlap. d. We are 95% confident that the mean difference in appraised value (Jill – Anya) for all houses currently for sale is between -$8,532 and -$1,928. e. Yes, since the confidence interval does not contain 0. f. Use the part d. analysis since it accounts for the fact that both Jill and Anya appraised the same houses and the part a. analysis does not. **33.** a. about individual bags and not a population mean, and no language of confidence b. about individual bags in the sample and not a population mean, and no language of confidence c. no context d. about individual bags and not a population mean e. no language of confidence f. best g. about the sample mean, not the population mean **35.** If we'd collect many, many SRS's of 10 locations from this field, and construct a 95%

confidence interval for each, about 95% of those intervals would contain the mean nitrate level for all locations in the field and about 5% would not. **37.** If we'd collect many, many SRS's of 10 8th grade classes in this state, and construct a 95% confidence interval for each, about 95% of those intervals would contain the mean percent proficient in ELA for all 8th grade classes in this state and about 5% would not.

Section 7.3

1. $\hat{p}$ **3.** SRS **5.** a. 1.34 b. 2.05 c. 3.89 d. 1.44 **7.** z^* gets closer to 0 **9.** Yes, since the data are from a SRS. **11.** No, since the data are not from a SRS. **13.** Both require a SRS. In addition, the one-proportion z-test requires that np_0 and $n(1 - p_0)$ are both at least 15. **15.** The only condition required is a SRS from a fairly large population. **17.** a. $\hat{p} = 0.21$ b. $(0.195, 0.226)$ c. The 99% confidence interval would be wider. **19.** a. $\hat{p} = 0.756$ b. $(0.637, 0.845)$ c. The 95% confidence interval would be narrower. **21.** a. subjects: seniors at this college; variable: career fair attendance status (categorical) b. i. p = proportion of all seniors at this college who have attended a career fair ii. SRS? Yes, given. iii. $(0.501, 0.759)$ iv. We are 95% confident that between 50.1% and 75.9% of all seniors at this college have attended a career fair. c. No, since the confidence interval does not lie entirely above 0.6. **23.** a. objects: recently-shipped orders; variable: on-time status (categorical) b. i. p = proportion of all recently-shipped orders that were shipped on time ii. SRS? Yes, given. iii. $(0.902, 0.951)$ iv. We are 90% confident that between 90.2% and 95.1% of all recently-shipped orders were shipped on time. c. There is marginal evidence since the confidence interval lies almost entirely below 0.95. **25.** $(0.906, 0.948)$ This interval does lie entirely below 0.95, but just barely. Technically the decision about whether $p < 0.95$ is different but just marginally. **27.** a. subjects: 18 – 20 year-olds at this school; variable: binge drinking status (categorical) b. SRS? Yes, given, so use the one-proportion score interval: $(0.332, 0.423)$ We are 95% confident that between 33.2% and 42.3% of all 18 – 20 year-olds at this school have engaged in binge drinking in the past month. **29.** a. subjects: current subscribers; variable: online version interest (categorical) b. SRS? Yes, given, so use the one-proportion score interval: $(0.133, 0.239)$ We are 95% confident that between 13.3% and 23.9% of all current subscribers are interested in an online version. **31.** Since we have the data on all the mutual funds available at Fidelity and can calculate the value of p (0.285), there is no need for the one-proportion score interval. **33.** a. $(0.226, 0.284)$ We are 95% confident that between 22.6% and 28.4% of all U.S. mothers think they spend too little time with their child/children. b. $(0.444, 0.508)$ We are 95% confident that between 44.4% and 50.8% of all U.S. fathers think they spend too

little time with their child/children. c. Yes, since the confidence intervals aren't even close to overlapping. **35.** a. no context b. about the sample, not the population and no language of confidence c. about the sample, not the population d. best e. about a probability, not a population proportion and no language of confidence f. about the wrong variable (percent of time using streaming), not whether or not streaming is used and no language of confidence g. about the sample, not the population and no context **37.** If we'd collect many, many SRS's of 5,000 U.S. adults, and construct a 95% confidence interval for each, about 95% of those intervals would contain the proportion of all U.S. adults who were raised in interfaith homes and about 5% would not. **39.** If we'd collect many, many SRS's of 60 adults in this county, and construct a 95% confidence interval for each, about 95% of those intervals would contain the proportion of all adults in the county who are white and about 5% would not. **41.** a. $(0.004, 0.897)$ b. 100% c. $(0.027, 0.061)$ d. 95% e. Even though we lose some confidence due to the uncertainty induced by a SRS, the gain in precision more than compensates! The value of the SRS is that it permits much more precise estimation compared to non-random samples. **43.** a. $(0.00433, 0.99996)$ b. 100% c. $(0.946, 0.998)$ d. 95% e. Even though we lose some confidence due to the uncertainty induced by a SRS, the gain in precision more than compensates! The value of the SRS is that it permits much more precise estimation compared to non-random samples.

Section 7.4

1. lower **3.** narrow **5.** 133 **7.** 34 **9.** 1,842 **11.** 271 **13.** 2,401 **15.** 20 **17.** a. 1,068 b. This required sample size is much smaller. c. There is much more information about a population in a SRS than in non-random samples of 10 or 100 times as many subjects. d. While it would be statistically optimal to obtain data on a random sample of people, the 4th amendment to the U.S. Constitution (which protects citizens from unlawful search and seizure) generally prohibits requiring compliance with such testing. **19.** 2,401 **21.** a.

MOE	n
0.005	38,416
0.01	9,604
0.02	2,401
0.05	385
0.10	97

b.

MOE	80%	90%	95%	99%
0.005	16,384	27,061	38,416	66,307
0.01	4,096	6,766	9,604	16,577
0.02	1,024	1,692	2,401	4,145
0.05	164	271	385	664
0.10	41	68	97	166

Section 8.1

1. requires a generalization inference **a.** schools **b.** percent of faculty that are full-time (numeric) **c.** all Bachelor's schools and all schools granting graduate degrees in the U.S. **d.** the 6 Bachelor's schools and the 13 schools granting graduate degrees in the top 20 ranked by tuition **e.** μ_1 = average full-time faculty percent for all Bachelor's schools, μ_2 = average full-time faculty percent for all schools granting graduate degrees **f.** A generalization inference is not permitted because the samples were not obtained randomly from the populations. The top 20 schools ranked by tuition are very different (by definition!) from the rest of the schools that didn't make the top 20. **3.** requires a generalization inference **a.** U.S. parents of children under 18 **b.** feeling about time spent with child(ren) (categorical) **c.** all U.S. mothers with children under 18 and all U.S. fathers with children under 18 **d.** the 870 mothers who responded and the 937 fathers who responded **e.** p_1 = proportion of all U.S. mothers with children under 18 who think they spend too little time with their child(ren), p_2 = proportion of all U.S. fathers with children under 18 who think they spend too little time with their child(ren) **f.** A generalization inference is not permitted because although the original sample of 12,047 parents was selected randomly, the data here are only for the 1,807 respondents. The data on the responders may not represent the populations of all mothers and fathers. The non-responders may feel differently! **5.** requires a causal inference **a.** students **b.** improvement in test score (numeric) **c.** homework assistance type **d.** no assistance, complete solution **e.** A causal inference is permitted because the homework assistance types were randomly assigned to the students. The random assignment should have made the two treatment groups look similar prior to the treatment so that differences after the treatment could be attributed to the differences in the treatments themselves. **7.** requires a generalization inference **a.** U.S. colleges and universities **b.** median student loan debt (numeric) **c.** all predominantly Associate's degree institutions in the U.S. and all predominantly Bachelor's degree institutions in the U.S. **d.** the 11 Associate's institutions and the 9 Bachelor's institutions **e.** μ_1 = average median student loan debt for all predominantly Associate's degree institutions in the U.S., μ_2 = average median student loan debt for all predominantly Bachelor's degree institutions in the U.S **f.** A generalization inference is permitted because the samples were obtained randomly. **9.** requires a causal inference **a.** days **b.** daily sales (numeric) **c.** name of flavor **d.** Chocolate Mud, Chocolate Dream **e.** A causal inference is not permitted because the flavor names were not randomly assigned to the days. It's likely that outside temperature is a confounding variable: sales may be higher on warmer June days so that the effect of temperature and the effect of flavor name cannot be separated. **11. a.** Helen's appraisals and Samara's appraisals **b.** appraised value (numeric) **c.** paired **13. a.** company's regular gas and the competitor's regular gas **b.** gas mileage (numeric) **c.** independent **15. a.** blue/green color palette and orange/yellow color palette **b.** click status (categorical) **c.** independent **17. a.** new drug and restricted diet **b.** memory test score (numeric) **c.** independent **19. a.** before debate and after debate **b.** opinion of candidate X (categorical) **c.** paired

Section 8.2

1. No. The population means are known, so no inference is needed. **3.** No. Welch's t-test requires a SRS. **5.** Yes. We have two independent SRS's and $n_1\hat{p}_1 = 99, n_1(1 - \hat{p}_1) = 276, n_2\hat{p}_2 = 132, n_2(1 - \hat{p}_2) = 243$ are all at least 15. **7.** No. The population mean difference is known, so no inference is needed. **9.** Yes. We have a SRS of pairs and, although the differences are right skewed, 1200 is considered sufficiently large. **11. a.** 0.0035 **b.** (0.014, 0.086) **13. a.** H_0: $p_1 = p_2$ vs. H_a: $p_1 \neq p_2$ where p_1 = proportion of all U.S. mothers with children under 18 who think they spend too little time with their child(ren) and p_2 = proportion of all U.S. fathers with children under 18 who think they spend too little time with their child(ren) **b.** We have 2 independent SRS's of parents and $n_1\hat{p}_1 = 221$, $n_1(1 - \hat{p}_1) = 649, n_2\hat{p}_2 = 446, n_2(1 - \hat{p}_2) = 491$ are all at least 15. No assumptions are necessary as all conditions are met. **c.** p-value < 0.000066 If there is no difference in the proportions of all U.S. mothers and all U.S. fathers with children under 18 who think they spend too little time with their child(ren), then there's less than a 0.0066% chance of observing data like these (an SRS of 870 and one of 937 having sample proportions at least 22.2% apart). **d.** (-0.265, -0.179) **e.** There is strong statistical evidence in these data that there is a difference in the proportions of all U.S. mothers and all U.S. fathers with children under 18 who think they spend too little time with their child(ren). The confidence interval suggests that between 17.9% and 26.5% more fathers think they spend too little time with their child(ren). **15. a.** H_0: $\mu = 0$ vs. H_a: $\mu > 0$ where μ = mean difference in tip percentage (Jill – Jack) **b.** We have a SRS of days and a normal probability plot of the differences shows a fairly linear pattern, so we'll consider the differences to be nearly-normal. No assumptions are necessary as all conditions are met. **c.** 0.0046 If there is no difference between Jack's and Jill's tip percentages, there's only a 0.46% chance of observing data like these (a SRS of 7 days where the difference in tip percentages is at least 2.37). **d.** (0.83, 3.91) **e.** We have strong statistical evidence in these data that there is an average difference between Jack's and Jill's tip percentage. The data suggest that Jill's tip percentage is between 0.83 and 3.91 higher than Jack's

f. No! Since the treatments (introduction, no introduction) were not randomly assigned to Jack and Jill, a causal inference like this is not permitted. It could be that patrons at restaurant B tend to tip more so that the effect of introductions and the effect of restaurant cannot be separated. **17.** a. No, the data are inconclusive about a difference between the two measurement methods: p-value = 0.2845; 95% CI for μ is (-92.6, 271.4). b. All newborns at this certain hospital during the time of the study. **19.** a. 0.0478; (-16.86, -0.16) Note that now, the evidence for a difference is only marginal. All else equal, larger sample sizes provide more evidence of a difference between population means. b. True c. 0.1174; (-5.37, 0.66) Note that now, the evidence for a difference is only marginal. All else equal, a larger difference between the sample means provides more evidence of a difference between population means.

Section 8.3

1. Yes. We have independent SRA to the two treatments and the response data in each treatment are nearly-normal. **3.** No. The kindergarteners are assigned to treatments using CRA, not SRA. **5.** Yes. There is a SRA of treatments within each object and, although the differences are mildly skewed, 150 is considered sufficiently large. **7.** No. $n_1\hat{p}_1 = 5$, $n_1(1 - \hat{p}_1) = 45, n_2\hat{p}_2 = 6, n_2(1 - \hat{p}_2) = 44$ are not all at least 15. **9.** a. 0.2984 b. (-0.064, 0.020) **11.** a. macular degeneration status; eat at least one serving of oranges a day, not b. observational study c. No. Since the treatments (at least one serving of oranges/day or not) was not randomly assigned to the adults, a causal inference like this is not permitted. It could be that those who eat oranges are different in other ways that actually affect the risk of developing macular degeneration. **13.** a. diabetes remission status; gastric bypass surgery, major dietary and lifestyle changes b. experiment c. Yes. Since the treatments (surgery or diet/lifestyle changes) were randomly assigned to the subjects, a causal inference like this is permitted as long as the effect of the random assignment can be ruled out. **15.** a. H_0: $\mu_1 = \mu_2$ vs. H_a: $\mu_1 > \mu_2$ where μ_1 = average gas mileage for cars like these on the company's gas and μ_2 = average gas mileage for cars like these on the competitor's gas b. We have independent SRA to gas type and both normal probability plots of the gas mileage data are quite linear, so we'll consider both data distributions nearly-normal. Use Welch's t-method. c. 0.003 If there is no difference in the effects of the two types of gas on gas mileage, then there's only a 0.3% chance of observing data like these (SRA of 40 cars where the mean gas mileages are at least 1.79mpg different). d. Since the p-value is quite small, the data provide convincing statistical evidence that the company's gas results in higher gas mileage compared to the competitor's gas for cars like these. **17.** a. H_0: $\mu = 0$ vs. H_a: $\mu \neq 0$ where μ = average difference in yield between the organic and conventional fertilizers (organic – conventional) for potato plants like these b. We have SRA to fertilizer type within each plot and the normal probability plot of the differences in yields is quite linear, so we'll consider the distribution of the differences nearly-normal. Use the paired t-method. c. 0.846 If there is no difference in the effects of the two types of fertilizer on yield, then there's an 84.6% chance of observing data like these (SRA of plants to fertilizer within each plot where the mean difference is at least 0.50 pounds). d. Since the p-value is not small, the data are inconclusive. The data do not provide convincing statistical evidence of an average difference in the effects of the two fertilizer types on the yield for potato plants like these. **19.** No, the data are inconclusive about the coupon effect. The data do not provide convincing statistical evidence that a $20 coupon would affect the average percent change in stress hormone levels for students like these: p-value = 0.183. **21.** Yes, these data provide convincing statistical evidence that including the word "good" would change the proportion of adults like these would definitely or somewhat agree with the statement: p-value = 0.0001.

CPSIA information can be obtained
at www.ICGtesting.com
Printed in the USA
LVHW022137131220
674072LV00011B/516